This book describes the research that has been done on the problems of Bénard convection – as well as its modern offspring, the Rayleigh–Bénard problem – and Taylor vortices. Bénard convection differs from Rayleigh–Bénard convection in that Bénard convection is characterized by the presence of surface tension and hexagonal cells, while Rayleigh–Bénard convection is characterized by buoyancy and parallel rolls. Toroidal vortices characterize Taylor vortex flow. Convection and Taylor vortex flow deal with the consequences of the presence of infinitesimal disturbances in an unstable fluid layer. Both problems are classical examples in the theory of hydrodynamic stability and share many features. Linear theory describing the onset of instability for both problems is practically complete; nonlinear problems have been at the forefront of research during the past 30 years. Professor Koschmieder describes the impressive progress that has been made in the theoretical and experimental investigation of the nonlinear problems and outlines the remaining basic problems.

CAMBRIDGE MONOGRAPHS ON
MECHANICS AND APPLIED MATHEMATICS

General Editors

G. K. BATCHELOR
Emeritus Professor of Applied Mathematics, University of Cambridge

L. B. FREUND
Professor of Engineering, Brown University

BÉNARD CELLS AND TAYLOR VORTICES

Bénard Cells and Taylor Vortices

E. L. KOSCHMIEDER

University of Texas at Austin

CAMBRIDGE
UNIVERSITY PRESS

Published by the Press Syndicate of the University of Cambridge
The Pitt Building, Trumpington Street, Cambridge CB2 1RP
40 West 20th Street, New York, NY 10011-4211, USA
10 Stamford Road, Oakleigh, Victoria 3166, Australia

First published 1993

Printed in the United States of America

Library of Congress Cataloging-in-Publication Data
Koschmieder, E. L.
Bénard cells and Taylor vortices / E. L. Koschmieder.
  p.   cm. – (Cambridge monographs on mechanics and mathematics)
Includes bibliographical references (p.   ) and index.
ISBN 0-521-40204-2
1. Bénard cells. 2. Rayleigh–Bénard convection. 3. Taylor
vortices. 4. Nonlinear theories. I. Title. II. Series.
QC330.2.K68   1992
536′.25 – dc20   92-11009
                                                              CIP

A catalog record for this book is available from the British Library.

ISBN 0-521-40204-2 hardback

For Kate,
Thomas, and Stefan

CONTENTS

PREFACE

The motivation to write this book originates from my own experience. When I started to work on Bénard convection about 30 years ago, it was possible to learn practically all that was known about convection caused by heating from below from Chapter II of Chandrasekhar's (1961) book. And when I became interested in the Taylor vortex instability, I found again practically everything concerning the linear theory of Taylor vortex flow in Chapter VII of Chandrasekhar's book. It was most helpful to have a summary of the work that had been done before on both topics. Today, 30 years after Chandrasekhar's book appeared, there are hundreds of papers in the literature concerned with Rayleigh–Bénard convection and Taylor vortices. A newcomer to the field will find it difficult to sort out these papers and to learn what progress has been made since 1960. Now, at a time when nonlinear problems dominate the field, theory is no longer as tidy as it was 30 years ago, nor are the experiments as simple as they were then. It is not easy to find out where we actually stand in the pursuit of the Bénard problem and the Taylor vortex instability. Therefore it seems desirable to summarize in a book of reasonable length the progress that has been made since 1960. An encyclopedic listing of all papers with some added comments would not be helpful and would soon wear down the interest of the reader. There must be some system in a presentation of hundreds of publications. What the reader will find in this book are the essential facts and problems concerning Bénard convection and Taylor vortex flow.

How, then, has it been established what is essential? I have deemed essential everything that has been verified. Verification is the hallmark of modern science and seems to be one of the prime causes of the extraordinary success of science since the time dogma was ruled out. Specifically, verification means that the validity of theories is tested by experiments. If the results of a theory do not agree with the experimental results, the theory is invalid.

Experiments are, of course, verified too. Experiments have to reproduce satisfactorily the results of accepted theories, or if a relevant theory does not yet exist, then the results of an experiment have to be corroborated by other, independent experiments. What the reader will find in this book is a comparison of the theoretical and experimental results concerning the two topics of interest. The self-imposed limitation of approximately three hundred pages does not permit formal derivations of the theoretical results or descriptions of experimental methods. Yet it is always stated from which equations a theoretical result originates, and the basic features of the experiments are described as well. The emphasis is on the results, not on the methods. I have not hesitated to point out problems. I do not think that it is good to pretend that we have solved all problems in convection or Taylor vortex flow. As a matter of fact, there are a number of problems whose solutions are long overdue, and there are problems that invite exploration for the first time. These problems are not easy, but therein lies the challenge.

I acknowledge with gratitude the help that I have received in writing this book. I am particularly indebted to Professor G. K. Batchelor, Professor A. Davey, Professor S. H. Davis, Professor G. M. Homsy, and Professor R. S. Schechter, who have read parts or chapters of the manuscript and helped me with constructive criticism. I am grateful to the many colleagues who have permitted me to reproduce figures from their publications. It is said that a good figure is worth a thousand words. All the figures in this book amount then to about as much as I have had to say. I am also grateful to the publishers that hold the copyright for these figures for their permission to reproduce them. I thank many other colleagues who have helped me with information about their work. I am grateful to my department for granting me the H. M. Alharthy Centennial Professorship to work for a semester uninterruptedly on the completion of the manuscript. And I thank Dr. T. H. Koschmieder for insisting that I learn to avail myself of a Macintosh personal computer in writing the manuscript. It was wonderful advice.

Part I

BÉNARD CONVECTION AND RAYLEIGH–BÉNARD CONVECTION

Bénard convection, or its modern offspring, called Rayleigh–Bénard convection, has been the subject of several reviews and has also been discussed in a number of books. The very first review of Bénard convection was written by Pellew and Southwell (1940) and served until 1960 as an excellent summary of the beginnings of the investigation of Bénard convection, in particular of the theoretical aspects of the problem. The first book in which a summary of the Bénard problem was presented was written by Chandrasekhar (1961). His book dealt with hydrodynamic and hydromagnetic stability in general, and Bénard convection and the Taylor vortex instability were described in all the detail known at that time. Chandrasekhar's book appeared when the first studies of nonlinear Bénard convection were being made and when the consequences of surface tension effects had just been outlined. The 44 references at the end of Chapter II of Chandrasekhar's book were a nearly complete list of the publications concerned with Bénard convection up to that time.

After three decades of research on the nonlinear aspects of Bénard convection, the number of relevant publications stands at more than 500, counting only papers whose principal topic is Bénard or Rayleigh–Bénard convection. Only about half of these papers are listed in the references at the end of this book. It is simply impossible and would be distracting to refer to each and every paper that deals with Bénard or Rayleigh–Bénard convection. However I have referred to any additional independent publication which supports a previously reported result, because that is an essential part of verification. I have, instead of going into innumerable detail, chosen to focus attention on what I consider to be the centerpiece of the Bénard and Rayleigh–Bénard convection problem. This is for me the instability of a Boussinesq fluid layer on an infinite plane of excellent thermal conductivity heated uniformly from below and cooled uniformly from above, or since the infinite plane is an abstraction, in containers of large aspect ratio. I stick very closely to the principal problem throughout Chapters 1–7, without pursuing many of the possible nuances which the framework of the principal problem still permits. Since science proceeds from the simple to the complex, the nuances will have to wait until we have reached agreement on the main problems. Only in Chapter 8 are variations of the main theme of the problem discussed; only four of them are examined, although actually there are many more. One of the possible items under miscellaneous topics was convection in porous media. This topic has just been covered in a book by Nield and Bejan (1991). This shows that, if we pursued in detail all topics related to convection, we would have to write a couple of volumes. Of the four variations of the main theme that we cover, two deal with obvious questions resulting from the simplifying assumptions usually made in the theory of convection; the other two variations deal with problems for which linear theories exist, and these theories have been discussed extensively in Chandrasekhar's book. It is, of course, also challenging to discuss topics of very recent interest, for example, binary convection. But I have abstained from discussing binary convection because I lack personal experience in this field and because I believe that independent work remains to be done to verify many of the results presented in this field.

I have focused attention on those papers which contribute most to the clarification of the principal problems. There are, naturally, conflicting views about which papers are convincing and which are not. Care has been taken to provide readers with sufficient sources so that they can sort out things by themselves and form their own opinions. It very quickly becomes evident that it is much easier to write about convincing results. They can be dealt with in much less space than controversial results, which require inordinate explanations because the criticism has to be justified. This is, of course, not the desired order of things, but is in line with the observation that problems of seemingly insurmountable difficulty, once they are solved, afterward appear to be simple. In the references preference has been given to regular journal articles because they are readily available in most university libraries. If, as is now quite common, the results of a particular study have been published piecemeal, as reports, letters, short communications, conference contributions, preliminary results, preprints, etc., only the most comprehensive of these papers has been quoted. Publications which are not in regular journals have been referred to only in exceptional cases. The latest papers referred to are from the printed material available to me at the end of December 1991.

Since 1960 several reviews on convection have been written, as have a number of books in which convection is discussed. The reviews most often used are those of Segel (1966) about early nonlinear theories, of Koschmieder (1974) about early nonlinear experiments, of Palm (1975) about nonlinear theory, of Normand et al. (1977) about a physicist's approach to convection, of Busse (1978) about nonlinear properties of convection on an infinite plane, and of Davis (1987) about thermocapillary instabilities. Books which have chapters on Bénard convection or deal with hydrodynamic stability and refer to convection are those of Saltzman (1962a), which contains a very practical collection of the classical papers on the theory of convection, of Turner (1973) about buoyancy effects in general, of Gershuni and Zhukovitskii (1976) about convection in general, of Joseph (1976) about the application of the energy theory to instability, of Drazin and Reid (1981) about hydrodynamic stability with an extensive discussion of nonlinear stability theory, of Legros and Platten (1983) about the application of numerical techniques to convection, and of Georgescu (1985) about the mathematical aspects of stability theory. Various mathematical and experimental aspects of convection and Taylor vortex flow are discussed in Barenblatt et al. (1983) in the context of the formation of turbulence.

Finally I would like to define the term ''convection,'' which is used so often in this book. Convection in general means fluid motions caused by temperature differences with the temperature gradient pointing in any direction. In Bénard convection and Rayleigh–Bénard convection the temperature differences are applied in the vertical direction. Instead of using the term ''convection caused by heating from below'' time and again, we shall, for the sake of brevity, just use the word ''convection.''

1

BÉNARD'S EXPERIMENTS

Convective motions in shallow fluid layers heated from below or cooled from above by evaporation occur naturally under many circumstances. Various kinds of material floating in the fluid frequently make the flow visible. Convection in shallow fluid layers must therefore have been noticed many times before 1900, and was described by several observers. These early observations are only of historical interest because they were not scientific in the sense that they did not take place under controlled conditions and were not made reproducible. It seems that an accidental observation made in 1897 by A. Guébhard, namely the observation of polygonal vortex motions in an abandoned bath of film developer, induced Bénard* to take a thorough look at convection. Bénard then performed the first systematic investigation of convection in a shallow fluid layer heated from below. The results of his studies, actually the results of his doctoral thesis, were published in the paper "Les tourbillons cellulaires dans une nappe liquide" (Bénard, 1900) and a subsequent paper (Bénard, 1901). Both papers cover pretty much the same ground. Bénard's work is usually dealt with in the form of a reference; a thorough review of his studies is not available. So, as a matter of fairness, and in order to provide a basis for Chapter 2, we shall describe his experiments here. Several results of his experiments are easy to understand with hindsight. Therefore in the following we shall often assume knowledge of basic concepts of the theoretical explanation of convection.

Bénard went at his experiments methodically. From earlier measurements of the thermal conductivity of fluids he knew that a fluid layer heated from above remains in hydrostatic equilibrium, i.e. at rest. Bénard, on the other hand, set out to study convection by heating from below. He chose to investigate the "most simple case" with uniformity of the conditions in the hor-

---

* Henri Bénard, 1874–1939, professor at the University of Paris.

izontal plane. He realized uniform horizontal conditions in his experiments by reducing as much as possible the influence of the necessarily present lateral wall of the fluid. In other words, Bénard strove to work with an approximation of an infinite horizontal fluid layer. And in order to ensure the uniformity of the temperature at the bottom of the fluid he made the bottom plates out of good thermal conductors, brass or cast iron. The uniformity of the conditions seems to have been of particular importance for him, because he was interested in the consequences of, as we would say, infinitesimal disturbances of the temperature, or in his words, in "très léger excès fortuit et local de température." He wondered whether a center of ascending motion above such a disturbance would remain immobile or whether it would be replaced without regularity. He said that only the experiment could answer this question. By posing this question in the context of an infinite fluid layer he had not only laid the foundation for a successful experiment but also opened the door to a successful theoretical analysis of the problem.

Bénard chose a free surface for the fluid layer, which introduced, as he noted, an asymmetry of the (boundary) conditions, but there was not much choice if he wanted to observe the convective motions in full detail. In his papers there is a noncommittal reference to a special artifice which could be used as a rigid upper surface and which is said to have produced the same results as the free surface. Be that as it may, all results presented by Bénard were obviously obtained with the free surface. The surface of the fluid was cooled by ambient air, which means that the cooling was not really uniform, since the ambient air convects itself when heated by the warm fluid layer. The fluid used for the entire series of steady-state experiments was spermaceti, a whale oil, which is a waxy, rigid substance at room temperature, but melts at 46°C. Molten spermaceti is a nonvolatile, viscous fluid of poor thermal conductivity; its thermal conductivity is about a thousandth of the thermal conductivity of the bottom plates. That ensures the uniformity of the temperature at the bottom.

Bénard's apparatus, which was quite simple, is shown in Fig. 1.1. The depth of the fluid was of the order of 1 mm, which is for future considerations very significant. The fluid layer was circular and about 20 cm in diameter, so that the aspect ratio, i.e. the ratio of the diameter to the depth of the fluid, was 200, thereby approximating a layer of infinite horizontal extent. The bottom plate was either brass (0.5 cm thick) or cast iron (7.5 cm thick) and was heated uniformly from below by steam from boiling water. The temperature at the bottom of the fluid was therefore 100°C. Assuming that the temperature of the ambient air was 20°C, there was then a temperature difference across the fluid layer of about 80°C, which is

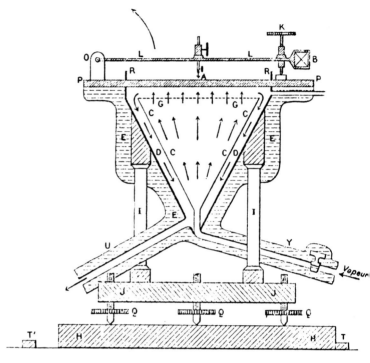

Fig. 1.1. Bénard's convection apparatus. After Bénard (1901).

much higher than the temperature difference usually applied in modern experiments. Since the depth of the fluid in Bénard's experiments was much smaller than the depth of the fluid layers in modern experiments, the temperature gradients across the fluid in Bénard's experiments were about a hundred times or more larger than the temperature gradients in most present-day convection experiments.

The initial conditions of Bénard's experiments are not spelled out. Heating of the fluid layer probably started by circulating steam underneath the bottom plate. That meant a very rapid increase of the bottom temperature to 100°C. This caused the appearance of a transient state of the fluid, which Bénard described but which we shall not consider. The transient state was followed by a steady state (''le régime permanent''), which produced the fundamental results of his experiments.

The fluid motions were made visible in three ways. Most often visualization was accomplished by the suspension of fine particles in the fluid. The particles were either aluminum or graphite powder. Occasionally Bénard

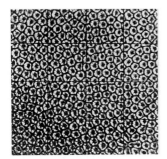

Fig. 1.2. Original photograph showing Bénard cells in a layer of spermaceti 0.810 mm deep. Visualization with graphite powder. Actual size. After Bénard (1900).

made the surface flow visible by fine floating particles, for which he used grains of lycopodium. He also applied various optical methods to make the flow visible; the optical method used most often is what we now call the shadowgraph technique.

The prime result of Bénard's experiments was the discovery of a stable, steady-state, regular pattern of hexagonal convection cells. He was obviously fascinated by the cells and believed that perfectly regular patterns existed. There are several figures in his papers showing patterns of hexagonal cells. Some of these patterns appear probably more regular than they were, because the outline of the cells was traced on photographs of shadowgraphs which seem to have been slightly out of focus. We reproduce in Fig. 1.2 a photograph which shows a less regular but perhaps more common cellular pattern. The cells in Fig. 1.2 are often polygonal; only a few are genuine hexagons. However assuming that one can realize perfect experimental conditions (which is not so easy), then one can believe in the existence of perfect patterns of hexagonal cells.

Investigation of the flow in the interior of the cells revealed that the fluid was ascending in the centers of the cells and descending along the hexagonal outline. Further optical measurements showed that the fluid surface was depressed over the centers of the cells. The depression was very small, of the order of $1 \mu m = 10^{-3}$ mm for a 1-mm-deep layer of spermaceti at 100°C. A section through a hexagonal cell with the circulation in the fluid and an exaggerated depression of the surface is shown in Fig. 1.3.

Bénard discussed possible reasons for the depression of the surface and mentioned in this context surface tension. He stated that ''surface tension by itself causes already a depression above the center of the cells'' (1901, p. 92). For future reference we note that Bénard also considered the variation of

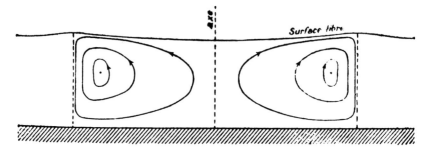

Fig. 1.3. Section through a hexagonal convection cell. The depression of the fluid surface
over the center of the cell is exaggerated by a factor of 100. After Bénard (1900).

surface tension with temperature, but not in the context of the cause of the
formation of the cells.

Bénard tried to determine the velocity of the flow in the interior of the
hexagonal cells, as well as to find a characteristic measure of the size of the
cells. The velocity measurements seem, from a modern point of view, to be
useless because it was not specified at which location in the cell the measurements of the period of the circulation of tracer particles was made. The
determination of the cell size, however, is still relevant because Bénard normalized the cell size with the depth of the fluid layer. As characteristic length
he used the distance between the centers of two neighboring cells, which he
designated by the letter λ. He searched for the laws of the variation of λ as a
function of the fluid depth, the heat flux, and the temperature.

Concerning the variation of the cell dimensions as a function of the depth
of the fluid, Bénard formulated a law according to which, in a first approximation, "the cell prisms remain geometrically similar, if the depth is varied"
(1900, p. 1321). In a formula, $\lambda/d = $ const. This is what, with hindsight, we
expect under critical conditions. Bénard added almost immediately after the
quoted law that "precise measurements reveal a systematic deviation" from
this law. He found that in spermaceti at 100°C bottom temperature, the ratio
of λ/d increased if the depth of the layer increased, namely from $\lambda/d = 3.378$
at $d = 0.440$ mm to $\lambda/d = 4.049$ at $d = 0.853$ mm. This observation is consistent with our present knowledge that the wavelength of convective motions
in a fluid layer between two rigid parallel plates increases with increased supercritical conditions. Since Bénard measured the cell sizes in the different
fluid layers usually at the same temperature difference (100°C at the bottom
minus the temperature of the ambient air), the Rayleigh numbers of the different fluid layers differ, being large for the deep layers and smaller for the

less deep layers. Our empirical knowledge from modern experiments tells us that the cell size of supercritical convective motions between two parallel plates is larger for large Rayleigh numbers (in this case for deep fluid layers) and smaller when the Rayleigh number is smaller (in this case in less deep layers), as Bénard observed.

Bénard could also vary the temperature difference across the fluid by permitting the big cast iron block on the bottom of the fluid to cool down. The effect of the variation of the temperature difference on the size of the cells is shown in various graphs, e.g. in Fig. 23 in Bénard (1900). It can be seen there that the cell size for one and the same fluid depth decreased overall as the temperature difference across the fluid decreased, in agreement with what we have since learned about the variation of the wavelength of convective motions as a function of the temperature difference. There was, however, a puzzling aspect of these cell size measurements. Bénard found that, near the point where the pattern disappeared, the cell size reached a minimum and increased a little when the temperature difference was decreased further, until the pattern disappeared. His observations of the variation of the cell size, in particular the puzzling minimum cell size, were correct but were confirmed only 90 years later.

The observations of the disappearance of the pattern at small temperature differences bear on the question of whether Bénard had detected the existence of a critical temperature difference in his experiments. He observed for four different fluid depths that the ratio λ/d was very nearly the same ($\lambda/d = 3.584 \pm 0.039$) whenever the cell pattern disappeared. Disappearance of the pattern means in modern terms that the layer had cooled down to the critical temperature difference. We know now that at the critical temperature difference the pattern has a unique wavelength $(\lambda/d)_c$, which is independent of the depth of the fluid layer. This is just what Bénard saw. But he did not realize that this was the consequence of a critical temperature difference, a concept introduced later by Rayleigh (1916a). Bénard attributed the disappearance of the pattern to the beginning solidification of the spermaceti. Much later on, Bénard (1930) was very skeptical about the concept of a critical temperature difference, although he had actually observed it when the pattern disappeared in his experiments.

All in all, Bénard's experiments provided a good basis for further experimental and theoretical work. His discovery of the hexagonal convection cells has been a lasting contribution to science. It is an irony of fate that, in spite of the quality of his work, his experiments created a monumental misconception by focusing the interest primarily on the hexagonal cells, which have, as was learned only 50 years later, little to do with buoyancy.

However at that time heating from below appeared to be the cause of the formation of the cells.

In the conclusion of his first paper Bénard referred to the interest that the hexagonal cells had found among the "naturalists." This interest has continued until today. However the obvious similarity between Bénard cells and living cells has yet to prove to be a fruitful connection. In view of the complexity of living cells and the difficulties which had to be overcome in order to explain even the "simple" Bénard cells, it is not surprising that a real connection between both phenomena has not yet been made. Practically the same can be said about the importance of convection for the "atmospheric circulation," to which Bénard refers in the introduction of his first paper.

We shall now see how the results of Bénard's experiments have been explained, and what else we have learned about convection in a fluid layer heated from below.

2

LINEAR THEORY OF RAYLEIGH–BÉNARD CONVECTION

2.1 Rayleigh's Work

Rayleigh* wrote his paper "On Convection Currents in a Horizontal Layer of Fluid, When the Higher Temperature Is on the Under Side" near the end of his brilliant career in science. In the first sentence of this paper (Rayleigh, 1916a) he refers to the "interesting results obtained by Bénard." Rayleigh's explanation of the essence of Bénard's experiments was guided by the experience he had gained from his studies of the stability of a liquid jet breaking up into droplets under the action of surface tension, a problem Rayleigh had solved theoretically in 1879. Rayleigh treated Bénard's problem similarly as a stability problem, looking for modes and their growth rates. He began his analysis with Euler's equation and the equation of thermal conduction, neglecting nonlinear terms and using the Boussinesq approximation, which permits, as he stated, that "even the variation with temperature may be disregarded except in so far as it modifies the operation of gravity" (1916a, p. 533). It was of great practical importance that he neglected the nonlinear terms, because only in this way was it possible to solve the problem analytically. Rayleigh considered a fluid contained between two infinite, plane, parallel, horizontal plates where the temperatures are maintained constant. His boundary conditions were that the vertical velocity component w and the temperature disturbance θ had to vanish at the top and bottom plates. In modern terminology that means that both boundaries were considered free surfaces and perfect thermal conductors. Using free boundary conditions was also a mathematical necessity, but seemed to be justified by the fact that the upper surface of the fluid in Bénard's experiments was in contact with air,

---

* Lord Rayleigh (John William Strutt), 1842–1919, professor at Cambridge University, chancellor of Cambridge University, 1904 Nobel laureate.

which seemed to make the surface free. However, by choosing a free boundary condition Rayleigh had, unknowingly, lost connection with Bénard's experiments because they were affected, in a crucial way, by surface tractions originating from surface tension gradients. Using the free boundary conditions, Rayleigh had introduced what we now call the Rayleigh–Bénard problem, which refers to convection caused by heating from below which is *not* affected by surface tension gradients.

Viscosity, an essential ingredient of the convection problem, was introduced only halfway through the paper. This was probably so because Rayleigh expected the fluid always to be unstable when the higher temperature is on the underside, in other words, when warm light fluid is under cold heavy fluid. As Rayleigh put it: "When the fluid is inviscid and the higher temperature is below, all modes of disturbance are instable, even when we include the conduction of heat during the disturbance" (1916a, p. 532). The consequence of the introduction of viscosity was the discovery, "a little unexpectedly," of what we now call the critical temperature difference, the minimal vertical temperature difference required for the onset of instability. At that temperature difference the buoyancy of the fluid is able to overcome the dissipation caused by the viscosity and the thermal diffusivity of the fluid. The discovery of the existence of the critical temperature difference was a fundamental step for the understanding of Rayleigh–Bénard convection, a result going beyond what Bénard had reported. The discovery of the critical temperature difference was also a fundamental step for stability theory in general.

Only the last two pages of Rayleigh's paper are devoted to the patterns of the convective motions. He showed that the flow in the cells is described by the solutions of the two-dimensional wave equation or, as he called it, the membrane equation. Rayleigh considered square cells, equilateral triangular cells, and regular hexagonal cells. He seems to have overlooked the possibility of the existence of cells in the form of two-dimensional straight, parallel rolls, although he had referred in the introduction of his paper to "infinitely long strips," which he had noted in the initial phases of Bénard's experiments. Rayleigh showed what the analytical solution of the membrane equation is for the case of the square cells. Since he did not know the analytical solutions of the membrane equation for either the equilateral triangle or the hexagons, he did not pursue these topics other than to discuss qualitatively an approximation of the hexagonal cells by circular cells. We shall present an outline of the modern form of Rayleigh's theory in the following two sections.

2.2 The Onset of Convection

Rayleigh's paper is the pioneering work on the theory of convection caused by heating from below. It has been said, with good reason, that the many subsequent theoretical studies of convection are variations on a theme of Rayleigh. The theme was first picked up by Jeffreys (1926, 1928), who considered rigid–rigid and rigid–free boundary conditions and obtained values of the critical Rayleigh numbers for these boundary conditions. Convective motions in a fluid layer heated from below were reviewed for the first time (referring to nine papers) and brought into sharper focus by Pellew and Southwell (1940). The topic was then summarized in the second chapter of Chandrasekhar's (1961) book, where a nearly complete picture of the linear analysis of what is now called Rayleigh–Bénard convection is presented. Rayleigh–Bénard convection can, in practice, be realized only with rigid–rigid boundary conditions, but in theory surface-traction-free surfaces are also considered for rigid–free and free–free boundaries, although the latter case is obviously academic.

Before we develop the theory describing Rayleigh–Bénard convection, we should outline the problem at hand. We are going to investigate the stability of a fluid layer of infinite horizontal extent contained between two plane parallel horizontal boundaries, the lower one of which is heated uniformly, whereas the upper one of which is cooled uniformly.

The analysis of convection caused by heating from below is based on the equations for conservation of mass, momentum, and energy, and on the equation of state of the fluid. The continuity equation is

$$\frac{\partial \rho}{\partial t} + \nabla \cdot (\rho \mathbf{v}) = 0, \tag{2.1}$$

the Navier–Stokes equation is

$$\rho \frac{d\mathbf{v}}{dt} = -\nabla p + \rho \mathbf{g} + \mu \nabla^2 \mathbf{v}, \tag{2.2}$$

and the equation of thermal conduction, neglecting the contribution of dissipation occurring in the viscous fluid, is

$$\frac{dT}{dt} = \kappa \nabla^2 T, \tag{2.3}$$

where ρ is the density of the fluid, $\mathbf{g}$ the acceleration of gravity, μ the dynamic viscosity, and κ the thermal diffusivity of the fluid. The equation of state of the fluid is

$$\rho = \rho_0(1 + \alpha \, \Delta T), \tag{2.4}$$

where $\alpha = 1/\rho_0 \, \partial\rho/\partial T$ is the volume expansion coefficient of the fluid. Since the density of fluids usually decreases with temperature, the expansion coefficient is usually negative. Now the so-called Boussinesq approximation (Boussinesq, 1903) is introduced, which was actually first formulated by Oberbeck (1879). A brief review of the history of the Boussinesq approximation can be found on p. 4 of vol. 2 of Joseph's (1976) book. The implications of the Boussinesq approximation have been studied by Spiegel and Veronis (1960), Mihaljan (1962), Pérez-Cordón and Velarde (1975), and Gray and Giorgini (1976). It suffices to state that the Boussinesq approximation means for our topic that the variation with temperature of the material properties of the fluid such as the kinematic viscosity, the thermal diffusivity, and the volume expansion coefficient, are negligible provided that the temperature variation is small. The density of the fluid is also considered constant, except of course for the variation of the density in the vertical direction by the imposed vertical temperature difference. The vertical density variation actually provides the driving mechanism of the convective motions.

We now look at the perturbation equations which describe the reaction of the fluid to small, or rather infinitesimal, disturbances of the basic steady state, in which the fluid is at rest. The fluid at rest is described by

$$\mathbf{v} = 0, \qquad \frac{d^2T}{dz^2} = 0, \qquad \frac{dp}{dz} = \rho_0 \, g[1 - \alpha(T(z) - T_0)] \, ,$$

where ρ_0 is the density of the fluid at the temperature T_0. A fluid at rest is the most simple basic state imaginable and should be realizable with near perfection and is therefore ideally suited for a study of the consequences of infinitesimal disturbances of an initial state. Since the perturbations of the basic state are considered to be infinitesimal, we can *linearize* equations (2.1)–(2.3) because the velocity of the motions caused by the infinitesimal disturbances will be so small that the squares of the velocity components, or the products of the velocity components, or the products of the velocity components with the temperature disturbance θ are negligible. That means, in mathematical terms, that we neglect $(\mathbf{v} \cdot \nabla)\mathbf{v}$ in the Navier–Stokes equation and $\mathbf{v} \cdot \nabla\theta$ in the energy equation. We also make the variables in equations (2.1)–(2.3) nondimensional by setting

$$x_i' = x_i/d, \qquad t' = \kappa t/d^2, \qquad u' = du/\kappa,$$
$$\theta' = \alpha g d^3 \theta/\nu\kappa, \qquad p' = pd^2/\rho_0\kappa^2,$$

where d is the depth of the fluid and θ an infinitesimal deviation of the temperature from the originally linear vertical temperature distribution in the fluid, brought about by the disturbances. The way θ is nondimensionalized anticipates the Rayleigh number, which we shall introduce shortly. There are different ways to nondimensionalize the variables; the choice made here seems to be the one preferred most often. For the rest of this section we shall drop the prime on the nondimensional variables and deal with nondimensional equations only.

The convection equations are then

$$\nabla \cdot \mathbf{v} = 0, \tag{2.5}$$

$$\frac{\partial \mathbf{v}}{\partial t} = -\nabla p + \mathscr{P}\theta\mathbf{e}_z + \mathscr{P}\nabla^2\mathbf{v}, \tag{2.6}$$

$$\frac{\partial \theta}{\partial t} = \nabla^2\theta + \mathscr{R}w, \tag{2.7}$$

where w is the vertical velocity component and $\mathbf{e}_z$ the unit vector in the vertical direction. $\mathscr{R}$ is the Rayleigh number given by

$$\mathscr{R} = \alpha g \, \Delta T d^3/\nu\kappa, \tag{2.8}$$

where ΔT is the temperature difference between the top and bottom of the fluid layer, ν the kinematic viscosity, and κ the thermal diffusivity. The Rayleigh number is a nondimensional measure of the vertical temperature difference applied to the fluid layer. Since the value of the volume expansion coefficient α is usually negative, the temperature difference ΔT must be negative in order to have a positive Rayleigh number $\mathscr{R}$. A negative ΔT means that the temperature decreases with height, which means that the fluid must be heated from below (or cooled from above). The name "Rayleigh number" was introduced by Sutton (1951). $\mathscr{P}$ is the Prandtl number given by

$$\mathscr{P} = \nu/\kappa. \tag{2.9}$$

From equations (2.5)–(2.7) we eliminate the pressure, the horizontal velocity components u and v, and the temperature θ (for details see Chandrasekhar, 1961), and obtain an equation for w only,

$$\left[\frac{\partial}{\partial t} - \nabla^2\right]\left[\frac{1}{\mathscr{P}}\frac{\partial}{\partial t} - \nabla^2\right]\nabla^2 w = \mathscr{R}\nabla_2^2 w, \tag{2.10}$$

where ∇_2^2 is the two-dimensional Laplace operator $\partial^2/\partial x^2 + \partial^2/\partial y^2$. The disturbance temperature θ satisfies the same equation as w.

Next we consider the boundary conditions. For free horizontal boundaries we require that the vertical velocity component w disappear at the boundaries, as well as the tangential stresses. We also require that the temperature disturbance θ disappear at a free horizontal boundary. This implies that the thermal conductivity in the medium adjacent to the fluid is excellent as compared with the thermal conductivity of the fluid itself, so that a temperature disturbance advected by the fluid is smoothed out by thermal conduction in the bounding medium. A free boundary is usually approximated by an air layer, but air is a very poor thermal conductor, so the boundary condition $\theta = 0$ is rather unrealistic for the so-called free surfaces. One can, of course, also treat analytically the case of finite conductivity at the boundaries, but we shall not pursue this topic other than by referring to Proctor (1981) for further reading. The boundary conditions imposed on a free surface are

$$\theta = 0, \qquad w = \frac{\partial^2 w}{\partial z^2} = 0 \qquad \text{(free)}. \tag{2.11}$$

On a rigid boundary the no-slip condition requires that not only w but also u and v disappear. From the continuity equation it follows that $\partial w/\partial z = 0$ at the boundary. We require again that the temperature disturbance disappears at the boundary, in other words that the medium adjacent to the fluid is a perfect conductor. This can be approximated very well with metallic boundaries. We have

$$\theta = 0, \qquad w = \frac{\partial w}{\partial z} = 0 \qquad \text{(rigid)}. \tag{2.12}$$

We now solve equation (2.10) with a normal mode solution for infinitesimal disturbances of the velocity of the form

$$w = W_0 W(z) e^{i(k_x x + k_y y)} e^{st} \tag{2.13}$$

and make a corresponding ansatz for the infinitesimal temperature disturbance:

$$\theta = \theta_0 \theta(z) e^{i(k_x x + k_y y)} e^{st}. \tag{2.14}$$

In (2.13) W_0 is the amplitude of the vertical velocity component, which by its sign also determines the direction of the flow. W_0 cannot be determined, because (2.10) is linear. $W(z)$ is the function which describes the variation with z of the vertical velocity component. k_x and k_y are the wave numbers in the x and y direction, respectively. The so-called growth rate s will be discussed later. The function

$$f(x,y) = e^{i(k_x x + k_y y)} \tag{2.15}$$

is periodic in x and y, as we must expect, because the cellular structure of the flow is periodic in the horizontal plane. The dimensionless wave numbers k_x and k_y determine the wave number a by

$$a^2 = k_x^2 + k_y^2. \tag{2.16}$$

We pause to consider the meaning of the wave number a. The fluid is certainly not perturbed by a systematic disturbance with a fixed wave number over the entire infinite plane of the fluid. Rather the disturbance with wave number a refers to one Fourier component of the spectrum of Fourier components of the actually present disturbances. The smallest disturbances one can hope for in a convection apparatus originate from white thermal noise, which will, however, always be smaller than the disturbances caused by other, ever so small imperfections of the apparatus.

From (2.15) it follows that

$$\nabla_2^2 f(x,y) + a^2 f(x,y) = 0. \tag{2.17}$$

This is the two-dimensional wave equation, or the membrane equation, as we shall refer to it following Rayleigh's example. The membrane equation determines the cellular structure of the fluid motion and will be discussed in detail in Section 2.3. The horizontal velocity components of the flow follow from the equations

$$u = \frac{1}{a^2}\frac{\partial^2 w}{\partial x\, \partial z}, \qquad v = \frac{1}{a^2}\frac{\partial^2 w}{\partial y\, \partial z}. \tag{2.18}$$

The term e^{st} in (2.13) describes the amplification of the velocity with time by the growth rate s. We shall postpone a discussion of the growth rate to the end of this section. Instead we shall now proceed with the case $s = 0$, i.e. a flow without amplification or dampening, or in mathematical terms the time-independent case of (2.10), which is referred to as the *marginal state* or the neutral state.

Equation (2.10) then reduces to

$$\nabla^6 w = \mathcal{R}\nabla_2^2 w. \tag{2.19a}$$

With the help of the membrane equation this can be easily transformed into

$$(D^2 - a^2)^3 w = -a^2 \mathfrak{R} w, \tag{2.19b}$$

with the nondimensional operator $D = d/dz$ and the wave number a.

The six boundary conditions for free–free surfaces are

$$w = \frac{\partial^2 w}{\partial z^2} = \frac{\partial^4 w}{\partial z^4} = 0 \tag{2.20}$$

at the top and bottom boundaries. The condition $\partial^4 w/\partial z^4 = 0$ is a consequence of the thermal boundary condition $\theta = 0$ at both boundaries. Equation (2.19b) is solved making the assumption that $W(z)$ in (2.13) is of the form

$$W(z) = \sin n\pi z, \tag{2.21}$$

where n is an integer number. It follows immediately that the eigenvalues are

$$\mathfrak{R}_n = \frac{(n^2\pi^2 + a^2)^3}{a^2}. \tag{2.22}$$

Equation (2.22) is, on a graph of $\mathfrak{R}$ versus the wave number a, represented by curves which have, for each value of n, a minimum which can be obtained from (2.22) by differentiation with respect to a. A graph $\mathfrak{R}$ versus a for rigid–rigid boundary conditions and $n = 1$ will be shown later in Fig. 2.1. The minima of (2.22) are the critical Rayleigh numbers given by

$$\mathfrak{R}_c = 27(n\pi)^4/4 \qquad \text{(free–free)}. \tag{2.23}$$

The so-called critical wave numbers a_c, which are the values of a at which the minima of the curves $\mathfrak{R}(a)$ occur, are given by

$$a_c = n\pi/\sqrt{2} \qquad \text{(free–free)}. \tag{2.24}$$

For the lowest mode ($n = 1$) we find

$$\mathfrak{R}_c = 27\pi^4/4 = 657.5 \qquad \text{(free–free)}, \tag{2.25}$$

$$a_c = \pi/\sqrt{2} = 2.221 \qquad \text{(free–free)}. \tag{2.26}$$

A different way to describe the horizontal periodicity of the flow is via the so-called wavelength, which is often more lucid than the wave number, because the wavelength can be seen with the naked eye. The dimensionless wavelength follows from the wave number through

$$\lambda = 2\pi/a \tag{2.27}$$

and is the horizontal distance between two consecutive maxima (or minima) of the vertical velocity w, i.e. the distance between two roll boundaries, divided by the fluid depth (see Fig. 2.2). In other words, the wavelength is a nondimensional measure of the cell size.

The critical Rayleigh number determines, via the definition of the Rayleigh number (2.8), the critical temperature difference ΔT_c required for a particular fluid layer perturbed with disturbances of wave number a to be in the marginal state, which means as much as being just on the verge of instability. At ΔT just above ΔT_c the entire infinite fluid layer should change spontaneously from the state of rest to cellular convective motions, if the temperatures on the top and bottom of the layer are indeed uniform.

Equations (2.25) and (2.26) express the two fundamental results of linear theory of Rayleigh–Bénard convection in the free–free case. The first is the fact that in a viscous medium with thermal conductivity the onset of convection does not take place at any arbitrarily small negative vertical temperature difference but rather that due to the action of viscous dampening and thermal dissipation a minimal vertical temperature difference has to be exceeded in order to commence convection. The second point is that the onset of convection at $\mathcal{R}_c$ takes place in the form of a pattern which is characterized by a specific wave number a_c or wavelength λ_c, a fundamental aspect of the formation of the pattern. We shall learn from the so-called balance theorem why the fluid picks just the critical wavelength.

The onset of convection in a fluid layer with rigid–rigid boundaries is likewise determined by (2.19b). The boundary conditions on the top and bottom of the fluid are then

$$w = \frac{\partial w}{\partial z} = \frac{\partial^4 w}{\partial z^4} - 2a^2 \frac{\partial^2 w}{\partial z^2} = 0. \tag{2.28}$$

Equation (2.19b) has been solved with these boundary conditions by several authors; the most thorough study is that of Reid and Harris (1958). The eigenvalues follow from a transcendental equation, a first approximation of which is given, according to Chandrasekhar (1961, p. 58), by

$$\mathcal{R} = \frac{(\pi^2 + a^2)^3}{a^2\{1 - 16a\pi^2 \cosh^2(a/2) \,/\, [(\pi^2 + a^2)^2(\sinh a + a)]\}} . \tag{2.29}$$

The marginal or neutral curve is plotted in Fig. 2.1. This curve is based on a short table of values of $\mathcal{R}$ and a in Reid and Harris (1958), which can also be found in Chandrasekhar (1961).

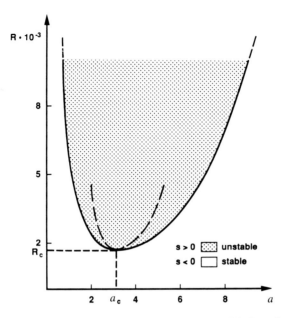

Fig. 2.1. Stability diagram of linear theory for Rayleigh–Bénard convection in a fluid layer
of infinite horizontal extent with rigid–rigid boundaries. The dashed line marks
the boundary of the so-called Eckhaus instability.

The minimum in Fig. 2.1 determines the critical Rayleigh number and the unique critical wave number. There is, expressed in mathematical terminology, a supercritical bifurcation at $\mathfrak{R}_c$. Note also that, according to linear theory, supercritical flow $(\mathfrak{R} > \mathfrak{R}_c)$ is *nonunique*. According to linear theory a *continuum* of unstable wave numbers exists for any supercritical Rayleigh number. This feature is of major importance for supercritical convection. The Eckhaus instability indicated by the dashed line in Fig. 2.1 will be discussed in Section 7.1. The lowest Rayleigh number of the first so-called odd mode (odd with respect to the midlayer), for which $n = 2$, is at $\mathfrak{R} = 17,610$ and at a wave number $a = 5.365$. This Rayleigh number is far above the range of Rayleigh numbers in Fig. 2.1. We shall see later that the higher modes are not realistic.

The critical Rayleigh number, the critical wave number, and the critical wavelength for the three possible combinations of the horizontal boundary conditions are listed in Table 2.1 for the lowest mode $n = 1$. As far as the wavelength is concerned, this means that in the rigid–rigid case a single convection roll is almost as wide horizontally as the fluid is deep, whereas in the free–free case a roll is flatter, being 1.41 times as wide as it is deep.

Table 2.1. *Critical parameters for the onset of convection*

Boundaries	$\mathcal{R}_c$	a_c	λ_c
Rigid–rigid	1707.8	3.117	2.016
Rigid–free	1100.7	2.682	2.342
Free–free	657.5	2.221	2.828

We shall now turn our attention to the growth rate of the disturbances. The growth rate s was introduced with the normal mode solution (2.13) and determines the rate of increase with time of an infinitesimal disturbance of the velocity field or the temperature field in the fluid. The value of the nondimensional growth rate s follows from equation (2.10), into which we insert (2.13), using only the lowest mode of the vertical velocity distribution (2.21), because $n = 1$ gives the minimal value of the critical Rayleigh number. We obtain a quadratic equation for s,

$$F(F + s) (F + s/\mathcal{P}) = a^2\mathcal{R}, \tag{2.30}$$

where $F = a^2 + \pi^2$. Equation (2.30) shows that the growth rate is a function of the wave number of the disturbance, of the Rayleigh number, and of the Prandtl number of the fluid. From (2.30) it follows that

$$2s = -(1 + \mathcal{P})F \pm [(1 - \mathcal{P})^2F^2 + 4a^2\mathcal{P} \; \mathcal{R}/F]^{1/2}. \tag{2.31a}$$

For slightly supercritical conditions, where s is small and s^2 can be neglected, we find for the growth rate in the free–free case

$$s = \frac{(a^2 + \pi^2) (\mathcal{R} - \mathcal{R}_c)}{(1 + 1/\mathcal{P})\mathcal{R}_c} . \tag{2.31b}$$

From (2.31a) it follows immediately that s is real if $\mathcal{R} > 0$. That means that the amplitude of the velocity is a monotonic function of time. The case of a complex s is called overstability; the amplitude still increases exponentially with time, but the increase varies periodically. Overstability occurs in the theory of convection with rotation. General proof that s is real in resting fluid layers will be provided soon. It also follows after some algebra from (2.31a) that $s > 0$ if $\mathcal{R} > \mathcal{R}_c$; i.e. the disturbances amplify exponentially if the Rayleigh number is larger than critical, meaning that the fluid is unstable. If for some values of $\mathcal{R}$ (or a) the value of s is negative, then the disturbances are dampened exponentially and will decay. The fluid is then stable in agreement with the definition of asymptotic stability, which requires that

$$|v(t) - v'(t)| \to 0 \qquad \text{for } t \to \infty, \tag{2.32}$$

where $v'(t)$ is the disturbed velocity.

The growth rates for a given set of parameters a, $\mathcal{R}$, $\mathcal{P}$ can be calculated from (2.31a). The growth rate as a function of a given $\mathcal{R}$ and $\mathcal{P}$ has a maximum at a particular wave number a_m. The maximum growth rate shifts to larger a_m as $\mathcal{R}$ increases. It was believed for some time that the maximum growth rate might provide a clue to the question of which wave number the flow would utilize at supercritical Rayleigh numbers. Experiments show that steady supercritical wave numbers decrease with increased $\mathcal{R}$, contradicting the expectation that the flow would choose the wave number with the maximum growth rate. The reason for this discrepancy seems to be that the concept of an exponential growth of the disturbances in Rayleigh–Bénard convection is unrealistic. The disturbances cannot grow ad infinitum; rather they must reach an equilibrium. The exponential form for the increase of the amplitude in equation (2.13) must be changed, as is done in the Landau equation to be discussed later.

The general proof that the growth rate s is real in the marginal state is provided as follows: One starts with the perturbation equation (2.10), which we write in the form

$$(D^2 - a^2 - s)F = -a^2 \mathcal{R} W(z), \tag{2.33}$$

using the operator $D = d/dz$ and the relation $\nabla^2 = D^2 - a^2$, which is obtained with the membrane equation (2.17). The function F is given by

$$F = (D^2 - a^2 - s/\mathcal{P})(D^2 - a^2)W(z).$$

We multiply (2.33) by the complex conjugate F^* and integrate over z. The result is

$$\int_0^1 F^*(D^2 - a^2 - s)F \, dz = -a^2 \mathcal{R} \int_0^1 F^*W(z) \, dz. \tag{2.34}$$

Integration by parts and the boundary conditions $F = 0$ at $z = 0$ and $z = 1$ yields for the left hand side (lhs) of equation (2.34)

$$\text{lhs} = \int_0^1 [|DF|^2 + (a^2 + s)\,|F|^2]\, dz. \tag{2.35}$$

The integral on the right hand side (rhs) of (2.34) is

$$\int_0^1 F^*W(z)\, dz = \int_0^1 G^*(D^2 - a^2 - s/\mathcal{P})W(z)\, dz, \tag{2.36}$$

with $G^* = (D^2 - a^2)W^*(z)$. Integration by parts together with the boundary conditions reduces this to

$$\text{rhs} = \int_0^1 [|G|^2 + s^*(|DW|^2 + a^2|W|^2)/\mathcal{P}]\, dz. \tag{2.37}$$

Since the real and imaginary parts of the difference between (2.35) and (2.37) must vanish separately, it follows for the imaginary part of s that

$$\text{im}(s)\left[\int_0^1 |F|^2\, dz + \frac{a^2\mathcal{R}}{\mathcal{P}}\int_0^1 [|DW|^2 + a^2|W|^2]\, dz\right] = 0. \tag{2.38}$$

Since all the terms under the integrals are positive, it follows that the growth rate s is real because

$$\text{im}(s) = 0, \tag{2.39}$$

if $\mathcal{R} > 0$. The principle that the onset of convection in the marginal state takes place with monotonically increasing amplitude carries the awkward name "the principle of the exchange of stabilities." It was derived first by Pellew and Southwell (1940); we have followed the presentation of Chandrasekhar (1961).

We shall finally discuss a theorem introduced by Chandrasekhar (1961) which we shall call the *balance theorem*. This theorem is important for understanding the value of the wavelength under critical, as well as supercritical, conditions. Suppose that the fluid is in a state of steady convection just beyond marginal stability with velocities u_i. We make the assumptions that the averages in the horizontal plane of the velocities u_i, the temperature disturbances δT caused by the convective motions, and the pressure disturbances δp are zero, i.e.

$$\langle u_i \rangle = 0, \qquad \langle \delta T \rangle = 0, \qquad \langle \delta p \rangle = 0. \tag{2.40}$$

The dimensional steady-state Navier–Stokes equations are in this case

$$0 = -\frac{1}{\rho}\frac{\partial p}{\partial x_i} - g\{1 - \alpha(T(z) - T_0 + \delta T)\} e_i + \nu \nabla^2 u_i. \quad (2.41)$$

We multiply (2.41) by u_i, average over the horizontal plane, integrate over z, and obtain

$$g\alpha \int_0^d \langle \delta T w \rangle \, dz + \nu \int_0^d \langle u_i \nabla^2 u_i \rangle \, dz = 0. \quad (2.42)$$

The average rate of viscous dissipation by a unit column of fluid is

$$E_\nu = -\rho\nu \int_0^d [\langle u\nabla^2 u \rangle + \langle v\nabla^2 v \rangle + \langle w\nabla^2 w \rangle] \, dz, \quad (2.43)$$

and the average rate of energy release by buoyancy per unit column is given by

$$E_b = \rho g\alpha \int_0^d \langle \delta T w \rangle \, dz. \quad (2.44)$$

With these two relations it follows from (2.42) that (in convective motions under the stated assumptions) the kinetic energy dissipated by viscosity is *balanced* by the internal energy released by buoyancy,

$$E_\nu = E_b. \quad (2.45)$$

We shall refer to equation (2.45) as the balance theorem, which holds in the nonlinear regime under the stated conditions. The balance theorem means that, to quote Chandrasekhar, "instability occurs at the minimum temperature gradient at which a balance can be steadily maintained between the kinetic energy dissipated by viscosity and the internal energy released by the buoyancy force" (1961, p. 34).

From very general thermodynamic stability considerations Glansdorff and Prigogine (1971) arrived at a similar balance principle according to which "instability occurs at the minimum temperature gradient at which a balance can be steadily maintained between the entropy generated through heat conduction by the temperature fluctuations and the corresponding entropy flow carried away by the velocity fluctuations'' (p. 159). The two mentioned variational principles have seen little use in later studies in spite of their heuristic value. Minimum entropy production was used, among other extremum assumptions, by Inoue and Ito (1984) in order to calculate the supercritical wave numbers of the convective motions, but the wave numbers found were larger than the critical wave number, in contradiction to the experimental findings.

From the balance theorem we understand qualitatively why the critical wavelengths are different when the boundary conditions are either rigid–rigid, rigid–free, or free–free, as follows from Table 2.1. It has been suggested by Davis (1967) that the differences in the size of the rolls in the three cases can be explained with the balance theorem. Narrow, tall cells dissipate more energy than wide, flat cells. The onset of convection in the free–free case takes place at the smallest critical temperature difference of the three cases. Therefore the minimum amount of energy is released by buoyancy in the free–free case. The small amount of energy released must be balanced by cells with little dissipation, i.e. by wide, flat cells, in agreement with a large critical wavelength ($\lambda_c = 2.83$) in the free–free case. The onset of convection in the rigid–rigid case occurs at a much higher critical temperature difference. Therefore more energy is released by buoyancy, which must be balanced by increased dissipation, i.e. by narrower cells. The wavelength in the rigid–rigid case must therefore be shorter than in the free–free case, and indeed $\lambda_c = 2.016$ (rigid–rigid), as compared with $\lambda_c = 2.83$ (free–free).

2.3 The Cellular Patterns

We shall now be concerned with the solutions of the membrane equation (2.17) which determine the various patterns of flow of the convective motions. We note first that the eigenvalues of (2.10), which determine the onset of convection, are degenerate; this means that to the eigenvalue $\mathcal{R}_c$ belongs an infinite number of patterns which all solve the membrane equation with the same wave number a_c. Degeneracy is the price we pay for using the simplifying assumption that the fluid layer is of infinite horizontal extent. If we would consider laterally bounded fluid layers, degeneracy would not occur, but the problem would be much more difficult and less general. Linear theory of convection on an infinite horizontal plane does not provide a criterion according to which a particular pattern is selected by a fluid which is heated with mathematical uniformity from below and cooled with the same uniformity from above. There is, according to the input into the theory, no preferred direction, nor are there any boundaries which could provide an orientation for the fluid. The reasons for which a particular pattern is actually selected by the fluid are still inadequately understood or are in dispute. The obvious preference for the hexagonal pattern in Bénard's experiments is not relevant to the Rayleigh–Bénard convection problem considered here, because the hexagonal Bénard cells are primarily a consequence of surface tension gradients.

The possible cell patterns are all solutions of the membrane equation (2.17) which cover an infinite plane regularly. These patterns are the parallel rolls, equilateral triangular cells, square cells, rectangular cells with all side ratios, the hexagonal cells, the axisymmetric ring cells, as well as ring cells with any number of azimuthal nodal lines. We require that no fluid particles pass through the cell boundaries, and that along the boundary

$$\frac{\partial w}{\partial n} = 0, \tag{2.46}$$

where n is the direction normal to the cell boundary. The simplest solution of the membrane equation is rolls. It is then

$$w = W_0 W(z) \cos ax \tag{2.47}$$

and with (2.18)

$$u = -\frac{W_0}{a} \frac{\partial W(z)}{\partial z} \sin ax, \qquad v = 0. \tag{2.48}$$

The direction x is chosen to be perpendicular to the axis of the rolls. A schematic representation of rolls is shown in Fig. 2.2.

Rolls are of prime importance in Rayleigh–Bénard convection. Although they are obviously a solution to the membrane equation, the word "rolls" did not appear in the literature before the paper of Malkus and Veronis (1958). On the other hand, rolls were in a way discovered by Bénard (1928) when he became aware of Rayleigh's theory, apparently through Jeffreys' (1926) paper. In his 1928 paper Bénard discusses "les tourbillons en bandes," realizing that some of his original observations could be interpreted as rolls and that rolls were a solution of the membrane equation. These rolls were, however, obviously not a steady-state solution in his experiments, and no notice was taken of Bénard's 1928 paper. Rolls pose a conceptional problem because they establish by the direction of their axes a preferred direction in the fluid, although the fluid does not possess a preferred direction according to the theoretical input.

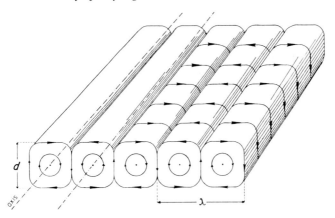

Fig. 2.2.   Plane, parallel convection rolls; λ, the dimensional wavelength. After Avsec (1939).

Square cells were described by Rayleigh as ''composed by superposition of two parts,'' each part being in modern terminology a roll, both sets of rolls being perpendicular to each other. So we have

$$w = W_0 W(z)(\cos ax + \cos ay). \tag{2.49}$$

Another description of the square cells was introduced by Pellew and Southwell (1940), according to which the flow in square cells is described by

$$w = W_0 W(z)\cos ax \cos ay. \tag{2.50}$$

Equation (2.50) rather than (2.49) is commonly used to describe square cells, although (2.49) is more lucid and leads easily to a picture of the velocity field (Fig. 2.3). Note that the sign of the vertical velocity in (2.49)–(2.50) remains arbitrary, since W_0 cannot be determined. Equation (2.50) can lead to confusion with regard to the cell boundaries, as was noted by Stuart (1964). Cell boundaries are a matter of definition, but as Stuart pointed out experimentalists observe as cell boundaries curves upon which the vertical velocity has one and the same sign. According to this definition the square cell shown in Chandrasekhar (1961) is not a genuine square cell. The vertical velocity field corresponding to equation (2.50) is shown in Fig. 2.4. This figure shows that (2.50) does indeed describe square cells, but that the axes of the rolls which make up the cells are oriented at angles of 45° and 135° from the x-axis. The square cell discussed by Chandrasekhar (1961) is marked by ABCD in Fig. 2.4.

A superposition of two perpendicular sets of rolls of unequal wave numbers produces, in theory, rectangular cells. We have

$$w = W_0 W(z)(\cos a_1 x + \cos a_2 y), \tag{2.51}$$

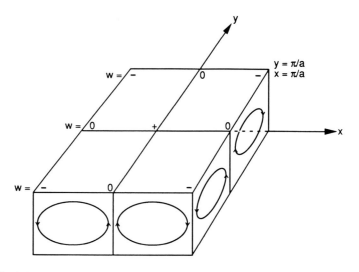

Fig. 2.3. Schematic of the cell boundaries of a square cell resulting from the superposition of two pairs of rolls with perpendicular axes, equal wave numbers, and equal amplitudes. The cell boundaries are the lines where the vertical velocity is of the same sign or zero. The rolls indicated on the vertical sides of the square surface are actually as deep as they are wide.

with $a_1^2 + a_2^2 = a_c^2$. According to Pellew and Southwell (1940) we have

$$w = W_0 W(z) \cos a_1 x \cos a_2 y. \qquad (2.52)$$

There is an infinite number of combinations of wave numbers a_1, a_2 such that $a_1^2 + a_2^2 = a_c^2$, so theoretically rectangular cells can exist with an infinite number of side ratios. One of the wave numbers making up a rectangular cell can, for example, be very small, which means according to (2.27) that its wavelength is very large, and that means that one side of the rectangular cell is very long compared with the other side. This, however, is unrealistic because there is no reason why, under the assumed uniform conditions, the fluid should develop in one direction a wavelength significantly different from the wavelength in the perpendicular direction. For rectangular cells to exist, infinite fields of parallel rolls with a wavelength either shorter or longer than λ_c would have to be present, but that is not compatible with the balance theorem. And indeed rectangular cells have never been observed experimentally, even in rectangular containers, whereas square cells which result from the superposition of two perpendicular fields of rolls with the same wavelength have been observed in square containers.

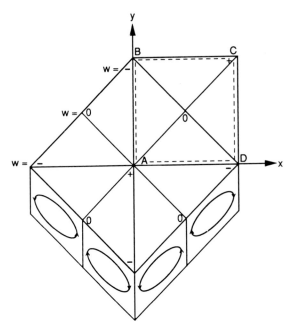

Fig. 2.4. Schematic of the cell boundaries of a square cell according to formula (2.52). The square marked ABCD shows the square cell discussed by Chandrasekhar (1961).

The solution of the membrane equation for the hexagonal cells was found by Christopherson (1940). It is

$$w = W_0 W(z)[\cos \sqrt{3} \, ax \cos ay + \tfrac{1}{2}\cos 2ay]. \tag{2.53}$$

Writing this in polar coordinates we obtain

$$w = W_0 W(z)[\cos(ar \sin \phi) + \cos(ar \sin(\phi + 60°))$$
$$+ \cos(ar \sin(\phi + 120°))], \tag{2.54}$$

with $r = \sqrt{x^2 + y^2}$. Equation (2.54) shows that hexagonal cells can be obtained by superposition of three pairs of rolls whose axes differ in orientation by angles of 60° and 120°. According to (2.53) hexagonal cells can also be considered as the superposition of rectangular cells and rolls. Streamlines of the flow are in Reid and Harris (1958) and Chandrasekhar (1961). A schematic picture of the flow in a hexagonal cell is shown in Fig. 2.5. If the medium is a liquid, the flow is upward in the center and downward along the circumference of the cells, as has been observed experimentally; in gases it is

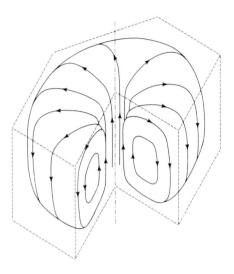

Fig. 2.5.   Schematic of the circulation in a hexagonal convection cell in a fluid. After Avsec (1939).

the opposite. The reason for this is the variation of viscosity with temperature, as will be discussed in Section 8.1. At the sides and in the center of hexagonal cells there is only a vertical velocity component.

Since hexagonal cells consist of six equilateral triangular cells, the triangular cells are part of Christopherson's solution of the membrane equation and do not need to be discussed separately.

The membrane equation (2.17) also has a solution in the axisymmetric case. The membrane equation in polar coordinates is solved with Bessel functions. Bessel functions in connection with convective motions were mentioned by Rayleigh (1916a) and Pellew and Southwell (1940), but they did not realize the form of motion which follows from the Bessel function. This was understood by Zierep (1958), who showed with the membrane equation that, in the axisymmetric case, the convective motions form a system of circular concentric rolls if

$$w = W_0 W(z) I_0(ar), \tag{2.55}$$

where $I_0(ar)$ is the Bessel function of zeroth order. The ring cells are shown schematically in Fig. 2.6. On an infinite plane with uniform conditions on top and bottom, there is no pole from which one can measure r. However the ring cells will pop up around any pointlike finite disturbance in an otherwise uniform critical temperature field. The existence of the ring cells was verified this way experimentally by v. Tippelskirch (1959).

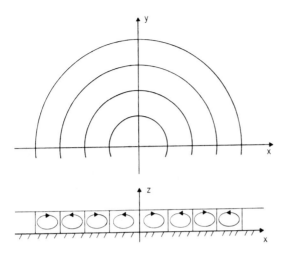

Fig. 2.6. The ring cells. After Zierep (1958).

The membrane equation can also be solved in a field which depends on r and an azimuth angle ϕ. It is then according to Zierep (1959)

$$w = W_0 W(z) I_n(ar) \cos n\phi, \tag{2.56}$$

where $I_n(ar)$ is the Bessel function of nth order. The radial dependence in (2.56) is only slightly changed as compared with (2.55), but the velocity now has n nodal lines $\cos n\phi = 0$, or $\sin n\phi = 0$. The patterns described by (2.56) are unrealistic, because the azimuthal wavelength, extending from one nodal line to the second following azimuthal nodal line, is a function of r and increases rapidly going outward. In the most simple case, $n = 1$, the azimuthal wavelength is equal to the circumference of the circular rolls. Since the circumference of circular rolls increases as r, the azimuthal wavelength in theory soon becomes much larger than the radial section through a pair of rolls, i.e. the radial wavelength. However it has been observed experimentally that the fluid prefers azimuthal wavelengths which are close to or of the same length as the wavelength of the radial component of the flow. If that is so the azimuthal cell boundaries can, in general, not be on common nodal lines through the origin.

So far we have discussed motions which extend in the vertical direction through the entire depth of the fluid. They are, in mathematical terms, described by the lowest mode ($n = 1$) of equation (2.21), which describes the vertical variation of w. The so-called higher modes with $n \geq 2$ were discussed first by Low (1929). The motions in the higher modes consist of cells stacked in the vertical direction. Their wave numbers or the corresponding

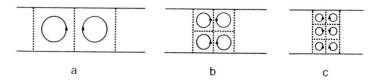

Fig. 2.7. Schematic of higher modes of rolls. (a) $n = 1$, (b) $n = 2$, (c) $n = 3$.

wavelengths are given by (2.24) for the free–free case and can be similarly determined for the rigid–rigid case. The horizontal dimensions of the cells of higher modes are smaller than those of the lowest mode because the wavelength of the modes is inversely proportional to n; see (2.24). For a schematic diagram of the cells in the higher modes see Fig. 2.7.

Higher modes require a critical temperature gradient much larger than the critical temperature gradient of the lowest mode. As follows from (2.23) the critical temperature gradient is proportional to n^4. The higher modes should therefore appear under highly supercritical conditions, in other words, under conditions which are not adequately described by linear theory. Higher modes are therefore not realistic and stacked cells have never been observed. As experience shows, in the lowest mode the fluid is not able to realize stable wavelengths shorter than the critical wavelength under supercritical conditions; it is therefore unlikely that the fluid will choose the higher modes, because they have wavelengths much shorter than the critical wavelength. Even if one could, perhaps through time-dependent operation, force stacked cells in a fluid layer, they would be ultimately unstable.

We mention finally that different solutions of the membrane equation can be superposed in theory. In this way solutions other than those discussed here have been constructed. There is no evidence that any one of these solutions is realistic. We also note that so far we have been concerned exclusively with convection in fluid layers of infinite horizontal extent. Infinite fluid layers are an abstraction; they are not present in reality. All experiments on Rayleigh–Bénard convection are made in laterally bounded containers. The consequences of the presence of lateral boundaries will be discussed in Section 5.2.

2.4 Subcritical Convection

The critical Rayleigh number dictates that, within the range of validity of the linear approximation of the Navier–Stokes equations, the onset of convection takes place spontaneously over the entire fluid layer at precisely the critical temperature difference ΔT_c. The critical temperature difference can have very different values, from very small for deep layers to very large for thin layers, because of the dependence of ΔT_c on d^{-3} [equation (2.8)]. Large changes of the value of ΔT_c can also be brought about by using either nearly inviscid or very viscous fluids. One wonders whether there no are limits to the range over which the value of ΔT_c can be varied without encountering problems with the concept of an immutable $\mathcal{R}_c$. As a matter of fact, for very large ΔT_c the Boussinesq approximation breaks down, as we shall discuss in Section 8.1. For very small ΔT_c we have to consider the possible consequences of the presence of finite disturbances on the onset of convection, which may result in subcritical instability. By subcritical motions we mean motions which occur in a fluid layer of large aspect ratio at $\mathcal{R} < \mathcal{R}_c$. This definition of the term "subcritical" differs from what is meant by a subcritical bifurcation, for which the onset of convection occurs with finite amplitude spontaneously in the entire layer at $\mathcal{R}_c$. Subcritical motions caused by the presence of lateral walls or by thermal noise are characterized by imperfect bifurcations for which the motions begin with infinitesimal amplitude at values of $\mathcal{R}$ below $\mathcal{R}_c$.

The spontaneous onset of convection in the entire fluid layer at the critical Rayleigh number which linear theory predicts does not seem to occur in reality. In experiments with (necessarily) bounded fluid layers the onset of convection usually proceeds from subcritical motions at the walls. Pictures of subcritical rolls along a circular wall can be found in Koschmieder and Pallas (1974). The critical condition in bounded fluid layers is reached when the convective motions fill the entire fluid layer. This is accompanied by a sudden increase of the heat transfer through the fluid caused by the onset of convection. Almost none of the descriptions of Rayleigh–Bénard convection experiments mention subcritical motions, nor do they show that the onset of convection was spontaneous in the entire layer. According to the studies of Brown and Stewartson (1978) one can actually expect the onset of convection to occur in the form of an imperfect bifurcation. In the Taylor vortex problem, spontaneous onset of the instability is also predicted by linear theory for fluid columns of infinite length. It is, however, an undisputed fact that the necessarily present ends of the column introduce finite disturbances which cause the formation of subcritical motions and make the bifurcation imperfect in this case as well.

Even for a fluid layer of infinite extent, subcritical motions may be possible because of finite disturbances of the fluid layer originating from causes other than the walls. The theoretical assumption of an absolutely uniform temperature on the top and bottom of the fluid layer is, strictly speaking, unrealistic. There will always be temperature fluctuations in the horizontal boundaries. Less obvious, but just as likely, are the always existing finite deviations of the horizontal surfaces from mathematically plane surfaces. The theoretical investigation of the consequences of finite disturbances of the fluid layer began with studies of the consequences of thermal noise. The equations of motion in the presence of thermal fluctuations were given by Landau and Lifshitz (1957). These equations were applied by Zaitsev and Shliomis (1970) to the Rayleigh–Bénard problem. Further theoretical investigations of this topic were made by Newell et al. (1970), R. Graham (1974), Swift and Hohenberg (1977), and Jhaveri and Homsy (1980).

The effect of thermal fluctuations on the critical Rayleigh number is much too small to be detected with current experimental techniques. On the other hand, it has been learned theoretically that the relaxation time and the correlation length of the fluctuations are inversely proportional to $|\mathscr{R} - \mathscr{R}_c|$. This means that the thermal relaxation time should become very large near the critical Rayleigh number, which effect is referred to as the "critical slowing down." Critical slowing down is not an obvious feature of the many convection experiments that have been made near the critical Rayleigh number. Verification of critical slowing down requires exceptional, long-time control of the temperature difference across the fluid very near the critical temperature difference. Any slight drift of ΔT can push the fluid into the unstable regime. A direct verification of critical slowing down was reported by Wesfreid et al. (1978). The relaxation time measured there was, e.g. in the case of Fig. 2 therein, only twice as long as the thermal relaxation time d^2/κ. So it might well be that the increase with time of the horizontal velocity maximum of the rolls observed in this experiment was caused by normal thermal relaxation. One also wonders whether the concept of critical slowing down, which was derived in studies dealing with fluid layers of infinite horizontal extent, holds in bounded fluid layers where the onset tends to be an imperfect bifurcation with subcritical motions starting at the walls.

The energy theory offers a different way to study the consequences of infinitesimal as well as finite disturbances on the onset of convection. Energy theory leads to a sufficient condition for stability of the fluid layer, whereas linear stability theory leads to a sufficient condition for instability with regard to infinitesimal disturbances. The results of the energy method concerning Rayleigh–Bénard convection have been discussed in detail in Joseph's (1976) book. It therefore suffices to state here that these studies end up with the statement that

$$\mathscr{R}_c \text{ (linear)} = \mathscr{R}_c \text{ (energy)}. \qquad (2.57)$$

Since, in order for a subcritical instability to exist, $\mathfrak{R}_c$(energy) would have to be smaller than $\mathfrak{R}_c$(linear), there can be no subcritical instability for finite disturbances of the classical Rayleigh–Bénard convection problem according to the energy theory.

The theoretical investigation of the existence of subcritical instabilities was begun by Sorokin (1953, 1954) and continued by Ukhovskii and Iudovich (1963), Sani (1964), and Joseph (1966). Subcritical instabilities can occur, according to the results of the energy method, if the disturbances are not self-adjoint (Davis, 1971). A practical case has been examined by Davis (1969a) in connection with surface-tension-driven convection.

Finally we look for experimental evidence for the existence of subcritical instability, as documented by convection in the *entire* layer at values of the Rayleigh number below the critical Rayleigh number for fluid layers of large aspect ratio. The onset of convection in this form has been reported by Chandra (1938) in shallow layers of air of depth less than 1 cm, whereas for layers deeper than 1 cm Chandra's data fit the conventional critical Rayleigh number. The results of these experiments were not confirmed by Thompson and Sogin (1966). Subcritical instability in very shallow layers of silicone oil under an air surface was observed with a modern apparatus by Koschmieder and Biggerstaff (1986). It appears that non-Boussinesq effects are involved in these experiments. A theoretical explanation of the results of these experiments has not yet been given. We conclude that it has, until now, not been established positively that subcritical instabilities, characterized by motions in the entire layer at Rayleigh numbers below $\mathfrak{R}_c$, really occur in a Boussinesq fluid layer.

3

THEORY OF SURFACE-TENSION-DRIVEN BÉNARD CONVECTION

3.1 Linear Theory of Bénard Convection

Let us now look at convection in a plane fluid layer heated uniformly from below and cooled from above by a gas at rest and of uniform temperature. In this case convection can be caused by surface tension gradients and can form a pattern of hexagonal cells. The classical example of such a flow appeared in Bénard's experiments, and consequently we shall refer to surface-tension-driven convection as Bénard convection. The name used here for this problem differs from the designation "Marangoni convection" used frequently in the literature. We do, however, believe that the name for this problem should honor the scientist who discovered the problem, and not someone who did not make a theoretical or an experimental contribution to this topic. Marangoni's work is done justice by the naming of the Marangoni number after him. If the designation "Bénard convection" is not sufficiently different from "Rayleigh–Bénard convection," then it may be appropriate to call surface-tension-driven convection the "Bénard–Pearson problem," as Davis (1969a) once proposed.

Consider the plane surface of a homogeneous fluid with a "free" surface that has a temperature-dependent surface tension. An infinitesimal disturbance of the temperature on the surface, say a warm spot, creates surface tension traction on the surface if the surface tension coefficient of the fluid is a function of temperature. Since surface tension usually decreases with temperature, a warm spot will be a soft spot at the surface from which the fluid will be pulled away laterally. The fluid underneath the warm spot must consequently rise. The fluid motion at the surface will be transmitted by viscosity to the interior of the fluid layer, but viscosity will also dampen the motion. The motion in the fluid can be sustained only if energy is provided to overcome the frictional losses. The required energy can be provided by a vertical temperature difference across the fluid layer, which can be created by heating from below. If sufficient energy is supplied, and if a critical vertical temperature difference is reached, convection caused by surface tension gradients will commence and will be sustained.

The first theoretical study of convection caused by surface tension gradients was made by Pearson (1958). The motivation for his investigation was the observation that polygonal cellular patterns in paint layers appear even if the paint is on the underside of a plane, horizontal surface. In this case the stratification of temperature in the paint, caused by evaporation at the free lower surface, must be stable, so buoyancy cannot be the cause of the formation of the cells. The cause of the formation of the cells suggested by Pearson was surface tension. His analysis was based on the Navier–Stokes equations, the equation of thermal conduction, and the continuity equation, just as in Rayleigh–Bénard convection. However the equation of state is now

$$S = S_0 + \frac{\partial S}{\partial T} \Delta T, \tag{3.1}$$

which describes the dependence of the surface tension coefficient S on temperature. Since surface tension is a monotonically decreasing function of temperature in most fluids, $\partial S/\partial T$ is usually negative. We apply the three conservation equations and the equation of state to an infinite homogeneous fluid layer of uniform thickness on a uniformly heated rigid bottom. The upper surface of the fluid is plane and free, "free" except for the surface tension force. The acceleration of gravity is neglected (g is set equal to zero); hence there is no buoyancy in the fluid and the Rayleigh number is zero. This assumption simplifies the analysis considerably and focuses attention on the effects of surface tension. We shall study the combined surface-tension-driven and buoyancy-driven convection problem later.

The initial state is a fluid layer at rest on an infinite plane with a constant linear vertical temperature gradient $\beta = dT/dz$. We disturb this layer with infinitesimal disturbances, which are so small that the nonlinear terms in the Navier–Stokes equations and the energy equation can be neglected. We use a modified Boussinesq approximation, which means we assume all material properties, including the density ρ, to be constant, exempting the surface tension coefficient S. After elimination of the pressure and the velocity components u and v, we arrive in the linear approximation at the perturbation equations

$$\left(\frac{\partial}{\partial t} - \nu\nabla^2 \right) \nabla^2 w = 0, \tag{3.2}$$

$$\left(\frac{\partial}{\partial t} - \kappa \nabla^2\right)\theta = \beta w, \tag{3.3}$$

where w is the vertical velocity component of the disturbance and θ the temperature disturbance.

There are four boundary conditions for the velocity. The no-slip condition at the bottom, at $z = 0$,

$$w = \frac{\partial w}{\partial z} = 0, \tag{3.4}$$

and

$$w = 0, \qquad \rho v \frac{\partial^2 w}{\partial z^2} = \sigma \nabla_2^2 \theta \tag{3.5}$$

at the free surface (at $z = d$), with $\sigma = -\partial S/\partial T$, taken at the temperature of the top surface. The second equation (3.5) equates the shear stress at the surface to the change of surface tension due to the temperature variations brought about by the disturbance. We see that surface tension, or the variation of surface tension with temperature, enters the equation for the velocity of the flow only via a boundary condition.

The two boundary conditions for the equation of thermal conduction are

$$\theta = Y \frac{\partial \theta}{\partial z} \tag{3.6}$$

at the bottom and

$$-\lambda \frac{\partial \theta}{\partial z} = q\theta \tag{3.7}$$

at the free surface. The boundary condition (3.6) takes the different thermal conductivities of the bottom plate into account through the factor Y. If the bottom plate is an excellent thermal conductor as compared with the fluid, then $Y = 0$, and the boundary condition $\theta = 0$ used in Rayleigh–Bénard convection is recovered. If, on the other hand, the bottom plate is insulating, then $Y \to \infty$. The parameter λ in the boundary condition (3.7) is the thermal conductivity of the fluid, and the parameter $q = \partial Q/\partial T$ represents the rate of change with temperature of the heat flux Q from the upper surface to the environment above. The particular value ascribed to q will depend on circumstances, e.g. on the thickness of the layer of air separating the fluid surface from a lid of uniform temperature covering the air. $q \to \infty$ corresponds to a conducting upper thermal boundary, whereas $q \to 0$ corresponds to an insulating thermal boundary, as exemplified, e.g., by a deep layer of air. We note that in contrast to the Rayleigh–Bénard problem the temperature disturbance θ is not zero at the top boundary. This must be so, since without a temperature disturbance the surface tension mechanism of Bénard convection cannot be triggered.

To solve (3.2)–(3.3) we try product solutions of the form

$$w = -\frac{\kappa}{d} W(z) f(x,y) e^{st}, \tag{3.8}$$

$$\theta = \Delta T \theta(z) f(x,y) e^{st}, \tag{3.9}$$

with the growth rate s. We pause to consider the disturbances in equations (3.8)–(3.9). These are the infinitesimal disturbances whose growth is said to be responsible for the pattern that is ultimately formed when the fluid is unstable. These disturbances correspond to the disturbances used in Rayleigh–Bénard convection, except for the functions $W(z)$ and $\theta(z)$ being different on account of the different boundary conditions, in particular θ being $\neq 0$ at the upper surface. We note that surface tension gradients play no direct role in these disturbances. This means that the pattern which forms from these disturbances grows out of fluctuations in the fluid, prior to the action of surface tension gradients. I wonder whether these disturbances do justice to the problem. Disturbances in the surface, in particular pointlike temperature disturbances, should play an important role in surface-tension-driven convection. We must keep in mind that surface tension is the only force present; gravity is assumed to be zero. We shall see in the following that the disturbances (3.8)–(3.9) lead to completely satisfactory critical conditions, the critical Marangoni number, but lead to an ambiguous statement about the pattern, whereas in reality the pattern seems to be unique, namely hexagonal, already in the linear case.

We assume that the function $f(x,y)$ in (3.8)–(3.9) can be described by normal modes because the pattern is periodic in x and y. Then $f(x,y)$ satisfies the nondimensional membrane equation

$$\nabla_2^2 f(x,y) + a^2 f(x,y) = 0, \tag{3.10}$$

where a is again the nondimensional wave number.

As in Rayleigh–Bénard convection there are an infinite number of possible patterns satisfying (3.10). However, the hexagonal pattern seems to be the unique solution for surface-tension-driven convection according to the experimental evidence, whereas in Rayleigh–Bénard convection rolls are predominant, although not unique. The solution of (3.10) for hexagonal surface-tension-driven cells is the same as the one found by Christopherson (1940) for the Rayleigh–Bénard problem [equation (2.53)]. We note, however, that in many photographs of surface-tension-driven hexagonal cells there are six azimuthal nodal lines of the vertical velocity component leading from the center to the vertices of the cell. These nodal lines are not present in Christopherson's solution.

We nondimensionalize with

$$x_i' = x_i/d \quad \text{and} \quad t' = \kappa t/d^2,$$

and replace the ∇ operator by $D^2 - a^2$, using the membrane equation and the operator $D = d/dz$. We drop the primes, and find for (3.2) and (3.3) the nondimensional equations

$$(s - \mathcal{P}(D^2 - a^2))(D^2 - a^2)W(z) = 0, \tag{3.11a}$$

$$(s - (D^2 - a^2))\theta(z) = -W(z). \tag{3.11b}$$

The marginal or neutral case, i.e. the time-independent case $s = 0$, is solved using nondimensional boundary conditions, $W = dW/dz = \theta = 0$ at the conducting bottom and at the top surface $W = 0$, as well as

$$\frac{d^2W}{dz^2} = a^2 \frac{\partial S}{\partial T} \frac{\Delta T d}{\rho \nu \kappa} \theta = a^2 \mathcal{M}\theta \tag{3.12a}$$

and

$$\frac{d\theta}{dz} = -\frac{qd}{\lambda}\theta = -\mathcal{B}\theta, \tag{3.12b}$$

with the Marangoni number

$$\mathcal{M} = \frac{\partial S}{\partial T} \frac{\Delta T d}{\rho \nu \kappa}. \tag{3.13}$$

The name was introduced by Scriven and Sternling (1964) and Nield (1964). The parameter $\mathcal{B}$ in equation (3.12b) is the surface Biot number, with λ being in this case the thermal conductivity of the fluid.

The Marangoni number is a nondimensional measure of the temperature difference applied to the fluid layer; it corresponds to the Rayleigh number in

Rayleigh–Bénard convection and is the most important parameter in Bénard convection. Since $\partial S/\partial T$ is usually negative, ΔT must be negative in order to arrive at $\mathcal{M} > 0$. $\Delta T < 0$ means that the fluid is heated from below. Smith (1966) has predicted the occurrence of instability with critical wave numbers of about 2 for negative values of $\mathcal{M}$, that means for heating from above. This applies to the case when the depth and the properties of the two layers are similar. This is, however, not the situation in what we usually call Bénard convection, where the upper layer is often very deep and is a gas, which means the upper layer has properties quite different from those of the lower layer. If the surface tension coefficient of the fluid should increase with temperature, then instability can also be caused by heating from above. The reality of this possibility is indicated by the cellular patterns in paint on the underside of surfaces where the lower free surface cools by evaporation.

The functions W and θ (3.8) and (3.9) can be constructed as sums of hyperbolic sines and cosines so that they satisfy the boundary conditions. From the boundary condition (3.12a), which correlates W and θ, follows a formula for the Marangoni number as a function of the wave number and the Biot number. It is, in the case of a conducting bottom boundary,

$$\mathcal{M} = \frac{8a(a \cosh a + \mathcal{B} \sinh a)\,(a - \sinh a \cosh a)}{a^3 \cosh a - \sinh^3 a}. \qquad (3.14)$$

The Marangoni number as a function of the wave number at a given value of $\mathcal{B}$ can therefore be calculated from (3.14), and from a similar formula for the insulating bottom boundary as well. The marginal (or neutral) stability curves for a conducting bottom are shown in Fig. 3.1. Vidal and Acrivos (1966) showed that the so-called principle of exchange of stabilities that we discussed in the Rayleigh–Bénard case is valid for Pearson's problem too; i.e. the amplitude of the motions at the critical Marangoni number is a monotonically increasing function of time. If the fluid surface is deformable (a condition to be discussed soon), the onset of convection may be oscillatory, as the theoretical studies of Takashima (1981), Garcia-Ybarra and Velarde (1987), Benguria and Depassier (1987, 1989), Gouesbet et al. (1990), and Pérez-Garcia and Caneiro (1991) indicate. Even solitary waves may form under certain circumstances on the surface, provided the surface is deformable, as a recent study of Garazo and Velarde (1991) indicates.

The curves in Fig. 3.1 are similar to the marginal curve of Rayleigh–Bénard convection (Fig. 2.1). There is, for a given value of $\mathcal{B}$, a minimal value of the Marangoni number, which is the critical Marangoni number $\mathcal{M}_c$. The onset of Bénard convection will occur spontaneously in the entire infinite

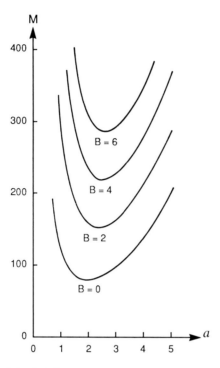

Fig. 3.1. Stability diagram of linear theory for surface-tension-driven convection in a fluid
layer of infinite horizontal extent, for different values of the surface Biot number.
After Pearson (1958).

fluid layer if $\mathcal{M}_c$ is reached or, in other words, if the corresponding critical
vertical temperature difference is reached. If the Biot number is greater than
0, then if more heat is transferred from the upper surface, a larger vertical
temperature difference is required for the onset of convection. The motions at
the onset of convection are unique and are characterized by the critical wave
number a_c. The values of the critical Marangoni number and of the critical
wave number are, for the conducting bottom boundary and $\mathcal{B} = 0$,

$$\mathcal{M}_c = 79.607, \qquad a_c = 1.993, \tag{3.15}$$

according to Table 1 in Nield (1964), where the critical Marangoni numbers
and critical wave numbers for different values of $\mathcal{B}$ are also given.

There are significant differences between the critical Rayleigh and the critical Marangoni numbers. These differences follow from the dependence of $\mathcal{R}$

on the third power of the depth d of the fluid, whereas $\mathcal{M}$ depends only on the first power of d. From

$$\Delta T_{c,R\text{-}B} = \frac{1}{d^2} \frac{\mathcal{R}_c}{\mathcal{M}_c} \frac{dS/dT}{\alpha \rho g} \Delta T_{c,M}$$

it follows that the onset of buoyancy-driven Rayleigh–Bénard convection takes place at a smaller ΔT than for surface-tension-driven Bénard convection if the depth of the fluid is large (for the same fluid). On the other hand, the onset of Bénard convection occurs at a ΔT_c smaller than the ΔT_c required for the onset of Rayleigh–Bénard convection in thin fluid layers ($d < 1$). This difference increases as d^{-2} and easily becomes so large that one can consider the surface-tension-driven instability being dominant in thin fluid layers. Remembering that the depth of the fluid layers in Bénard's experiments was of the order of 1 mm, it follows that, to quote Pearson, "somewhat surprisingly, it seems almost certain that many of the cells observed in molten spermaceti by Bénard were in fact caused by surface tension forces" (1958, p. 490). Block (1956) had arrived at the same conclusion independently two years earlier from experimental observations.

Finally we compare the critical wave number of Bénard convection, $a_c = 1.993$, with the critical wave number $a_c = 2.342$ of Rayleigh–Bénard convection in the rigid–free case under a perfectly conducting surface. According to these numbers hexagonal cells in surface-tension-driven convection should be noticeably larger than buoyancy-driven hexagonal cells in fluid layers of the same depth.

Pearson's (1958) theory tacitly assumed that the upper surface of the fluid is plane and nondeformable, a condition expressed by the requirement that the capillary number

$$\mathcal{C} = \frac{\mu \kappa}{dS_0} \tag{3.16}$$

be zero, or that the surface tension coefficient S_0 tend to infinity. It is, on the other hand, known from Bénard's experiments that the surface of the fluid in hexagonal convection cells is deformed (Fig. 1.3). If the deformation of the fluid surface is permitted, the boundary conditions on top of the layer change. For the boundary conditions with a deformable surface we refer to Davis (1987). The consequences of a deformable surface ($\mathcal{C} \neq 0$) on the onset of surface-tension-driven convection without gravity were investigated by Scriven and Sternling (1964), in the linear approximation. For wave numbers $a \cong 2$ and $\mathcal{C} < 10^{-2}$ the stability results of Scriven and Sternling are similar to those of Pearson (1958). On the other hand, the fluid layer on a conducting bottom was found always to be unstable with respect to disturbances of small wave number ($a < 0.1$), i.e. to very long wavelength disturbances. Under these circumstances a critical Marangoni number does not exist, because the onset of convection will occur at arbitrarily small temperature differences for sufficiently long disturbances. An instance of such an instability has not been observed so far. The consequences of surface deflections of very long wavelength have been clarified by Davis (1987). He showed that in the absence of gravity the fluid layer is unconditionally unstable if $a^2 < \frac{3}{2}\mathcal{M}\mathcal{C}$, and he showed that long-wave instabilities can always be made significant if the layer is made thin enough. However one has to exercise caution with the extension of the results of the theory to very thin fluid layers, because in such a case the critical temperature differences tend to become so large that the Boussinesq approximation may not hold. Smith (1966) has tried to reconcile the results of Scriven and Sternling with the apparent fact that a critical Marangoni number *is* required for the onset of surface-tension-driven convection. He considered the consequences of gravity waves on the surface of the fluid on the stability of a layer without buoyancy and found that gravity waves stabilize the long-wavelength instabilities. Usually surface waves are important only for very long disturbances.

Returning to Scriven and Sternling's paper, we note that they found that "in steady cellular convection driven by surface tension there is upflow beneath depressions" (1964, p. 339) of the surface of the fluid. This explains Bénard's (1900) observations and solves the puzzle posed by Jeffreys' (1951) result that a truly free surface over a cell center with rising motion should be elevated, in obvious contradiction to the experimental fact that in surface-tension-driven hexagonal convection cells the fluid surface is depressed although there is upflow at the centers of the cells. Measurements of surface elevations and depressions will be discussed in Chapter 4.

The investigation of surface-tension-driven convection was carried an essential step further when Nield (1964) considered the combined action of surface tension and gravity on a fluid layer heated uniformly from below. Since gravity was present in Bénard's experiments as well as in all other experiments involving convection in a fluid layer with a free surface in a laboratory on the earth, an extension of Pearson's theory was necessary. Using the same linear perturbation analysis that was originally introduced by Rayleigh (1916a), which we have summarized in Chapter 2, and trying the same product solutions as in (2.13) and (2.14), Nield arrived, after elimination of the pressure and the velocity components u and v, at the nondimensional equations

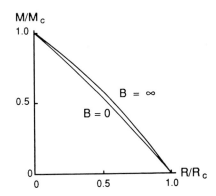

Fig. 3.2. Stability diagram of linear theory for surface-tension-driven convection in the presence of gravity. $\mathscr{B} = 0$ and $\mathscr{B} = \infty$ refer to insulating and conducting free surfaces. After Nield (1964).

$$(D^2 - a^2)(D^2 - a^2 - s)W = a^2\mathscr{R}\theta, \tag{3.17}$$

$$(D^2 - a^2 - \mathscr{P}s)\theta = -W, \tag{3.18}$$

with the nondimensional operator $D = d/dz$. Equation (3.17) is, in the case $g = 0$, i.e. $\mathscr{R} = 0$, the same as equation (3.2) used in Pearson's theory. The surface tension gradients come into consideration through the boundary conditions on the top surface. The boundary conditions are the same as in (3.4)–(3.7). The equations (3.17) and (3.18), subject to the boundary conditions, were solved in the marginal case with a Fourier series method. The eigenvalue parameters are the wave number a, the Rayleigh number $\mathscr{R}$, the Marangoni number $\mathscr{M}$, and the surface Biot number $\mathscr{B}$. Fixing $\mathscr{M}$ and $\mathscr{B}$, one can find numerically the critical Rayleigh number by varying the wave number a. Fixing $\mathscr{R}$ and $\mathscr{B}$, one can find the critical Marangoni number by varying a. One can then plot curves $\mathscr{M}$ versus $\mathscr{R}$ for given values of $\mathscr{B}$. The results so obtained are shown in Fig. 3.2.

As Fig. 3.2 shows, the onset of convection in the absence of surface tension effects ($\mathscr{M} = 0$) takes place at the critical Rayleigh number. In the case of an upper insulating boundary $\mathscr{R}_c = 669.0$ and $a_c = 2.086$, according to Table 1 in Nield. The onset of surface-tension-driven convection in the absence of gravity, meaning $\mathscr{R} \to 0$, takes place at the critical Marangoni number, which has the value $\mathscr{M}_c = 79.6$ [equation (3.15)] for the insulating upper surface. The critical wave number in this case is $a_c = 1.993$; this means that the wave number then is 4.6% smaller than in the corresponding buoyancy-driven case ($\mathscr{M} = 0$). The upper surface in surface-tension-driven convection will practically always be nearly insulating because of the extremely poor thermal conductivity of gases. In general, the Marangoni and Rayleigh numbers at which the onset of convection occurs are coupled through the equation $\mathcal{R}/\mathcal{R}_c + \mathcal{M}/\mathcal{M}_c \cong 1$ (Fig. 3.2). At values of $\mathcal{R} \approx \mathcal{R}_c$ buoyancy is dominant, which means that the corresponding Marangoni number is much smaller than the critical Marangoni number. On the other hand, if $\mathcal{R}$ is small, the onset of convection takes place at $\mathcal{M} \approx \mathcal{M}_c$, and surface tension is dominant. Changes in the ratio $\mathcal{R}/\mathcal{R}_c$ are accompanied by small changes in the critical wave number, which were calculated by Nield and are plotted in Fig. 2 of his paper. The patterns of Bénard convection in the presence of gravity follow from the membrane equation, (3.10). There is, as before, an infinite number of possible patterns, including the hexagonal cells. The relative streamlines in hexagonal Bénard cells induced by surface tension in the presence of gravity were calculated by Nield (1966).

Concerning Bénard's experiments, in which the fluid depth was of the order of 1 mm, Nield's results confirm that the onset of convection must have taken place essentially at the critical Marangoni number and was therefore in essence driven by surface tension, in spite of the presence of gravity. This is in agreement with Pearson's (1958) conclusions and with the conclusions drawn by Block (1956) from his observations.

The stability of a fluid layer having a free, nondeformable surface subject to gravity and surface tension forces was investigated by Davis (1969a) with the energy method which studies the stability with regard to finite disturbances. The energy method provides a sufficient condition for stability. Studying a thin fluid layer of infinite horizontal extent heated uniformly from below, Davis found a possible subcritical instability in a small range of Marangoni numbers below the critical Marangoni number, which ranges from $56.77 < \mathcal{M} < 79.61$ if $\mathcal{R} = 0$ and $\mathcal{B} = 0$, i.e. in the exclusively surface-tension-driven case with an insulating surface. The subcritical interval shrinks to zero as $\mathcal{R}$ goes to $\mathcal{R}_c = 669.0$, i.e. in the case of Rayleigh–Bénard convection. This range of subcritical $\mathcal{M}$ is much larger than the range of subcritical instability predicted by the three nonlinear theories to be discussed later. Subcritical instability was studied again with the energy method by Davis and Homsy (1980) in the two-dimensional case, by Castillo and Velarde (1982) in the two-dimensional case with a deformable surface, and by Lebon and Cloot (1982) when negative Rayleigh and Marangoni numbers were considered, where a negative $\mathcal{R}$ means heating from above and a negative $\mathcal{M}$ means a positive $\partial S/\partial T$ with heating from below. All these studies using the energy method arrive essentially at the same range of subcritical instability. Only one example of subcritical surface-tension-driven convection has been observed experimentally (Koschmieder and Biggerstaff, 1986). It does not appear that the subcritical flow in this case is qualitatively the type of flow expected, and quantitatively the range of subcritical flow in this experiment extends deep into the range of stable $\mathcal{M}$ predicted by Davis (1969a).

3.2 Nonlinear Theory of Bénard Convection

So far we have discussed the results of linear analysis of surface-tension-driven convection on a plane of infinite horizontal extent, from which we have learned the values of the critical Marangoni number and the critical wave number. The most spectacular feature of the experimental observations, namely the hexagonal cell pattern, cannot be explained with linear theory because at $\mathcal{M}_c$ the membrane equation permits an infinite number of different solutions with the same a_c. In order to explain the apparent uniqueness of the hexagonal pattern we have to consider the nonlinear terms in the Navier–Stokes equations and the energy equation. A nonlinear investigation of surface-tension-driven convection should also elucidate what is the wavelength of slightly supercritical convection or the size of slightly supercritical cells. Since the exponential growth of the disturbances assumed in linear theory is unrealistic, we also want to learn how the amplitude of the hexagonal cells increases in reality. According to the results of linear theory as shown in Fig. 3.1 there should be a continuum of unstable supercritical wave numbers for a given supercritical $\mathcal{M}$, which makes it appear that supercritical Bénard convection is nonunique. Clarification of the question of the uniqueness of the supercritical wave numbers is the second principal objective of the nonlinear analysis of surface-tension-driven convection. Whereas in Rayleigh–Bénard convection we have postponed the discussion of nonlinear theory to Chapter 7 in order to look first at the relevant experimental results, we shall discuss the nonlinear theories of Bénard convection now, because there are only a few nonlinear theoretical investigations of supercritical surface-tension-driven convection.

The first attempt to tackle the nonlinear Bénard convection problem theoretically was made by Scanlon and Segel (1967). In order to reduce the difficulties of the problem they made a number of simplifying assumptions. Their model deals with a fluid layer of semi-infinite depth and infinite horizontal extent, having an infinite Prandtl number, a nondeformable surface, and no gravity. The assumption of an infinite Prandtl number was justified by the correct observation that in many convection experiments the Prandtl numbers are in the hundreds, or even thousands. Working with an infinite

Prandtl number modifies the nonlinear problem significantly by eliminating the nonlinear terms in the Navier–Stokes equations. The only nonlinear terms remaining are those in the energy equation. The basic equations are

$$\nabla^4 w = 0, \tag{3.19}$$

and

$$\frac{\partial \theta}{\partial t} + \mathbf{v}\nabla\theta = \nabla^2\theta + w. \tag{3.20}$$

The boundary conditions on the free upper surface are the same as in (3.5) and (3.7), and at the bottom at $z \rightarrow -\infty$ it is required that w and $\partial w/\partial z$ as well as $\partial T/\partial z$ be bounded.

Solutions of the nonlinear problem were obtained by an iteration scheme, with the eigenvectors of the linear problem being the first approximation. Attention was focused on the question of whether hexagonal cells are a stable secondary flow. Therefore the initial disturbance was chosen to be of the form

$$w = W(z)\left\{ Z(t) \cos ay + Y(t) \cos \frac{\sqrt{3}}{2}ax \cos \frac{1}{2}ay \right\}. \tag{3.21}$$

This is a modified form of Christopherson's solution of the membrane equation for hexagonal cells [equation (2.53)]; the linear solution is obtained by setting $Y = \pm 2Z$. With the amplitude $Y(t) = 0$, one has parallel rolls. From the existence conditions for the solution of the iteration scheme follows a pair of amplitude equations,

$$\frac{dy}{dt} = \varepsilon Y - \gamma YZ - RY^3 - PYZ^2 + \cdots, \tag{3.22a}$$

$$\frac{dZ}{dt} = \varepsilon Z - \frac{\gamma}{4}Y^2 - R_1 Z^3 - \frac{1}{2}PY^2 Z + \cdots, \tag{3.22b}$$

with ε being proportional to $(\mathcal{M} - \mathcal{M}_c)/\mathcal{M}_c$, and γ, P, R, and R_1 being constant coefficients whose values follow from the existence conditions. The amplitude equations for the three-dimensional flow in hexagonal cells differ significantly from the Landau equation for the amplitude of two-dimensional flow (rolls), which we shall discuss in Section 7.3.

From (3.22a) and (3.22b) one can determine the final equilibrium states by making a linear stability analysis for each equilibrium solution. The principal conclusion of the analysis of Scanlon and Segel is that the hexagonal pattern is stable in a range of Marangoni numbers from just below critical to 64 times critical. It seems, as Scanlon and Segel comment themselves, that the supercritical range of up to $64 \mathcal{M}_c$ is well beyond the limits of the analysis. We note that the theory predicts a small subcritical range $-0.023 < (\mathcal{M} - \mathcal{M}_c)/\mathcal{M}_c < 0$ in which hexagonal cells should appear. A small range of subcritical surface tension convection is a persistent feature of the nonlinear theories. In the experiments this subcritical range would become apparent in hysteresis at $\mathcal{M}_c$; however hysteresis at $\mathcal{M}_c$ has not yet been observed.

Although the analysis of Scanlon and Segel predicts unquestionably the correct stable solution, namely hexagons, it appears to be uncertain whether the result obtained is conclusive, in view of the highly simplifying assumptions about the fluid layer, as well as the very selective pick of the initial disturbance. Davis (1987) has also questioned whether higher order terms of the amplitude equation must not be included in the analysis.

Kraska and Sani (1979) extended the analysis of nonlinear surface-tension-driven convection to more realistic conditions. They considered a fluid layer of finite depth with surface tension gradients in the presence of gravity. They used the Boussinesq approximation, now in a form in which the density and the surface tension coefficient are linear functions of the temperature. The free surface was considered deformable. On the other hand, attention was restricted to disturbances and cellular flows of a single wave number. They posed the nonlinear stability problem as follows: Given disturbances whose x,y,t dependence is described by

$$\phi = \sum_{m=1}^{N} A_m(t) \, \phi_m \, (x,y), \tag{3.23}$$

where the ϕ_m obey the membrane equation

$$\nabla_2^2 \phi_m = -a^2 \phi_m,$$

what are the possible stable equilibrium amplitudes $A_m(\infty)$?

Actually Kraska and Sani considered only six of the infinite number of possible ϕ_m, namely the three presumably most dangerous disturbances already studied by Scanlon and Segel and three more disturbances of the same type, shifted in phase by 90°. A seventh rectangular disturbance was also mentioned, but in the end no definite conclusion could be reached concerning the stability of the system if the seventh disturbance was included in the analysis. Five equilibrium states of the nonlinear problem were found, including rolls and hexagons. The stability of these solutions was studied with a linear stability analysis, and only hexagons and rolls were found to be stable. Stable hexagons were found at subcritical as well as supercritical Marangoni numbers, just as was the case in Scanlon and Segel. Rolls were found to be stable only under supercritical conditions.

The range in which stable hexagons can occur according to Kraska and Sani is proportional to the wave number, and $\Delta \mathcal{M}$ or $\Delta \mathcal{R}$, and is an increasing function of the Prandtl number, and is maximal for a given ratio $\mathcal{R}/\mathcal{M}$ if the free surface is deformable. The stability of the finite amplitude solutions with regard to arbitrary disturbances [equation (3.23)] of the same wave number and to disturbances with a continuum of different wave numbers remained, however, unsolved. Davis (1987) has questioned the validity of the adjoint operator used by Kraska and Sani. We note also that it has still to be established experimentally that subcritical hexagonal patterns as well as supercritical surface-tension-driven rolls actually exist. Rolls seem to present a radical departure from the three-dimensional cellular character of surface-tension-driven convection, which has been so obvious so far.

Cloot and Lebon (1984) carried the analysis of surface-tension-driven convection an additional step further. They were able to make predictions about the cell form *and* the cell size of supercritical convection. Their analysis was concerned with a Boussinesq fluid layer of finite depth and infinite horizontal extent, having a flat nondeformable surface. Gravity and surface tension gradients act on the fluid, which is heated from below. They solved the nonlinear convection problem with a power series expansion technique introduced by Malkus and Veronis (1958) for the study of the Rayleigh–Bénard problem. We shall discuss this technique in more detail in Chapter 7. In the immediate vicinity of the critical Marangoni number the field variables and the Marangoni number were expanded as

$$\mathbf{u} = \sum_{i=1}^{N} \varepsilon^i \mathbf{u}^{(i)}, \qquad \theta = \sum_{i=1}^{N} \varepsilon^i \theta^{(i)}, \qquad \mathcal{M} = M^0 + \sum_{i=1}^{N} \varepsilon^i M^{(i)}, \qquad (3.24)$$

where ε is so small that terms beyond the second order approximation can be omitted. From the second order approximation it follows that the nonlinear convection problem has an infinite number of steady solutions, namely rolls, rectangular cells of all side ratios, and hexagonal cells – just as with the linear Rayleigh–Bénard problem or in Pearson's analysis.

In order to determine which of these solutions is actually preferred, Cloot and Lebon investigated the stability of the solutions with regard to infinitesimal disturbances. They found that the hexagonal pattern is stable with regard

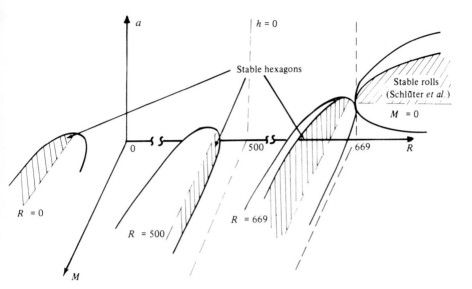

Fig. 3.3. Stability diagram of weakly nonlinear theory for surface-tension-driven convection
as well as Rayleigh–Bénard convection in $(\mathcal{R},\mathcal{M},a)$-space. The curves represent
the marginal stability curves obtained by projection either on the $(\mathcal{M},a)$-plane or on
the $(\mathcal{R},a)$-plane. After Cloot and Lebon (1984).

to disturbances in the form of rolls, whereas rolls cannot be a stable solution
because they are unstable with regard to hexagonal disturbances. Disturbances with wave numbers different from the wave number of the steady solution were also studied and, again, hexagons were found to be stable. The
results obtained by Cloot and Lebon are summarized in Fig. 3.3. This figure
shows that in the case $\mathcal{R} = 0$, i.e. in the total absence of buoyancy, and for
$\mathcal{M} \geq \mathcal{M}_{c} = 79.6$, stable cells of surface-tension-driven convection should be
hexagons, as well as in the case $\mathcal{R} = 669$, i.e. when the critical Rayleigh
number under an insulating surface has been reached but surface tension is
present. On the other hand, Bestehorn and Pérez-Garcia (1987) found with a
generalized Ginzburg–Landau equation that in thick fluid layers, in which
surface tension effects tend to be small as compared with buoyancy effects,
rolls and hexagons can coexist. Be that as it may, in all experiments to which
surface tension effects made a substantial contribution the pattern has always
been unambiguously hexagonal.

The wave numbers of the hexagonal cells for $\mathcal{R} = 0$ and $\mathcal{M} \geq 79.6$ were
found by Cloot and Lebon to be in a continuum of wave numbers *larger* than

a_c, actually larger than $a = 3.2$ for $\mathcal{R} = 0$ and $\mathcal{P} = 7$ according to Fig. 8 in Cloot and Lebon. In an intermediate range of Rayleigh numbers, $400 \leq \mathcal{R} \leq 500$, when surface tension effects and buoyancy act simultaneously, the supercritical wave numbers of the hexagonal cells should be in a continuum of wave numbers *smaller* than a_c, actually ≤ 1.57 at $\mathcal{R} = 400$ and ≤ 1.88 at $\mathcal{R} = 500$. On the other hand, at $\mathcal{R} = \mathcal{R}_c = 669$, with $\mathcal{P} = 7$ and $\mathcal{M} \geq 0$, the pattern should still be hexagonal but the continuum of supercritical wave numbers should again be *larger* than the a_c of linear theory, just as it was in the case $\mathcal{R} = 0$. Note that in all of these cases the stable supercritical wave numbers do not connect with the critical wave number of linear theory $a_c = 1.993$ (as it should be). In the absence of all surface tension effects ($\mathcal{M} = 0$), when the surface of the fluid is in contact with a rigid lid, rolls should be stable for $\mathcal{R} \geq 669$ and the supercritical continuum of wave numbers should have values *larger* than a_c, as was originally shown by Schlüter et al. (1965) for Rayleigh–Bénard convection. This is indicated on the far right side of Fig. 3.3. Cloot and Lebon mention that they cannot advance a conclusive argument for the perplexing behavior of the stable supercritical wave numbers in the different ranges of $\mathcal{R}$. We note that the predicted stable supercritical wave numbers of surface-tension-driven Bénard convection with Rayleigh numbers of ≈ 100 do not agree with the experimental observations (Koschmieder, 1991). The observed wave numbers of hexagonal cells at moderately supercritical $\mathcal{M}$ are between 1.85 and 2.05 and thus are much smaller than the minimum $a = 3.64$ for which stable hexagons are predicted at $\mathcal{R} = 100$. The trend of the theory toward wave numbers larger than a_c, i.e. toward hexagonal cells smaller than the critical cells, is, however, in the right direction. In the experiments there is also no continuum of supercritical wave numbers, but a preferred wave number at a given supercritical $\mathcal{M}$. Details will be discussed in Chapter 4.

We mention finally that Cloot and Lebon found a subcritical instability at values of $\mathcal{M}$ where linear theory predicts a state of rest. The subcritical instability occurs when surface tension effects are dominant, i.e. at low values of $\mathcal{R}$. This result confirms the findings of Scanlon and Segel (1967) and Kraska and Sani (1979). The range of subcritical instability is, according to Cloot and Lebon, very narrow; it is outlined by $(\mathcal{M} - \mathcal{M}_c)/\mathcal{M}_c = 0.003$, if $\mathcal{P} = 7$.

Let us summarize the results of the nonlinear investigations of surface-tension-driven convection. All three nonlinear theories duly find the hexagonal pattern to be preferred under moderately supercritical conditions. Nevertheless it does not appear that a convincing and conclusive explanation of the preference for the hexagons has emerged. Actually, the hexagonal cells already exist at the onset of convection, i.e. under linear conditions. It seems that, in order to be convincing, not only the cell pattern but also the size of the cells under supercritical conditions has to be explained satisfactorily. We have not yet succeeded in doing that. We face with this problem the same formidable difficulties which have prevented us from arriving at a solution of the corresponding problem of the wavelength of supercritical rolls in Rayleigh–Bénard convection.

4

SURFACE-TENSION-DRIVEN BÉNARD CONVECTION EXPERIMENTS

The pioneering experimental study identifying surface tension as the cause of the hexagonal Bénard cells was made by Block (1956). This paper marks a turning point in the history of Bénard convection. Until then Bénard's experiments as well as the other early convection experiments were believed to be driven by buoyancy. Surface tension, although occasionally mentioned, did not seem to play a significant role. This was completely changed by the one-page text which describes Block's qualitative experiments. He worked with thin hydrocarbon films less than 1 mm thick, which had a free surface and in which he established hexagonal Bénard cells. The method by which the hexagonal cells were "established" is not spelled out; the temperature at the base of the films is referred to only in passing, and temperature differences are not given. Two decisive observations were made. First, Block observed that when a convecting layer was covered with a spreading monolayer of silicone (which was insoluble in the hydrocarbon), the flow in the fluid layer stopped "wherever and as soon as the [mono]layer passed over the Bénard cell." Block concluded correctly, "Clearly the variation of vertical density could not be affected so rapidly by the covering by the [mono]layer" (1956, p. 650). That meant that the driving mechanism of the cells was not buoyancy, but rather had to be located at the surface of the fluid. Second, Block observed Bénard cells in films of a thickness as small as $50 \mu m = 5 \times 10^{-3}$ cm. He stated that this thickness was "at least an order of magnitude smaller" than the depth for which convective instability could be expected under his experimental conditions. In modern terminology, the critical temperature difference for a 50-μm-deep layer of, say, ethanol with a free surface is about 6×10^{4}°C, because of the d^{-3} dependence of the critical temperature difference. The cells that Block observed in the 50-μm-deep fluid layer at temperature differences very much smaller than the critical temperature difference were, therefore, incompatible with the theory of buoyancy-

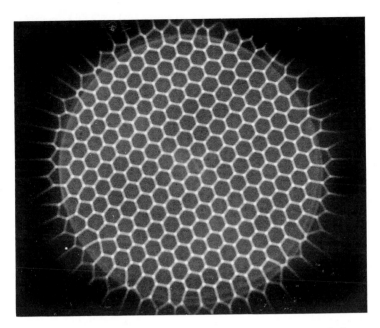

Fig. 4.1.   Shadowgraph picture of Bénard cells in a 1.9-mm-deep layer of silicone oil of 10.5 cm diameter. The fluid is under a 0.4-mm-deep layer of air, which is under a uniformly cooled sapphire lid. The bottom is a uniformly heated chromium-coated silicon crystal. Bright lines indicate cold fluid. After Koschmieder (1991).

driven convection. Block suggested that the phenomena which he observed were caused by surface tension. In view of the qualitative nature of his results one wonders whether his paper would not have been forgotten if the point that surface tension can be the cause of the hexagonal Bénard cells had not been made again two years later by Pearson (1958).

There was no rush to verify Block's results or Pearson's theory. Koschmieder (1967a) made a set of experiments studying the formation of hexagonal convection cells under an air layer. In order to bring the thermal conditions on top of the fluid under control, a uniformly cooled glass lid was placed about 1 mm above the fluid surface. Very regular patterns of hexagonal cells formed. For a modern shadowgraph picture of hexagonal cells from a similar experiment see Fig. 4.1. The patterns produced by the shadowgraph technique have been discussed in an informative article by Jenkins (1988). From the cell patterns follows the critical wavelength of the motions. In a 4.28-mm-deep layer of silicone oil, Koschmieder found $\lambda_c = 3.12 \pm 0.03$, or $a_c = 2.014 \pm 0.02$. This value is in striking contrast to the value of $\lambda_c = 2.34$

predicted by linear theory for buoyancy-driven flow with a free surface and a rigid bottom (Section 2.2). Nield (1964), on the other hand, predicted a critical wavelength $\lambda_c = 3.14$, or $a_c = 1.993$, for surface-tension-driven convection with insulating conditions on the fluid surface. Since air at rest is a very poor thermal conductor, the insulating boundary condition is well approximated in these experiments. The critical wavelength measured by Koschmieder confirmed the value of the wavelength predicted by Nield's theory of surface-tension-driven convection in the presence of gravity.

The value of the critical Marangoni number at the onset of surface-tension-driven convection was verified by Palmer and Berg (1971), using the break of the heat transfer curve as an indicator for the onset of convection as Schmidt and Milverton (1935) had done. The experiments of Palmer and Berg differed from those of Schmidt and Milverton by the gap of air which, for surface-tension-driven convection, has to separate the fluid layer and the lid. Since air is such a poor thermal conductor, much of the temperature difference between the top and bottom plates falls off in the air gap. The uncertainty of the temperature difference across the fluid layer and consequently the uncertainty of either the Marangoni number or the Rayleigh number is therefore comparatively large, but in general the predicted values of the critical Marangoni number as well as Nield's (1964) stability diagram were confirmed. Neither the flow pattern nor the onset of convection was observed visually.

An attempt to verify Pearson's (1958) theory was made by Koschmieder and Biggerstaff (1986). As discussed in Chapter 3 the consequences of surface tension effects should become dominant over buoyancy effects if the depth of the fluid is decreased sufficiently. It has, therefore, been believed that the results of Pearson's theory, which does not incorporate gravity and consequently does not consider buoyancy, can be approximated in laboratory experiments if the fluid layers are made about 1 mm deep or less, regardless of the specific fluid, since most fluids have similar values of $\partial S/\partial T$. Experiments made with such thin fluid layers are similar to Bénard's (1900) experiments, which were done with layers between 0.5 to 1 mm deep, the difference with modern experiments being mainly that the thermal conditions on top of the fluid are controlled, whereas in Bénard's experiments the fluid was cooled by ambient air. A surprising result of the experiments of Koschmieder and Biggerstaff was the appearance of subcritical motions before the onset of the instability which causes the hexagonal cells. By subcritical we mean that these motions appeared at $\mathcal{M} < \mathcal{M}_c$. These subcritical motions had the characteristics of an imperfect bifurcation, not of a subcritical bifurcation. If the layer of silicone oil was 2 mm deep or more, the onset of convection occurred out of the resting fluid in the form of hexagonal cells at the expected value of the critical Marangoni number. If, however, the depth of the fluid was less than 2 mm, subcritical motions visible with the naked eye formed first, the motions consisting of ill-defined small-scale cells. The interval ΔT over which the subcritical flow was observed increased very much with decreased fluid depth. The development of the subcritical motions seemed to be a continuous process from a barely perceptible to a weak flow, thereby differing markedly from a critical phenomenon. Hexagonal Bénard cells formed spontaneously out of the subcritical motions when the temperature difference was increased to the critical temperature difference given by the critical Marangoni number. Patterns with very large numbers of regular hexagonal cells were observed.

The subcritical motions are at variance with the basic result of linear theory of surface-tension-driven convection according to which, in a plane Boussinesq fluid layer with surface tension gradients, motions are dampened exponentially before the critical temperature difference is reached; i.e. sustained motions caused by heating from below are not possible. On the other hand, Davis (1969a) predicted with the energy method the possibility of a subcritical instability in surface-tension-driven convection. However the subcritical motions observed by Koschmieder and Biggerstaff extended, in the case of their thinnest fluid layer, deep into the stable range of Davis' (1969a) stability diagram. So the origin of the subcritical flow is a mystery, and poses a principal problem. By putting a lid on the fluid, one can test whether the observed subcritical motions are a consequence of surface tension gradients or other surface conditions.

The outcome of the experiments of Koschmieder and Biggerstaff (1986) is not an indication that Pearson's (1958) theory is not valid within the framework of its premises, but rather that Pearson's theory cannot be tested in a laboratory, in the presence of the acceleration of gravity, simply by reducing the depth of the fluid layer. Pearson's theory can be verified rigorously only in a microgravity environment, such as on the space shuttle. Two preliminary experiments undertaken to verify the formation of Bénard cells in microgravity were made on the *Apollo XIV* and *Apollo XVII* space flights, and were described by Grodzka and Bannister (1972, 1975). In spite of the admirable efforts of the astronauts performing these experiments, the results are inconclusive, because in both experiments the fluid surface was not plane and the fluid was stirred while it was heated. Preliminary microgravity Bénard convection experiments have also been described by Schwabe et al. (1990).

Besides by their pattern, Bénard cells are characterized by the deformation of the fluid surface over the cells. Bénard observed that the fluid surface was depressed by about 1 μm = 10^{-3} mm over the cell centers where the fluid

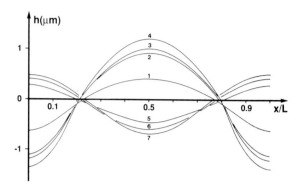

Fig. 4.2.   Relief of the surface of hexagonal Bénard convection cells in silicone oil layers for different supercritical conditions expressed by $\varepsilon = \mathcal{R}/\mathcal{R}_c + \mathcal{M}/\mathcal{M}_c - 1$. Upper part, fluid depth 4.04 mm, $\varepsilon = 0.32, 1.82, 2.41, 3.26$ in ascending order. Lower part, fluid depth 1.75 mm, $\varepsilon = 0.04, 0.18, 0.98$ in descending order. After Cerisier et al. (1984).

rises, in fluid layers about 1 mm deep (see Fig. 1.3). This problem did not receive attention for many years. Jeffreys (1951) showed theoretically that the fluid surface over the centers of hexagonal cells should be elevated, not depressed. Jeffreys' paper dealt with a free surface, free in the exact sense, and with buoyancy-driven flow; it was written at a time when the significance of surface tension for convection was not recognized yet. Jeffreys, of course, wondered about the discrepancy between theory and observation and called for more experiments. This puzzle was, as already mentioned, solved theoretically by Scriven and Sternling (1964). When in the theory of surface-tension-driven convection (Pearson, 1958) the condition of a nondeformable top surface is relaxed and surface deformation is permitted, there is upflow beneath depressions and downflow beneath elevations of the surface. Note that in the linear theory of Rayleigh–Bénard convection the direction of motion in the centers of hexagonal cells can be either upflow or downflow. This ambiguity disappears in linear theory if both small deflections and surface tension gradients on the upper surface are introduced. Scriven and Sternling's result explains Bénard's observation.

   The first modern quantitative experimental determination of the curvature of the fluid surface over hexagonal convection cells was made by Cerisier et al. (1984). Using various optical methods, they arrived at a profile of the relief of the surface, as shown in Fig. 4.2. From this figure it follows that for thin layers of silicone oil of depth <2 mm the surface over the cell centers is depressed and that the depression increases with increased Marangoni number. The effects of surface tension gradients are dominant in such thin layers.

On the other hand, for thicker fluid layers of depth >3 mm, the surface of the fluid is elevated over the centers of the hexagonal cells and the elevation increases with increased Rayleigh number. In such layers buoyancy is dominant. It must be observed that Fig. 4.2 is somewhat idealized; the statistical variations of the height of the fluid surface brought about by the irregularities of the hexagonal cells alone limit the accuracy of the measurement of the height of the fluid surface. Nevertheless the results of Cerisier et al. (1984) give a good picture of the variation of the fluid surface over hexagonal cells and clearly confirm Bénard's observations.

The size of the Bénard cells, or in other words the wavelength of the motions, is another characteristic of Bénard convection. As mentioned in Chapter 1, Fig. 23 of Bénard's (1900) paper shows that the dimension of the cells in his experiments had a minimum at temperatures slightly above the temperature at which the pattern disappeared, presumably at ΔT_c, when the fluid gradually cooled down. In modern terminology Bénard's observations indicate that after the onset of convection the wavelength first *decreases* with increased ΔT, reaches a minimum at moderately supercritical temperature differences, and then *increases* with further increases of ΔT. These observations have been overlooked for more than 80 years. The first modern study of the size of surface-tension-driven Bénard cells was made by Cerisier et al. (1987). In very large fluid layers of aspect ratio $\Gamma \approx 70$ they found a continuous increase of the cell size with increased Marangoni number in the range from $\mathcal{M}_c$ to around $5\mathcal{M}_c$. Their results confirm the observations of Dauzère (1912). The increase in the cell size of supercritical hexagonal cells was observed again in the experiments of Cerisier and Zouine (1989). In this paper measurements of the cell size as a function of the aspect ratio are also given. It was found that the critical cell size is, within the experimental error, the same up to aspect ratios of about 100.

The variation of the cell size of Bénard cells near the onset of convection was studied by Koschmieder (1991). Working with large fluid layers ($\Gamma \approx 55$) in thermal equilibrium and with practically perfect patterns, such as in Fig. 4.1, he found that the number of cells first increases if, after the onset of convection, the Marangoni number is increased from $\mathcal{M}_c$ to about $1.5\mathcal{M}_c$. An increase of the number of cells in a container of a given size amounts to a *decrease* of the cell size or of the wavelength of the motions. In this range of Marangoni numbers the Rayleigh number was only of the order of 100, so that the flow was certainly surface tension driven. The variation of the number of the cells is shown in Fig. 4.3.

This figure confirms Bénard's findings and means that, indeed, the wavelength of supercritical surface-tension-driven convection first *decreases* with increased ΔT. The decrease of the wavelength of surface-tension-driven

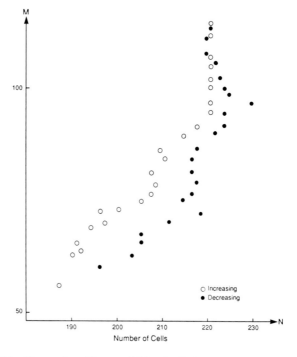

Fig. 4.3.   The number of hexagonal Bénard cells in a 1.9-mm-deep layer of silicone oil of 1 cm^2/sec viscosity under a 0.4-mm layer of air, as a function of either increasing or decreasing temperature difference. The number of cells is inversely proportional to the cell size or the wavelength of the motions. After Koschmieder (1991).

convection is in startling contrast to the continuous increase of the wavelength of buoyancy-driven Rayleigh–Bénard convection, which we shall discuss in Chapter 6. When the temperature difference across the fluid is increased, it is not only the Marangoni number that increases, but also the Rayleigh number, because both depend on the same ΔT. As the Rayleigh number increases with increased ΔT, at some point the Rayleigh number apparently reaches a value sufficient for the onset of buoyancy-driven convection. Thereafter buoyancy seems to be dominant, and from then on the usual increase of the wavelength of buoyancy-driven convection prevails. So when the temperature difference is increased, the wavelength first decreases, reaches a minimum, and then increases, in agreement with Bénard's observations and, as far as the final increase is concerned, in agreement with Cerisier et al. (1987). Another result apparent from Fig. 4.3 is that the wavelength of the motions is, within the experimental uncertainty introduced by the effects of the lateral wall, the same whether the temperature difference is very slowly increased or decreased, i.e. independent of initial conditions. That means that the wavelength of slightly supercritical surface-tension-driven convection is *unique*, in contradiction to the theoretical prediction of Cloot and Lebon (1984).

All investigations of surface-tension-driven convection discussed so far deal with fluid layers of practically infinite horizontal extent. The experimental observations indicate that the always present lateral walls of the fluid layers affect only the cells bordering the lateral boundary, whereas in the interior of the layer the hexagons do not seem to be affected by the lateral constraints; see, e.g., Fig. 4.1. This will, of course, no longer be so if the size of the fluid layer permits only the formation of a few cells. Nonlinear theoretical studies of surface-tension-driven convection in small circular and rectangular containers were made by Rosenblat et al. (1982a,b). They found that in sufficiently small circular layers the cells are no longer hexagonal but have the form of pieces of pie separated by the nodal lines of sin $n\phi$. The critical Marangoni numbers were found to be a sharply increasing function of the aspect ratio for $\Gamma = L/d < 1$, in agreement with the corresponding results for Rayleigh-Bénard convection in small containers obtained by Davis (1967) and Charlson and Sani (1970). The patterns of surface-tension-driven convection in small circular containers were confirmed experimentally by Koschmieder and Prahl (1990). Similar patterns were observed under similar conditions by Ezersky et al. (1991). Pattern formation in rectangular containers with a ''free'' surface is apparently more difficult. In the experiments of Koschmieder and Prahl two triangular cells formed if there was space for only two cells in a square container. On the other hand, with space for either one or four cells, either one or four perfect square cells formed. But with space for three, five, or six cells in a square container, three, five, or six cells of unorthodox shape appeared. The bifurcations in a two-dimensional rectangular cavity of aspect ratio $\Gamma = 2$ were studied theoretically by Winters et al. (1988). The consequences of the lateral boundary conditions on surface-tension-driven flow in rectangular containers were studied numerically by Dijkstra and van de Vooren (1989). An experiment on two-dimensional surface-tension-driven convection in an annulus with $\Gamma_r \cong 1$ and $\Gamma_\phi \cong 77$ was made by Bensimon (1988). In his experiment the lateral boundaries of the fluid dominated the flow; there were no hexagons but short parallel rolls whose axes point in the radial direction, just as in the experiments by Stork and Müller (1975) with Rayleigh–Bénard convection in an annulus. Actually, buoyancy may have a bearing on the outcome of the experiments of Bensimon, because he worked with $\mathcal{M}/\mathcal{M}_c \approx 0.5$ and $\mathcal{R}/\mathcal{R}_c \approx 0.5$. These investigations of surface-tension-driven convection in small containers complete our knowledge about Bénard convection.

There is, all in all, a dearth of good data about surface-tension-driven Bénard convection. Only one feature stands out, namely the hexagonal cell form, which is apparently unique. We need better data if we want to understand, finally, after 90 years, the reason for the preference of the hexagonal planform in Bénard convection and solve one of the most simple and clearcut pattern selection problems.

5

LINEAR RAYLEIGH–BÉNARD CONVECTION EXPERIMENTS

5.1 Early Experiments

Bénard seems to have lost interest in cellular convection after the completion of his thesis. He pursued other topics, the most notable result of which was the discovery (Bénard, 1908) of what later came to be known as the von Karman vortices. In 1912 Dauzère, a student of Bénard, picked up cellular convection again. In the absence of a theory to be guided by, the task was to find something that went beyond Bénard's principal results. No new avenues opened by sticking closely to Bénard's experimental methods. One result (Dauzère, 1912) deserves mention: the observation of an obvious decrease of the number of hexagonal cells in a fluid layer when the temperature of the bottom plate was increased beyond the 100°C Bénard had worked with. A decrease of the number of cells is equivalent to an increase of the wavelength of the convective motions. A systematic investigation of the dependence of the cell size on the temperature difference across the fluid layer would have been an early guide to a principal feature of supercritical convection, the variation of the wavelength of the motions with increased temperature difference.

The focus of experimental work on cellular convection then shifted to England, probably because of the theoretical work done by Jeffreys (1926) and Low (1929). The very first experiments with air as the convecting medium were performed by A. Graham (1933). Out of necessity, because the air had to be contained on top, a rigid lid was used as the upper boundary for the first time in these convection experiments. Although Graham's apparatus was very simple (cooling was by ambient air only), he made one basic observation. He found that, when the conditions had become steady, "polygons with descent in the centers" had formed, the flow thus being of a direction opposite to the direction of motion in the centers of hexagonal cells in fluids, as observed by Bénard. Graham suggested that the direction of the circulation in hexagonal cells in gases and in fluids is determined by the variation of the viscosity of the medium with temperature. As is well known, the viscosity of fluids decreases with temperature, whereas the viscosity of gases increases with temperature. Graham's suggestion was verified experimentally by v. Tippelskirch (1956), whose experiment will be discussed in Section 8.1.

The onset of convection was probed very convincingly by Schmidt and Milverton (1935). They measured the heat flux through a layer of water contained between two plane, parallel, horizontal brass plates. They expected and found a discontinuity in the rate of heat transfer at the critical temperature difference for the onset of convection. At the critical temperature difference, thermal conduction through the molecular conductivity of the fluid is supplemented by additional heat transfer by the bulk motion of the fluid. From the location of the discontinuity of the heat transfer curve they determined the critical Rayleigh number, which they found at $\Re_c = 1770 \pm 150$, in reasonable agreement with Jeffreys' (1928) prediction of the critical Rayleigh number. This was an important accomplishment which over the years has developed into the standard method for the experimental determination of the onset of convection. A modern heat transfer curve with the discontinuity of the heat transfer at $\Re_c$ is shown in Figure 5.8, Section 5.4.

In the 1930s Bénard became active again in cellular convection. One of his associates made a futile effort to develop a theory of thermal convection that was different from Rayleigh's theory (Vernotte, 1936). Vernotte's effort was apparently fueled by the correct observation that the temperature differences in Bénard's experiments were incompatible with the critical temperature difference expected from Jeffreys' calculations. This had been noticed earlier by Bénard (1930); the magnitude of the disagreement was, however, toned down by Vernotte. As we know now the discrepancy between the theoretically expected and experimentally observed critical temperature difference in Bénard's experiments is due not to a failure of linear theory, but rather to a misinterpretation of Bénard's experiments as being buoyancy driven (Rayleigh's theory) whereas they were actually surface tension driven (Pearson's theory). Much more successful than Vernotte's work was the experimental work done in Bénard's laboratory by Avsec. He investigated in detail convection caused by heating from below in an air layer with horizontal shear. This was careful work which has not been improved upon even today. No doubts are expressed in Avsec's (1939) thesis about the validity of Rayleigh's theory. The essential results of this study can be found in Bénard and Avsec (1938).

5.2   The Planform of Convection

Linear theory of convection has only three experimentally verifiable results: the cellular pattern, the critical wavelength, and the critical temperature difference for the onset of convection. We shall discuss each of these topics, beginning with the oldest feature, the cellular pattern or, as it is often referred to, the planform. We note that linear theory is ambiguous about the pattern except when sidewalls are taken into account. Although the determination of the pattern seems to be next to trivial – a pattern can easily be visualized – we shall find that no consensus has been reached among experimentalists as to what the correct pattern is at the onset of convection. The formation of the spatial structures in Rayleigh–Bénard convection was discussed recently by Getling (1991).

We must first discard one of the usual assumptions of linear theory of convection, namely that the fluid layer is of infinite horizontal extent. The notion of an infinite horizontal layer is a convenient theoretical simplification which allows us to disregard lateral boundaries of the fluid. This concept is, however, unrealistic since in experiments the fluid must be bounded. That does not mean that the fluid is thereby confined to a small layer; very many experiments study convection in large fluid layers for which insulating lateral walls appear to be of little dynamic significance. Conceptionally the width of the layer does not make much difference; the lateral boundaries introduce an orientation into the setup whether the walls are far away or not, whereas in an infinite layer there is no orientation at all, provided we retain the assumption of perfectly uniform temperatures on the top and bottom of the fluid. Otherwise an orientation is introduced by the nonuniformity of the temperature field. The width of the fluid layer is, however, important for the experiments because it becomes increasingly difficult to maintain uniform temperatures on the top and bottom of the layer as the width of the layer is increased. The width of a fluid layer is characterized by the nondimensional aspect ratio $\Gamma = L/d$, which we define as the ratio of the width L of the layer divided by the depth d of the fluid. We shall use this definition regardless of whether the fluid layer is rectangular or circular. It should be noted that in the literature the aspect ratio in circular vessels is commonly defined as the ratio of the *radius* to the depth, introducing thereby a difference of a factor of 2 in the aspect ratios of a circular layer and a square layer of the same diameter, which is confusing. In rectangular containers there are, of course, two aspect ratios, but usually only the ratio of the length of the longer side to the depth is given.

Lateral boundaries with their associated boundary conditions remove the degeneracy of the pattern that exists on an infinite plane in linear theory at $\mathcal{R}_c$. Even the most simple lateral boundary condition, the requirement that the velocity component normal to the wall of the container be zero at the wall, makes it impossible for most patterns to match a container of a given shape; e.g. straight parallel rolls do not match a circular container. For practical reasons only circular, square, and rectangular containers have been used in convection experiments. Even in the presence of lateral walls we expect *absolutely regular* patterns at the onset of convection, in response to the assumed *absolutely uniform* temperatures on the top and bottom of the fluid and correspondingly uniform conditions along the lateral boundaries. If we observe an irregular pattern or defects in a pattern, we must conclude that the irregularities are caused by finite imperfections of the temperature field or other deviations from the assumed ideal uniformity. Thermal noise cannot be the cause of local defects, because the magnitude of thermal noise is orders of magnitude below the nonuniformities of the temperatures in the horizontal boundaries of the fluid, in particular of the lid. Experiments made with low Prandtl number fluids often result in patterns of poor regularity. Nonlinear effects unquestionably play a larger role in low Prandtl number fluids, but in order to prove that irregularities in the pattern are the consequence of nonlinear effects it has to be proved that the temperature nonuniformities in the horizontal boundaries are much smaller than the critical temperature difference of the layer.

For a comparison we shall look at the Taylor vortex problem. There is very little talk about defects in Taylor vortex flow. This is not so because there is no noise in the velocity distribution of the subcritical azimuthal Couette flow, or because there are no nonlinearities involved. Actually the problem seems to be more susceptible to nonlinearities, but the absence of defects appears to be a result of the fact that the ideal boundary conditions used in theory can be much more easily realized in Taylor vortex experiments than in Rayleigh–Bénard convection experiments. As a consequence it is taken for granted that perfect two-dimensional toroidal vortices or regular wavy three-dimensional vortices are formed at the onset of instability in the Taylor vortex problem.

The first experimental investigation of the planform of Rayleigh–Bénard convection in bounded circular, as well as square and rectangular, containers was made by Koschmieder (1966a). In this paper convection of a Boussinesq fluid of Prandtl number $\mathcal{P} = 950$ was studied in circular containers of aspect ratio $\Gamma = 19.7$ and $\Gamma = 38.8$. The lateral walls were made of lucite, whose thermal conductivity differs by only 10% from the thermal conductivity of the silicone oil used for the experiments. The vertical temperature gradient in the wall was thereby made practically identical to the temperature gradient in the fluid. The temperature difference across the fluid layer was increased very slowly in order to establish a linear vertical temperature gradient in the fluid, in order to comply with the assumption of time independence in theory. The bottom of the apparatus was a 2.5-cm-thick copper plate. The excellent thermal conductivity of copper in combination with the very poor thermal conductivity of the silicone oil ensures a practically uniform temperature underneath the fluid. The lid of the fluid was either a 2.5-cm-thick copper plate or, in order to see through the lid, a 3-mm-thick glass plate. Both were held at a constant temperature by water circulated over the lid. Therefore the experimental conditions conform in a good approximation to the conditions set forth in linear theory, except for the fluid being bounded. The pattern observed in the circular containers consisted of circular, concentric rolls (Fig. 5.1). Hexagonal cells did not appear because the lid was in contact with the fluid, thereby eliminating surface tension effects, which are the prime cause of the formation of hexagonal cells. However without the lid a very regular pattern of hexagonal cells formed in the fluid when it was heated from below.

The observed circular concentric pattern reflects the axisymmetry of the setup. Axisymmetric lateral boundary conditions make the problem axisymmetric if the top and bottom temperatures of the fluid layer are perfectly uniform. We expect, therefore, an axisymmetric solution of the linear equations of motion at the critical Rayleigh number. Axisymmetric circular concentric rolls are the only solution which satisfies everywhere along the wall the elementary boundary condition $v_n = 0$. Circular concentric rolls in a circular container are the result of a superposition of ring cells [equation (2.55)] of different amplitudes and wave numbers which fits the size of the container. It appears that circular containers are particularly suited to the investigation of convective motions in bounded layers in the context of theories of convection in infinite fluid layers, because a circular lateral wall is the smallest boundary for a given area of the fluid layer.

Convection in rectangular containers was also studied by Koschmieder (1966a). For this purpose square or rectangular frames were inserted into the otherwise unchanged apparatus. A particular effort was made with square containers of aspect ratio $\Gamma = 12$. In some of the experiments square cells formed (Fig. 5.2). The square cells resulted from a superposition of rolls which formed parallel to the walls of the square frame, as is shown in a photograph in this paper. The superposition of rolls corresponds exactly to the theoretical description of square cells. Figure 5.2 is the only photograph in the literature showing square cells at $\mathcal{R}_c$ under the conditions of linear theory of convection. Photographs of another square cell pattern were published by

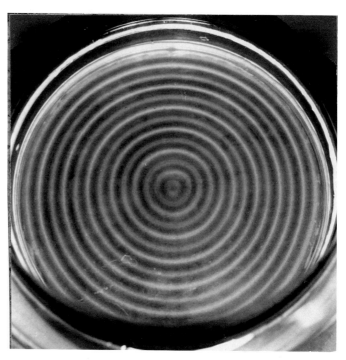

Fig. 5.1. Rayleigh–Bénard convection in the form of circular concentric rolls. A silicone oil
layer of 20 cm diameter and 0.765 cm depth is on a uniformly heated copper block
and under a uniformly cooled glass lid. The fluid is in contact with the lid. $\nu =$
1 cm^2/sec, $\mathcal{P} \approx 950$. Just critical. Visualization with aluminum powder. After
Koschmieder (1974).

LeGal et al. (1985). The convective motions in their experiment were, however, triggered by the controlled initial conditions method introduced by Chen
and Whitehead (1968). This method produces, as will be described in detail
later, a temporary, spatially periodic temperature distribution in the slightly
supercritical fluid. Many photographs of square cells in non-Boussinesq fluid
layers can be found in White (1988). These square cells have likewise been
induced by controlled initial conditions.

In the majority of Koschmieder's (1966a) experiments with the square container, convection started in the corners, in which arced rolls developed. The
center of the layer was finally filled with a patch of four square cells. This
unsatisfactory result points out the great sensitivity of the square cell pattern
to very small nonuniformities of the temperature field. On the other hand, in
a rectangular container of $(9 \times 12)d^2$ size straight rolls formed whose axes
were parallel to the short side of the container. For an example of the pattern
of convection in a rectangular container see Fig. 5.3.

Fig. 5.2. Square convection cells under a glass lid in a square container of $(10 \times 10)d^2$ size. Just critical. After Koschmieder (1966a).

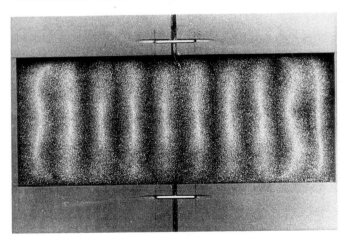

Fig. 5.3. Rayleigh–Bénard convection in a rectangular container. $\Gamma_x = 10$, $\Gamma_y = 4$. After Srulijes (1979).

From the experiments with bounded fluid layers Koschmieder concluded that the form of the pattern of Rayleigh–Bénard convection in a bounded (Boussinesq) fluid layer on a uniformly heated plate "is determined by the shape of the lateral wall" (1966a, p. 11).

The Rayleigh–Bénard instability is often referred to as an example of a symmetry-breaking instability, and a particularly simple example at that. It is said that the onset of convection breaks the original translational and rotational invariance of the static fluid layer before the onset of instability. Strictly speaking, symmetry breaking occurs only when the fluid layer is of infinite horizontal extent. In reality the fluid layers are always bounded. In a bounded static circular fluid layer there is no translational invariance. Rotational symmetry is preserved if the onset of convection takes place in the form of circular concentric rolls. In a static fluid layer in a rectangular container there is no translational invariance nor is there rotational invariance, before or after the onset of convection. Nevertheless rolls parallel to the short side of the container reflect the symmetry of the container, and so do square cells in a square container. If the pattern of motion after the onset of convection is irregular, symmetry clearly seems to be broken, but only if there was symmetry before the onset of convection, in particular uniform temperatures on the top and bottom of the layer. Usually there is no proof that this was really so. The only pattern that always seems to break symmetry is the hexagonal pattern of surface-tension-driven convection. Regardless of whether the container is circular or rectangular, genuinely hexagonal patterns form with ease in such layers and break the symmetry established by the shape of the container.

Convection in laterally bounded fluid layers is amenable to theoretical analysis, although realistic no-slip boundary conditions prevent the separation of the variables, as was first noted by Pellew and Southwell (1940) and Zierep (1963). They circumvented this difficulty by considering slip boundary conditions. The first theoretical investigation of convection in rectangular boxes with no-slip conditions at the sidewalls was made by Davis (1967). In this paper the linear equations of motion in boxes with dimensions $\frac{1}{4} \leq L/d \leq 6$ and with perfectly conducting lateral boundaries were solved with the Galerkin method. The principal result of this study was the prediction that rolls with axes parallel to the shorter side of the box should fill rectangular containers, instead of the rectangular cells one might expect from the solutions of the membrane equation. In the special case of a square box, square cells were predicted, which are the consequence of rolls which form along each of the equivalent sides and superimpose. The rolls in question here are so-called finite rolls. These are rolls with two nonzero velocity components which are perpendicular to the axis of the roll and depend on all three spatial variables, the third velocity component in the direction of the axis of the rolls being zero. This is different from the infinite rolls of linear theory, where the two nonzero velocity components depend only on the two coordinates

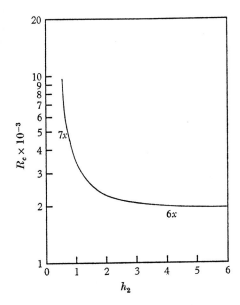

Fig. 5.4. Critical Rayleigh numbers in a rectangular box with a fixed length $x/d = 6$ as a function of the width $y/d = h_2$; nx denotes n finite x-rolls, i.e. rolls whose axes are parallel to the y-direction. After Davis (1967).

perpendicular to the axis of the roll. The finite rolls have been introduced so that all boundary conditions can be satisfied, in particular so that all three velocity components are zero at the lateral walls. Davies-Jones (1970) has shown that finite rolls aligned perpendicular to rigid sidewalls are not exact solutions of the linear convection equations, but they approximate the computed solutions for free top and bottom boundaries for a large range of aspect ratios. Finite rolls can be clearly recognized in a numerical investigation of convective motions in a 10 : 4 : 1 box made by Kirchartz and Oertel (1988, Fig. 7 therein). Similar computations for a 4 : 2 : 1 box were made by Kessler (1987).

   The upper bounds of the critical Rayleigh number as a function of the length and the width of the rectangular container were also determined by Davis. Characteristic of the curves of the critical Rayleigh number is a very sharp increase of $\mathcal{R}_c$ as the length of the container becomes shorter than the depth of the fluid, whereas in extended fluid layers the critical Rayleigh number decreases asymptotically to the conventional value 1708. A curve showing the variation of the critical Rayleigh number in rectangular boxes is shown in Fig. 5.4. The critical Rayleigh numbers were calculated again with other trial functions and insulating lateral boundaries by Catton (1970). Luijkx and Platten (1981) investigated the convective motions in an infinite channel of rectangular cross section with rigid top and bottom plates and insulating sidewalls. They found lower critical Rayleigh numbers for three-dimensional flow rather than for finite rolls. However the three-dimensional form of motion does not agree with the observations of Stork and Müller (1972), nor with what we see in Fig. 5.3. Edwards (1988) found crossed rolls at the onset of convection in boxes with small aspect ratio, in particular for square or near-square containers. Davis' prediction that the preferred mode at the onset of convection in a rectangular container consists of rolls parallel to the short side of the container was strongly supported by the papers of Chana and Daniels (1989) and Daniels and Ong (1990). Chana and Daniels studied the onset of convection in a channel with linear theory, and Daniels and Ong studied, with an amplitude equation, weakly nonlinear convection in an infinite horizontal channel with a rectangular cross section with conducting sidewalls. They found rolls with axes aligned perpendicular to the sidewalls, but these rolls have a nonzero velocity component perpendicular to the sidewalls, in contrast to the finite rolls of Davis' theory, for which $v_y = 0$. Chana and Daniels discussed details of the motions near the sidewalls and found that "close to the walls the motions remain fully three-dimensional and [there is] a reversal of the vertical flow" near the wall (1989, p. 257). The details of these motions are in qualitative agreement with the interferograms shown by Bühler et al. (1979) and are potentially of great value for the understanding of pattern selection in bounded layers.

As we see there is a strong theoretical foundation for the notion that Rayleigh–Bénard convection in rectangular containers forms a *regular* pattern of rolls parallel to the short side of the container. In other words, the form of the convective motions in a rectangular container has been found in theory to be determined by the shape of the lateral walls.

Why, then, do the rolls in rectangular boxes orient themselves parallel to the short sides of the boxes and not parallel to the long sides? This is, as suggested by Davis (1967), probably caused by different amounts of dissipation of the rolls at the walls when the axes of the rolls are either parallel or perpendicular to a wall. Finite rolls with axes perpendicular to a wall produce little additional dissipation at that wall, whereas rolls with their axes parallel to a wall produce additional dissipation at that wall. In order to maintain balance of dissipation and energy release as the balance theorem requires or, in practical terms, in order to maintain the value of the wavelength of infinite rolls for the finite rolls in a rectangular container, the fluid chooses that orientation of the rolls which reduces the amount of dissipation at the walls to a minimum. Another, but probably equivalent, way to explain the preference of the orientation of the rolls in a rectangular box was proposed by Segel (1969). Since according to his calculations the critical Rayleigh number in a finite container is equal to $\mathcal{R}_{c\infty}$ plus a term proportional to L^{-2}, where L is the extension of the layer, it follows that "the fluid will 'prefer' to convect in the form of rolls parallel to the shorter side of the rectangular dish so that L, the length of the side perpendicular to the roll axes, is as large as possible" (p. 208). Hence the critical Rayleigh number is smaller in the direction along the long side of the box, and the onset of convection occurs at the lowest value of the Rayleigh number.

The results of Davis' (1967) study are in agreement with the square pattern in the square container and the roll pattern in the rectangular container observed by Koschmieder (1966a). A specific test of the patterns predicted by Davis was made by Stork and Müller (1972), and general agreement was found. In large containers Stork and Müller observed somewhat irregular patterns, which are probably caused by the unorthodox stirring of the fluid under nearly critical conditions prior to the formation of the pattern. Support for the existence of rolls parallel to the short sides of a rectangular container comes also from the paper of Dubois and Bergé (1978). Innumerable convection experiments in rectangular containers performed since Davis' study seem to confirm that rolls which form in such containers are parallel to the short side of the box. One can, furthermore, investigate convection in narrow annuli, which are quasi-rectangular containers having no ends in the direction parallel to the walls. Stork and Müller (1975) have shown that the convective motions in annuli consist of rolls whose axes point in the radial direction; this means that they are parallel to the short section through the container and not along the long wall, confirming Davis' prediction. On many occasions the rolls observed in later experiments with rectangular containers were neither straight nor really parallel to the short side. One wonders whether that was due to nonuniformities of the temperatures in the walls or in the horizontal boundaries of the fluid, rather than to deficiencies in Davis' (1967) theory.

The geometry of narrow annuli has recently served as a means to study what is called one-dimensional convection, which refers to a periodic cellular state that depends on only one space variable and has periodic boundary conditions in that direction. One-dimensional convection is approximated by the rolls in a narrow annulus whose axes are oriented in the radial direction. One-dimensional Rayleigh–Bénard convection at fairly high Rayleigh numbers has recently been studied experimentally by Ciliberto (1991) and by Daviaud and Dubois (1991), and one-dimensional Bénard convection has been studied by Bensimon (1988).

The first theoretical investigation of convection in circular containers was made by Charlson and Sani (1970). They used a variational formulation and the Rayleigh–Ritz method in order to approximate the solution of the linear equations of motion. Assuming axisymmetry they determined the streamlines of the flow, which was in the form of circular concentric rolls, and they determined the wavelength of the circular rolls for aspect ratios ranging from 1 to about 26. Charlson and Sani also determined the critical Rayleigh numbers for various aspect ratios using conducting as well as insulating lateral boundaries. They found sharply rising values of $\Re_c$ for small aspect ratios, whereas the critical Rayleigh numbers were found to decrease rapidly to $\Re_c = 1708$ for aspect ratios $\Gamma > 10$. This corresponds to the results of Davis (1967) concerning convection in rectangular containers. The results of Charlson and Sani are in good agreement with the flow patterns observed by Koschmieder (1966a). The results of Charlson and Sani confirm the notion that the form of the convective motions in a bounded fluid layer is determined by the shape of the lateral wall.

Charlson and Sani (1971) extended their calculations to nonaxisymmetric flows in circular containers. In this paper they plotted the critical Rayleigh numbers versus the aspect ratio for different numbers of azimuthal nodes. From these curves it follows that axisymmetric flow is not necessarily the preferred mode of convection in small circular containers in the presence of nonaxisymmetric disturbances, in particular in the case with an insulating lateral wall. This result was confirmed by Rosenblat (1982), who investigated weakly nonlinear flow in circular containers of aspect ratio $\Gamma < 2$ with free boundary conditions on the sidewalls and found that the convective motions just above critical are not axisymmetric. Charlson and Sani found that the pattern with just one azimuthal node requires critical Rayleigh numbers clearly larger than the $\Re_c$ for patterns with two, three, and four azimuthal nodes if $\Gamma > 2$. Similar results were obtained by Buell and Catton (1983a) when they studied the same problem with different trial functions. For $\Gamma > 10$ the critical Rayleigh numbers calculated by Charlson and Sani for axisymmetric as well as nonaxisymmetric flows seem to converge to $\Re_c = 1708$, as one would expect. In other words, in a circular container with, say, $\Gamma > 20$ any pattern, whether axisymmetric or asymmetric, will appear at practically the same critical Rayleigh number. The degeneracy of the patterns predicted by linear theory for the infinite layer is, within the accuracy of our measurements, also present in bounded circular containers of large aspect ratio.

In general, the critical Rayleigh number in finite containers is the critical Rayleigh number $\Re_{c\infty}$ for an infinite layer plus a correction which is inversely proportional to the square of the (nondimensional) length L of the container, as was first shown by Segel (1969) and later by, e.g., Brown and Stewartson (1978). Segel's paper will be discussed in detail in Section 7.4. The dependence of the correction to the critical Rayleigh number on L^{-2} means that the influence of lateral walls fades rapidly as the area of the fluid layer increases. To give a specific example, the critical Rayleigh number of a circular fluid layer of aspect ratio 26 is 1.0023 times $\mathcal{R}_{c\infty}$, according to Charlson and Sani (1970). This shows that in convection experiments fluid layers of aspect ratios of 20 or more are, for all practical purposes, fairly close approximations of infinite fluid layers.

The existence of circular concentric rolls in circular containers was confirmed experimentally by Hoard et al. (1970) and by Koschmieder and Pallas (1974). Experiments in circular containers of small aspect ratio are described in Liang et al. (1969). Numerical studies of axisymmetric flow in small aspect ratio containers are described in Jones et al. (1976), and for moderate aspect ratio containers in Tuckerman and Barkley (1988). On the other hand, there are a number of experiments with circular apparatus in which axisymmetric patterns have not been found. These are described in the papers of Stork and Müller (1975), Kirchartz et al. (1981), Croquette et al. (1983), Steinberg et al. (1985), Heutmaker and Gollub (1987), and Croquette (1989b). The patterns found in these experiments vary, but can be summarized as fields of disordered rolls whose axes tend to end perpendicular to the circular wall; see, e.g., Fig. 5.5, which shows one of the more regular patterns of disordered rolls. It is often said that rolls in a circular container should be oriented perpendicular to the wall, just like the finite rolls in a rectangular container, because that would minimize friction at the wall. There is no theoretical support for this argument. Besides, the pattern in the interior of the circular container is then necessarily irregular, which does not match expectations from linear theory, and it is not known whether the total amount of dissipation of an irregular pattern is smaller or larger than that of a regular pattern. It is not difficult to make disordered rolls. The problem is that they are not strictly reproducible. To quote Croquette et al.: "The structure changes with each experiment" (1983, p. 294). On the other hand, experiments concerning linear Rayleigh–Bénard convection must be strictly reproducible, an elementary condition for a successful experiment.

The question of regular versus irregular patterns is most clearly apparent in the papers of Croquette (1989a,b). In Croquette (1989a) patterns in rectangular containers were studied. Croquette expects that "convection displays periodic and erratic solutions" (p. 121). Apparently in all of his experiments with rectangular containers the pattern was erratic. He stated, "The straight-roll pattern is more an exception than the rule, and if no special care is taken,

Fig. 5.5.   Quasi-steady disordered rolls in a circular container filled with water. Aspect ratio $\Gamma = 15.0$, $\mathcal{R} = 1.141\mathcal{R}_c$. Enhanced shadowgraph picture. After Ahlers et al. (1985).

no straight-roll pattern will be observed. Instead, a disordered and generally time-dependent pattern appears'' (p. 133). The special care referred to means the introduction of nonuniformities of the temperature at the shorter sides of the rectangular container by heating wires along these walls. All pictures of regular or nearly regular straight-roll patterns shown in this paper have been forced this way. Since the result of these investigations of the patterns in rectangular containers at the onset of convection and under moderately supercritical conditions is that the patterns are "disordered and generally time-dependent," this study contradicts the results of linear theory of convection in rectangular containers (Davis, 1967) and the subsequent theoretical studies discussed above, as well as the results of weakly nonlinear theory of convection (Schlüter et al., 1965), which predict parallel rolls, i.e. an ordered pattern, not an erratic solution. There is also no indication of time dependence in the theoretical studies. Croquette's observations also contradict earlier experiments which produced parallel rolls in rectangular containers, as e.g., in Koschmieder (1966a), Stork and Müller (1972), and Dubois and Bergé (1978), and they are at variance with the rolls shown in Fig. 5.3 as well as with the computations of Kirchartz and Oertel (1988).

In Croquette (1989b) patterns in circular containers are discussed. Croquette observed near the onset of convection a "fairly straight, defect-free roll pattern," i.e. nearly parallel rolls. He continued, "The orientation of the pattern is in principle a free parameter; however, in practice, the same orientation is chosen in each experiment. Evidently, a small breaking of the cylindrical symmetry, which is difficult to avoid, decides this orientation" (p. 154). In other words the fluid layer was not really axisymmetric, and the asymmetry created the roll pattern. How such a pattern can be formed from an initial disturbance was shown numerically by Bestehorn and Haken (1991). In order to create circular patterns Croquette used "the same trick as in the rectangular cell which consists of placing a small heating wire along the entire perimeter of the container" (1989b, p. 157). The patterns then observed consisted of nearly circular concentric rolls, but had defects, were off-center, or spiraled toward the center. These patterns, all of them forced, do not agree with the linear theory of convection in circular containers (Charlson and Sani, 1970, 1971), which predicts axisymmetric motions under axisymmetric conditions or *regular* nonaxisymmetric motions when small nonaxisymmetric disturbances are considered. The patterns observed by Croquette disagree also with the axisymmetric motions observed experimentally by, e.g., Koschmieder and Pallas (1974).

Recently Meyer et al. (1991) investigated whether the irregular convection patterns are the consequence of stochastic influences on pattern formation. In the experiments of Meyer et al. patterns of circular concentric rolls were observed in a container filled with water, having a circular polyethylene wall. This lateral wall was replaced by a circular wall made from a gel consisting of 95% water and 5% polyacrylamide. The thermal conductivity of the gel should therefore be very close to the thermal conductivity of water, much closer than that of polyethylene, which does not match the conductivity of water very well. Using this innovative lateral boundary Meyer et al. observed, under otherwise the same experimental conditions, the formation of irregular patterns which had no relationship to the geometry of the container and were not reproducible from one experiment to the next. From this observation it was concluded that the onset of convection in the irregular form "is provoked by spatially and temporally random perturbations, i.e. by stochastic forcing" (1991, p. 2515). On the other hand, the authors found that the measured convective heat flows required in theoretical models to explain the experimentally observed onset in terms of stochastic effects are 2×10^4 larger than the intensity of thermal noise. Meyer et al. state, "We unfortunately have to report at this time that we do not know the origin of this factor 10^4" (p. 2515). This enormous difference shows that the irregular patterns observed were caused not by stochastic noise, but by finite irregularities other than thermal noise. Since the gel matches the thermal conductivity of water so well, the influence of the wall was reduced to a minimum in these experiments. The next largest nonuniformity of the temperature field to which the fluid is exposed is on the lid, which in these experiments was cooled by water coming from eight nozzles placed around the lid; this arrangement makes the cooling of the lid necessarily irregular. The pattern of convection which forms under these circumstances reflects the nonuniformity of the temperature on the lid and is consequently irregular.

We have to discuss experimental detail in order to understand the discrepancy between the experiments producing axisymmetric and asymmetric patterns in circular vessels, or between regular and irregular patterns in general. First of all we have to recall the fundamental theoretical assumption that the temperatures on the top and bottom of the fluid are absolutely uniform. Then we must realize that this condition is not met in each and every convection experiment, except for experiments with helium at a few kelvins. The nonuniformity of the temperature on the top of the fluid is there for a principal reason. We have no other way of cooling the lid than by circulating a coolant over it, which means that the coolant is warmer after it has passed over the lid than where it entered. In all conventional convection experiments there is therefore a small but finite temperature nonuniformity on the lid and also on the bottom, if the bottom is heated by circulating water. It is easy to calculate the temperature difference in the coolant on the lid if the flow rate of the coolant is known, but the flow rate is rarely given. Usually the temperature difference in the coolant on the lid will be quite small; in Koschmieder's (1966a) experiment the radial temperature difference on the lid was 10^{-3}°C at the critical vertical temperature difference of 2°C. If, on the other hand, the bottom of the fluid is an electrically heated thick copper plate, there will be a practically uniform temperature on the bottom, as is the case in the majority of the experiments. It is elementary that the nonuniformity of the temperature on the top of the fluid must not interfere with the axisymmetry of the setup in a circular container. The cooling applied in the experiments of Koschmieder (1966a) and of Hoard et al. (1970) maintained the axisymmetry of the temperature on the top of the fluid by introducing the coolant at the center of the lid. In the experiments of Croquette et al. (1983), Steinberg et al. (1985), and Heutmaker and Gollub (1987) the cooling was not axisymmetric; the coolant moved horizontally on top of the lid from one side to the other side. A schematic drawing of the isotherms produced by this type of cooling is shown in Fig. 5.6. It does not appear likely that the fluid will form an axisymmetric pattern in an apparatus which is not axisymmetric to begin with. Any finite nonaxisymmetric disturbance will break the symmetry of

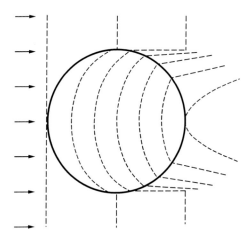

Fig. 5.6. Isotherms (dashed lines) on the lid of a shallow circular static fluid layer which is cooled from above by a uniform flow of a coolant coming from the left side, while the fluid and the surrounding container are heated uniformly from below. The thermal conductivity of the static fluid is three times larger than the thermal conductivity of the material surrounding the fluid, similar to water surrounded by lucite.

a uniformly heated fluid layer which is near the critical temperature difference and is axisymmetric by virtue of the presence of an insulating circular lateral wall.

The uniformity of the temperature field on the top and bottom of the fluid is usually not discussed in the papers describing conventional convection experiments, except in a very few cases. The consequences of cooling the lid with a coolant coming from the side (Fig. 5.6), for example, have never been mentioned, although this arrangement was used in many experiments. However it appears that the uniformity of the temperature on the horizontal boundaries of the fluid is fundamental for the success of an experiment. To express this as an empirical experimental rule: the more uniform are the temperatures on the top and bottom of the layer, the more regular is the pattern.

Sometimes there are other problems with the apparatus or procedure which contribute to the confusing multitude of asymmetric patterns in circular containers. Axisymmetry of the conditions in the experiments of Stork and Müller (1975), for example, was destroyed by the stirring of the fluid under nearly critical conditions. The experiments of Steinberg et al. (1985) were made in a container in which the fluid could not expand although it was heated. The fluid must have leaked at the wall and the leak destroyed the

symmetry. The unavoidable expansion of the fluid with increased temperature difference has to be taken care of with an axisymmetric outlet. Koschmieder (1966a) used for this purpose a very narrow gap between the upper
end of the lucite lateral wall and the lid. This outlet is axisymmetric and the
expansion of the fluid can be accommodated this way by minimal motion of
the fluid. The papers describing the recent experiments with circular containers criticize this arrangement on the ground that the gap under the lid causes
a deviation of the vertical temperature gradient in the wall from the vertical
temperature gradient in the fluid. This argument is correct, but the effect is
minimal because the thermal conductivities of lucite and silicone oil differ
very little. It can be calculated easily that the ratio of the temperature gra-
dients in the lucite wall and in the silicone oil differs from 1 by only 2.5×10^{-3} in Koschmieder's (1966a) experiment. Nevertheless the apparatus deviates from the assumptions of the theory because of the expansion of the
fluid, but the setup is still axisymmetric. The consequences of this way of
coping with the expansion of the fluid have to be compared with the consequences of the other ways of dealing with the expansion of the fluid in other
apparatus, in which expansion likewise introduces an imperfection differing
from the theoretical input. These problems are not addressed in the more recent papers on convection in bounded containers.

Differences of the vertical temperature gradients in a lateral wall and in the
fluid introduce horizontal temperature differences at the wall. The consequences of small horizontal temperature gradients at a lateral boundary on
the onset of convection in bounded layers have been investigated analytically by Daniels (1977), by Brown and Stewartson (1977) for rectangular
containers, by Brown and Stewartson (1978) for circular vessels, and by Hall
and Walton (1977). Common to these papers are the usual assumptions of
linear theory of convection, in particular uniform temperatures in the top and
bottom boundaries of the fluid. It is predicted in these papers that the supercritical bifurcation occurring at $\Re_c$ will be replaced by an imperfect bifurcation (Fig. 5.7) if there is a small horizontal heat flux through the insulating
lateral wall. The imperfect bifurcation becomes apparent from rolls which
spread from the circular wall through the fluid as the Rayleigh number is increased to $\Re_c$. This is the way the onset of convection occurred in the experiments of Koschmieder (1966a) and Koschmieder and Pallas (1974). This
is also the way convection patterns develop when a radial temperature gradient is introduced deliberately in the horizontal boundaries, as will be discussed in Section 8.2. If the lateral heat flux is vanishingly small, the
difference between the supercritical bifurcation and an imperfect bifurcation
becomes ever so small, so that it is only a matter of degree, not of principle.

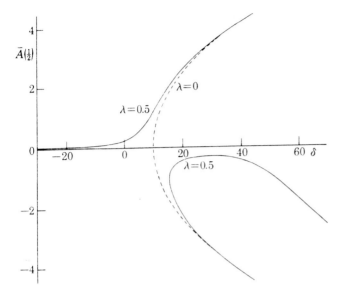

Fig. 5.7. Conditions for the onset of convection in a shallow circular container with a small horizontal heat flux through the lateral wall. $A(\frac{1}{2})$ is the steady amplitude at $r = L/2$, λ is the parameter characterizing the heat flux through the lateral wall, and δ is proportional to $\mathcal{R} - \mathcal{R}_c$. After Brown and Stewartson (1978).

how well ideal theory is satisfied. If there is no horizontal heat flux at all through the circular wall, the pattern is nevertheless axisymmetric, as Charlson and Sani (1970) showed theoretically, provided of course that the conditions are axisymmetric.

The consequences of idealized lateral walls on convection in bounded fluid layers have also been investigated. Drazin (1975) proposed a simple model, Kidachi (1982) studied the influence of stress-free, insulating lateral walls, Metzener (1986) studied the motions in a small rectangular container with three or four rolls bounded by rigid walls, and Nagata (1990) studied symmetric flows in a container with insulating, rigid sidewalls. It is much more difficult to confirm the results of these studies than those dealing with the more realistic lateral boundary conditions discussed above.

We note finally that in the Taylor vortex instability it is an accepted fact that, in the necessarily finite fluid columns, Taylor vortices form first at the ends of the column and spread from there through the column when the Taylor number is increased. The end effects destabilize the fluid column at the ends first, even if the upper boundary of the fluid column is free, i.e. open to air. Regardless of how cleverly one arranges the configuration of the ends, in an unstable fluid the ends will always be a steady finite disturbance and will be the only finite disturbance if the instability of the fluid is otherwise uniform.

Hexagonal cells have not been observed to occur in time-independent Rayleigh–Bénard convection experiments in Boussinesq fluid layers, although hexagonal cells are certainly a solution of the linear equations describing Rayleigh–Bénard convection on an infinite plane. But, as we shall see later, hexagonal cells are not a stable solution in the weakly nonlinear theory of convection. We note, however, that Krishnamurti (1968a) has shown theoretically that in fluid layers with slowly changing mean temperatures hexagons can be stable solutions near the critical Rayleigh number. She verified this prediction experimentally (Krishnamurti, 1968b) in a square apparatus with rigid–rigid boundaries, an aspect ratio $\Gamma = 15$, and a silicone oil of 2 cm^2/sec viscosity. Using a very elegant visualization technique she showed that hexagons appear for slow rates of change in the mean temperature of the order of a few degrees Centigrade per hour. The hexagons had either ascending or descending flow in the cell center, the direction of motion depending on the sign of the rate of change. In the stationary state she observed two-dimensional rolls, in agreement with the other experiments which investigated the planform of the convective motions on uniformly heated plates in the steady state.

To summarize the results about pattern selection, there seems to be little disagreement about the pattern in rectangular containers. Davis' (1967) and the following theoretical predictions of rolls parallel to the shorter side of the rectangular container are apparently accepted, although there is an increasing tendency to find ever more irregular patterns in the experiments. The results of the experiments on convection in circular containers are in dispute. Either perfect circular concentric rolls or disordered, irregular rolls have been found at the onset of convection. The theoretical study of Charlson and Sani (1970) predicts circular concentric rolls, assuming that the conditions are axisymmetric. If axisymmetry is dropped from the assumed conditions, theory predicts axisymmetric as well as regular nonaxisymmetric patterns. Axisymmetry of the experimental conditions is, of course, a crucial requirement if one wants to verify the results of theories studying axisymmetric convection. A convincing answer to the simple question of whether the pattern of convection in bounded layers is determined by the shape of the lateral walls or whether the pattern is simply irregular has not been found, because the issue of circular rolls versus disordered rolls has not been resolved. After 30 years of research on Rayleigh–Bénard convection, we still do not seem to know whether any one of the regular patterns predicted to exist by linear theory at the onset of convection does indeed appear and should appear exclusively, or whether the planform of convection in a bounded layer heated uniformly from below is actually irregular.

5.3 The Critical Wavelength

Although the critical wavelength λ_c is a fundamental feature of Rayleigh–Bénard convection, little work has been devoted to its measurement, in spite of the fact that the wavelength can, in some cases, be determined just by counting. If one has a regular pattern, the average dimensional wavelength is the width of the layer divided by the number of pairs of rolls. This determination of λ_{av} involves the quantization condition, which says that only integer numbers of rolls can fill a container. Any deviation of the value of the wavelength either $<\lambda/2$ smaller or $<\lambda/2$ larger than the true wavelength will therefore not be noticed. The principal uncertainty of the wavelength measurement is $\Delta\lambda = \pm \lambda_{av}/2N$, where N is the number of rolls. If the rolls are regular, the principal accuracy of the wavelength measurement obviously increases with increased aspect ratio, i.e. the number of rolls. If the pattern is irregular, the wavelength has to be determined from sections through the fluid layer in order to obtain a spatial average of λ. This procedure is likely to result in a large standard deviation for the measured wavelength. Picking selected rolls or particular areas of the flow field for the determination of λ, as has been done, is arbitrary.

There seem to be no reports in the literature about specific efforts to measure λ_c. One can extrapolate to the critical wavelength from several measurements of the wavelength of moderately supercritical convection, which we shall discuss later. From such extrapolations one can draw the qualitative conclusion that the experimental wavelengths at $\mathcal{R}_c$ are compatible with the theoretical value of λ_c. In order to end this discussion on a positive note, we calculate λ_c from an experiment with large aspect ratio, rigid–rigid boundaries, and regular (circular) rolls. As reported in Koschmieder (1966a) a pattern with 19 circular concentric rolls was established at $\mathcal{R}_c$ in a fluid layer of 20 cm diameter and 5.15 mm depth. (For a photograph of this pattern see Koschmieder, 1974.) From the number of rolls in this experiment follows a value of the critical wavelength which is $\lambda_c = 2.044 \pm 0.053$. This value is compatible with the theoretical value $\lambda_c = 2.016$ within the principal uncertainty of $\pm 2.6\%$ originating from the quantization condition, which is in this case given by one-half the cell width divided by the radius of the layer.

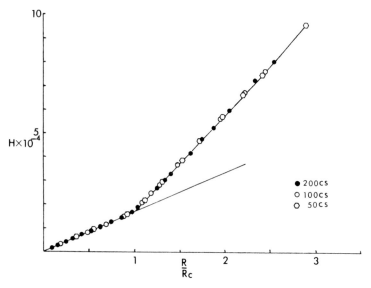

Fig. 5.8.   The break in the heat flux curve which accompanies the onset of convection at $\mathfrak{R}_c$. The experiments were done in a circular container with aspect ratio 26.6 with fluids of Prandtl number $\mathfrak{P} > 500$, the fluids being in contact with a sapphire lid. The straight line marks the heat flux expected from thermal conduction only. After Koschmieder and Pallas (1974).

5.4   The Critical Rayleigh Number

The critical Rayleigh number can, of course, be determined from the temperature difference at which one visually observes the onset of convection. This method is, however, fairly subjective. The objective way to determine the critical Rayleigh number employs the "break" in the heat transfer curve, which we mentioned in Section 5.1. This break occurs when molecular conduction of heat through the resting fluid is supplemented by heat transferred by the motion of the fluid after the onset of convection. An example of the break in the heat transfer curve is shown in Fig. 5.8. The value of $\mathfrak{R}_c$ found with this method by Schmidt and Milverton (1935) was $\mathfrak{R}_c = 1770 \pm 152$. This is, within the limits of the experimental error, in agreement with the theoretical value $\mathfrak{R}_c = 1708$. The first careful modern experiment measuring $\mathfrak{R}_c$ was made by Silveston (1958). He determined the break in the heat flux curve in five different fluids, and extended his measurements of the heat flux to highly supercritical temperature differences. Silveston found the critical Rayleigh number at $\mathfrak{R}_c = 1700 \pm 51$. A realistic error analysis showed that his experiments had a systematic uncertainty of 10–15%. It is worth noting that the accuracy of the measured value of $\mathfrak{R}_c$ depends on the accuracy of the value of the Nusselt number and on the accuracy of the material properties of the fluid, which are usually known to about $\pm 1\%$. The accuracy of the Nusselt number cannot be smaller than the uncertainty of the value of the thermal conductivity of the fluid, which is usually in the 1% range. Silveston also observed the onset of convection visually. The patterns photographed by him are, however, not the same patterns that existed in his heat transfer measurements, because the cooled copper lid used in the heat transfer measurements was replaced in the visualization experiments by a glass lid which was cooled by ambient air and therefore not necessarily of uniform temperature.

The critical Rayleigh number in gases was determined by Thompson and Sogin (1966) with an unusual method. In their experiments the gas was brought through the state of marginal stability by increasing the pressure, while the temperature difference was held constant. They found $\mathfrak{R}_c = 1793 \pm 80$, considering errors from all sources except fluid properties. The critical Rayleigh numbers for three silicone oils of different viscosity were measured by Koschmieder and Pallas (1974) in an apparatus in which simultaneously the onset of convection could be observed visually through a plate of sapphire, which is a very good thermal conductor. They found $\mathfrak{R}_c = 1682 \pm 56$, the uncertainty being $\pm 6\%$ and being due mainly to the uncertainty of the fluid properties. The onset of convection in helium at cryogenic temperatures was studied by Behringer and Ahlers (1982). These experiments seemed to hold great promise, because of the extraordinary accuracy with which temperatures can be measured in low-temperature physics. This should result in an improved accuracy of the heat flux measurements. Unfortunately this advantage is negated by the uncertainty of the fluid properties. The uncertainty of the viscosity of helium at these low temperatures (≈ 2 K) "is perhaps as large as 10%" (1982, p. 232); the value of the thermal diffusivity is also uncertain by "perhaps as much as 10%" (p. 234). Behringer and Ahlers found $\mathfrak{R}_c = 1599 \pm 240$ in one cell with $\Gamma = 9.44$, where according to Charlson and Sani (1970) $\mathfrak{R}_{ctheor} = 1734$, and $\mathfrak{R}_c = 1694 \pm 250$ in another cell with $\Gamma = 4.15$, where $\mathfrak{R}_{ctheor} = 1840$. The possible systematic errors were as large as $\pm 15\%$. Similar studies were made by Lucas et al. (1983) with an apparatus designed to investigate the influence of rotation on the onset of convection. With the rotation rate $\Omega = 0$ they found for $\Gamma = 15.6$ in a couple of experiments at different temperatures between 2 and 3 K an average $\mathfrak{R}_c = 1412 \pm 145$ with a principal uncertainty of $\pm 16\%$.

To summarize the results of the experiments concerning the onset of convection, we have found that the measurements of the critical Rayleigh number in large aspect ratio containers agree with an accuracy of not better than ±5% with the theoretical critical Rayleigh number for an infinite, plane fluid layer. We have also found agreement within a couple of percent of the measured critical wavelength with the theoretical critical wavelength. We have been less successful with the verification of the predicted patterns. There is still no consensus as to whether the patterns which appear at the onset of convection are regular, as linear theory predicts, or whether the patterns at the onset of convection in a bounded Boussinesq fluid layer are irregular. But in general, there is no reason to doubt that linear theory of Rayleigh–Bénard convection correctly describes the onset of convection. The contradictions between the patterns of absolute regularity predicted by linear theory in bounded and unbounded fluid layers and the irregular patterns observed in many experiments do not prove a failure of the rather unambiguous linear theory, but are the consequence of experimental problems.

6

SUPERCRITICAL RAYLEIGH–BÉNARD CONVECTION EXPERIMENTS

6.1   Steady, Moderately Supercritical Convection

It is possible, according to linear theory, that a significant change in convection will take place as soon as the critical Rayleigh number is exceeded and the Rayleigh–Bénard problem has become nonlinear. Potentially the most important feature is the continuum of nonunique wave numbers at supercritical Rayleigh numbers, which we have seen in the stability diagram of linear theory (Fig. 2.1). Also, it is believed that the degeneracy of the patterns at $\mathcal{R}_c$ in linear theory may disappear in favor of a preferred pattern, if the nonlinear terms in the equations of motion are taken into account. The theoretical investigation of just supercritical planforms culminates in the prediction that Rayleigh–Bénard convection on a uniformly heated infinite plane should form a pattern of straight, parallel rolls. This result was established by Schlüter et al. (1965), whose paper we shall return to in Section 7.1. We look now at the experimental facts concerning the patterns and wave numbers under moderately supercritical conditions. "Moderately supercritical" is here defined as being at Rayleigh numbers between, say, $1.2\mathcal{R}_c$ and about $10\mathcal{R}_c$. It is a matter of experience that for Rayleigh numbers higher than $10\mathcal{R}_c$ the flow tends to be time-dependent, although for fluids of small Prandtl number this may happen earlier. Time-dependent convection will be discussed separately in Section 6.2.

No significant change in the patterns has been observed to occur with the transition from critical to slightly supercritical $\mathcal{R}$ with Boussinesq fluids. If a pattern consisted of regular rolls at $\mathcal{R}_c$, it continued as regular rolls (see, e.g., Fig. 6.2); if a pattern consisted of disordered rolls at $\mathcal{R}_c$, it remained disordered. It appears that the consequences of the presence of lateral walls on the form of the pattern are more important than the weak nonlinear effects associated with the change from critical to just supercritical

Fig. 6.1.   Bimodal Rayleigh–Bénard convection in a layer of silicone oil of $\mathcal{P} = 450$ at
$\mathcal{R} = 2.5 \times 10^4$. Visualization with aluminum powder. Courtesy of G. E. Willis.

flow. Regular patterns with straight parallel rolls in rectangular containers and circular rolls in circular containers have their shape not because of nonlinear effects, but because of the configuration of the lateral boundaries. In regular patterns in high Prandtl number fluids there is a tendency to develop distinctly three-dimensional flow if $\mathcal{R}$ is greater than about $5\mathcal{R}_c$. This kind of motion has been aptly called bimodal convection (Busse and Whitehead, 1971); the most convincing demonstration of bimodal convection is shown in Fig. 6.1. Bimodal convection is steady. The change to bimodal flow is gradual and may depend on the configuration and conditions at the lateral boundaries, as well as on the Prandtl number. Very little is actually known about bimodal convection.

Whereas the increase to supercritical flow does not affect the patterns very much, the wave number or its more obvious equivalent, the wavelength, changes as soon as the critical Rayleigh number is exceeded. At $\mathcal{R}_c$ the wavelength is unique and equal to λ_c according to linear theory and, within the experimental error, also according to experiments. As soon as the Rayleigh number is supercritical, however, the fluid begins to utilize the nonunique range of wavelengths in order to *increase* the value of the wavelength of the motion. The increase of the wavelength of supercritical convective motions has a long history, but the fact was established by Koschmieder (1966a). His experiments were made with a circular container of aspect ratio 19.7 and silicone oil of Prandtl number $\mathcal{P} = 920$. As the Rayleigh number was slowly increased above $\mathcal{R}_c$ a consecutive disappearance of the center roll of a pattern of circular concentric rolls was observed. The circular geometry is particularly suited for this purpose, because the fluid is free to move upward or downward in the center of the layer, whereas the direction of the flow near a rigid lateral wall is fixed by the lateral boundary condition. As the number of rolls decreases the wavelength increases, because the average dimensional wavelength is equal to the diameter of the container divided by the number of the rolls. The increase of the wavelength can be observed easily with the naked eye (Fig. 6.2). At that time the prevalent theories predicted just the opposite. It was then believed that an increase of the Rayleigh number was accompanied by a decrease of the wavelength of the flow. Of the half a dozen theories making this prediction we shall refer here only to the two best-known papers, those of Malkus and Veronis (1958) and Schlüter et al. (1965).

Earlier indications of the increase in the wavelength of convective motions or in the size of the cells can be found in Bénard (1900), Dauzère (1912), Chandra (1938), Avsec (1939), and in particular Silveston (1958). The last paper contains a graph showing the variation of the wavelength as a function of the Rayleigh number, but no conclusion was drawn from this result. After 1966 the increase of the supercritical wavelength was observed by Leontiev and Kirdyashkin (1968), Krishnamurti (1970a), Willis et al. (1972), Farhadieh and Tankin (1974), and Koschmieder and Pallas (1974). More recently the increase was observed again in rectangular containers by Bühler et al. (1979), Martinet et al. (1984), and Kolodner et al. (1986). Note that in the last experiment the copper bottom plate of the fluid layer is so thin (4.7 mm) that the temperature at the bottom of the fluid will reflect the nonuniform heating by the heating wire beneath the copper plate, which may affect the form of the pattern and the wavelength of the motions. The observed variation of the wavelength according to various experiments is shown in Fig. 6.3. This figure portrays the consensus of all time-independent convection experiments that the wavelength increases with increased Rayleigh number. The accuracy of the wavelength measurements is not particularly good. A careful examination of the experimental errors in Koschmieder and Pallas (1974) shows that the uncertainty of λ ranges up to $\pm 5\%$, even with good axisymmetric patterns; in less regular patterns the standard deviation of λ is certainly larger.

In the study of Krishnamurti (1970a) a dependence of the increase of the wavelength on the Prandtl number of the fluids, water and various oils of large Prandtl numbers, was found in a large aspect ratio container ($\Gamma \approx 25$). The smaller the Prandtl number of the fluid the longer was the supercritical wavelength, at the same Rayleigh number. There was some hysteresis of the

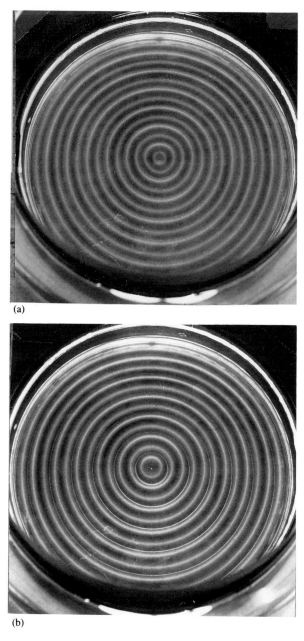

(a)

(b)

Fig. 6.2.   Demonstration of the increase in the wavelength of axisymmetric supercritical Rayleigh–Bénard convection in silicone oil ($\mathscr{P} \approx 950$) in a circular container. Quasi-steady states. (a) $\mathscr{R} = 1.05\mathscr{R}_c$, $\lambda = 2.01 \pm 0.08$. Thirteen slightly super-critical rolls. Compare with Fig. 5.1. (b) $\mathscr{R} = 2.18\mathscr{R}_c$, $\lambda = 2.18 \pm 0.09$. Twelve rolls just after the disappearance of the thirteenth roll. The bright circles at the

(c)

(d)

bottom of the layer originate from aluminum powder settled under the location of rising motion. (c) $\mathscr{R} = 5.29\mathscr{R}_c$, $\lambda = 2.61 \pm 0.13$. Ten rolls just after the disappearance of the eleventh roll. (d) $\mathscr{R} = 7.32\mathscr{R}_c$, $\lambda = 2.90 \pm 0.16$. Nine rolls just before the appearance of three-dimensional flow. After Koschmieder (1974). Copyright John Wiley & Sons, Inc.

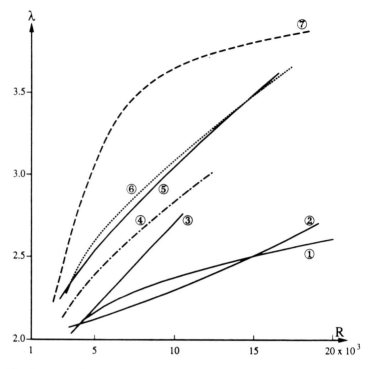

Fig. 6.3. Summary of measured steady supercritical wavelengths of Rayleigh–Bénard convection, with increasing ΔT only, in large aspect ratio containers as a function of the Rayleigh number. (1) Silveston (1958), silicone oil, $\mathscr{P} \approx 3000$. (2) Krishnamurti (1970a), silicone oil, $\mathscr{P} \approx 870$. (3) Koschmieder and Pallas (1974), silicone oil, $\mathscr{P} \approx 950$. (4) Martinet et al. (1984), air, $\mathscr{P} \approx 0.7$. (5) Willis et al. (1972), water, $\mathscr{P} \approx 7$. (6) Farhadieh and Tankin (1974), water, $\mathscr{P} \approx 7$. (7) Willis et al. (1972), air, $\mathscr{P} \approx 0.7$. The different values of the measured wavelengths for fluids of similar Prandtl number reflect the differences in the apparatus, different experimental procedures, and different methods for determining the wavelength. Of the two experiments measuring the wavelength in air, that of Martinet et al. employs an apparatus that is a better match of the theoretical assumptions; these authors also use the more advanced visualization technique.

wavelength in Krishnamurti's experiments; this seems, however, to be within the experimental uncertainty of the wavelength measurement, which was made from selected straight rolls. The experiments of Willis et al. (1972) were made with air ($\mathscr{P} = 0.71$), water ($\mathscr{P} = 7.16$), and silicone oil ($\mathscr{P} = 450$) in square containers with aspect ratios up to 80. They observed likewise an increase of the wavelength with increased Rayleigh number. Remembering that, according to Segel (1969), the critical Rayleigh number in bounded containers is the critical Rayleigh number in a layer of infinite horizontal extent $\mathcal{R}_{c\infty}$ plus a correction proportional to L^{-2}, where L is the width of the layer, we see that the experiments of Willis et al. approximate an infinite fluid layer very well. This consideration applies not only to the onset of convection, but also to supercritical flow. The fluid changes its stability and wavelength in response to the available energy. In the experiments of Willis et al. so much energy comes into the fluid through the large bottom plate that the energy which comes into the fluid through the (insulating) lateral walls is negligible. This means that the presence of the lateral walls in these experiments with very large aspect ratio was irrelevant to the changes of the wavelength. This is of importance, because it was believed that the presence of the lateral walls might be the reason for the (unexpected) increase of the wavelength, as will be discussed in Section 7.4. The hysteresis reported by Willis et al. is certainly the consequence of insufficient waiting times, because the horizontal relaxation time L^2/κ for their largest container filled with silicone oil was of the order of a month, but the experiments were done over a day.

Farhadieh and Tankin (1974) made experiments with water ($\mathcal{P} = 7$) in a rectangular container of aspect ratio $\Gamma \cong 21$, and an increase of λ was found likewise. This shows that the increase of the wavelength is not seriously affected by the shape of a container if the aspect ratio is larger than about 10. Bühler et al. (1979) observed a discontinuous increase of the wavelength in a rectangular container of size $(10 \times 4)d^2$ and filled with nitrogen ($\mathcal{P} = 0.71$). Martinet et al. (1984) made experiments with air ($\mathcal{P} = 0.71$) in a rectangular container with aspect ratio $\Gamma_x = 18$. Up to $\mathcal{R} = 7\mathcal{R}_c$ they observed an increase of the wavelength of the steady flow. The wavelength measurements made with gases are those with the lowest Prandtl number; they confirm that if $\mathcal{P}$ is small the wavelength at a given $\mathcal{R}$ is particularly large. Martinet et al. found some hysteresis, but they did not indicate whether the experiments were made in strict thermal equilibrium.

So far we have discussed the increase of the wavelength under moderately supercritical conditions. Very little is known about the value of the wavelength when $\mathcal{R} > 10\mathcal{R}_c$, because the cells are then usually irregular and also time-dependent. It appears that the wavelength continues to grow with increased $\mathcal{R}$; there is a curve in Fitzjarrald (1976) giving values of λ up to about $6\lambda_c$, but these data are, in the author's words, equivocal. It can also be considered that atmospheric convective motions have been observed with wavelengths of the order of $100\lambda_c$ or more, but it cannot be ruled out that anisotropy plays a role in these phenomena.

It is now a certainty that the wavelength of moderately supercritical Rayleigh–Bénard convection increases with increased supercritical Rayleigh number. The slope $d\lambda/d\mathscr{R}$ of the curves in Fig. 6.3 showing the variation of the wavelength is not known with an accuracy of better than about $\pm 10\%$. It is likely that $d\lambda/d\mathscr{R}$ depends on the Prandtl number. Although the various experiments measuring the wavelength have been made in very different apparatus under quite different conditions, they have one principal feature in common. In none of these time-independent experiments have convective motions been observed with wavelengths *smaller* than the critical wavelength; to use a phrase from Krishnamurti (1970a, p. 306), "The domain $a > a_c$ is conspicuously bare" (a being the wave number, $a = 2\pi/\lambda$). This feature is also obvious in Fig. 6.3, which shows that there exists, for a given $\mathscr{R}$ and $\mathscr{P}$, a preferred wavelength. It is one of the principal problems of the theory of nonlinear Rayleigh–Bénard convection to explain the increase of the wavelength of supercritical convective motions. The wavelength is the most discriminating feature of steady supercritical convection. The significance of the wavelength is apparent from the fact that the critical wavelength is one of only three features predicted by linear theory of convection. (The preferred wavelength of supercritical convection was also reviewed by Getling, 1991.)

It has, on the other hand, been maintained by Busse and Whitehead (1971) that *stable* supercritical convective motions with wavelengths $\lambda < \lambda_c$ can be produced with the so-called controlled initial conditions. The controlled initial conditions technique was introduced by Chen and Whitehead (1968). The experiments of Busse and Whitehead (1971) proceeded as follows: A silicone oil layer of either 80×80 cm^2 or 30×30 cm^2 in a state of pure conduction was heated rapidly with a temperature increase of $1°$C/min through the critical temperature difference ($\Delta T_c = 1.1°$C), while the fluid layer was illuminated from above through a grid by two 500-W heat lamps. The illumination through the parallel gaps of the grid produced spatially periodic temperature disturbances in the fluid of the order of $0.05°$C, or 5% of ΔT_c. At the onset of convection at ΔT_c, the fluid was therefore not in thermal equilibrium (the thermal relaxation time d^2/κ being 285 s), nor was there a uniform temperature in the horizontal direction. In other words, the controlled initial conditions procedure does not comply with the assumptions made in the theory of Rayleigh–Bénard convection. In these experiments the fluid formed a transient pattern of straight, parallel rolls, the size of the rolls being determined by the width of the gaps in the illuminated grid. If the distance of the gaps of the grid was small, rolls with wavelengths smaller than the critical wavelength could be produced in this way. According to Busse and Whitehead, "The question of stability was usually decided half an hour after the observation had started" (1971, p. 315). This time interval is insufficient to establish the steady state in this apparatus. If a roll forced upon the fluid wants to adjust its size it must change horizontally, so the horizontal relaxation time L^2/κ (L being again the width of the layer) determines equilibrium. The horizontal relaxation time in the 30 × 30 cm^2 container was 280 h. The rolls caused by the controlled initial conditions did not stay straight and parallel for such a long time; rather they changed within hours to a "patchy structure" of a few wavelength diameter. This patchy structure is "similar to the convection produced" in this apparatus after quasi-steady heating without controlled initial conditions. The rolls created by the controlled initial conditions are not stable – they are transient; the patchy structure is the steady-state solution in this apparatus.

The wavelength of the flow in the patchy structures was not discussed by Busse and Whitehead (1971), but one can conclude from Chen and Whitehead (1968, Fig. 6 therein) that the wavelengths in the patchy structures tend with time to wavelengths which are longer than λ_c. Experiments with controlled initial conditions made by Berdnikov et al. (1990), which specifically investigated the existence of an inherent optimal scale of rolls, confirmed that rolls created by initial conditions similar to those of Busse and Whitehead tend to change their wavelength in favor of longer wavelengths.

So the fact remains, the experiments with the controlled initial conditions notwithstanding, that stable supercritical convective motions have only wavelengths *longer* than the critical wavelength λ_c. This does not mean, however, that *all* wavelengths $\lambda > \lambda_c$ which could form according to linear theory are actually stable supercritical wavelengths. Figure 6.3 indicates that, for one given moderately supercritical Rayleigh number, and for a fluid with a given Prandtl number, there is only one value of λ plus or minus the experimental uncertainty, or at most a small range of steady supercritical wavelengths which are accessible by quasi-steady increases of the Rayleigh number, *not* a continuum which covers all wavelengths $>\lambda_c$ permitted by linear theory. This applies to fluid layers of large aspect ratio and is true within the experimental uncertainty of the wavelength measurements, which is between ±5 and ±10%. In small aspect ratio containers the wavelength can be forced by the geometry and the quantization condition to differ substantially from the wavelength in wide containers. That, however, amounts to nothing but an increase of the systematic uncertainty $\Delta\lambda = \pm\lambda_{av}/2N$ of the wavelength. Supercritical wavelengths *shorter* than the critical wavelength can be produced by rapid heating of the fluid layer, as was observed by Koschmieder (1969) and by Simpkins and Dudderar (1978), but these motions are not stable.

The experimental data as summarized in Fig. 6.3 make it appear that the wavelength of moderately supercritical Rayleigh–Bénard convection is *unique*, within the error of measurement. This observation is, however, not

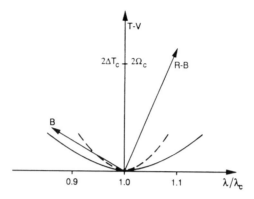

Fig. 6.4. Qualitative picture of the variation of the wavelength of supercritical Rayleigh–Bénard convection (R-B), axisymmetric Taylor vortex flow (T-V), and surface-tension-driven Bénard convection (B) after quasi-steady increases of the control parameters. Shown also are the marginal curve of linear theory and (dashed curve) the Eckhaus instability. After Koschmieder (1991).

absolutely certain; a definitive experimental proof of the uniqueness of the supercritical wavelength of Rayleigh–Bénard convection has still to be provided, and requires the measurement of λ with decreasing temperature differences. The apparent uniqueness is surprising, because it contradicts not only the expectations from the stability diagram of linear theory, but also the expectation that the nonlinear terms in supercritical flow foster nonuniqueness. In the corresponding supercritical Taylor vortex problem the flow is clearly nonunique. Figure 6.3 is the strongest indication of the existence of a unique solution for moderately supercritical convection. Some theoretical studies of supercritical convection to be discussed in Chapter 7 predict the existence of broad bands of nonunique supercritical wavelengths, including wavelengths with $\lambda < \lambda_c$. On the other hand, if the balance theorem discussed in Section 2.2 holds, one would expect uniqueness of moderately supercritical flow, because balance of buoyancy and dissipation can be accomplished in only one way.

Having discussed the variation of the wavelength of supercritical Rayleigh–Bénard convection, we should recall that we found in Chapter 4 that the wavelength of supercritical surface-tension-driven Bénard convection *decreases* with increased ΔT in the slightly supercritical range, as shown in Fig. 4.3. The astounding difference in the way the fluid reacts in Rayleigh–Bénard convection and in Bénard convection to an increase of the temperature difference across the fluid is illustrated in Fig. 6.4. This figure shows furthermore that the wavelength of axisymmetric supercritical Taylor vortices does not vary with the Taylor number (or the rotation rate), as will be discussed in Section 13.2. As Fig. 6.4 demonstrates each of these instabilities reacts differently to an increase of the control parameter in the nonlinear range, whereas they are so similar at the onset of instability when the conditions are linear.

Let us summarize what we have learned from the experiments about the wavelength of convective motions under moderately supercritical conditions. It has been found in a large number of experiments that the wavelength of supercritical convection increases with increased Rayleigh number in the steady state. The preferred wavelength seems to depend on the Prandtl number of the fluid. We know that stable supercritical wavelengths are only in a very small portion, if not on a curve $\lambda(\mathcal{R},\mathcal{P})$, in the range of nonunique supercritical wavelengths predicted to exist by the weakly nonlinear theory of Eckhaus (1965). There are no stable steady supercritical convective motions with wavelengths shorter than λ_c (or wave numbers $a > a_c$) at all.

We shall now discuss a new avenue for detailed investigations on moderately supercritical convection which was opened by the advent of laser–Doppler velocimetry. The first accurate measurements of the velocity of convective motions were made with this technique by Bergé (1975) and Dubois and Bergé (1978). They measured the x and z velocity components at different Rayleigh numbers in a set of parallel rolls in a rectangular container of 10×3 cm^2 size and 1 cm depth, filled with silicone oil of Prandtl number $\mathcal{P} = 926$. The top and bottom of the container were 1-cm-thick copper plates, in compliance with the usual assumption of perfect thermal conductivity in the horizontal boundaries made in the theory of Rayleigh–Bénard convection. Due to the geometry of the container, the wavelength of the rolls – or the number of the rolls – did not change in this apparatus with increased $\mathcal{R}$ over the range of Rayleigh numbers studied, as was also found in similar experiments by Bühler et al. (1979). In a rectangular container the wavelength increases discontinuously because the lateral boundary conditions at the short sides of the container fix the direction of the flow at these sides. The wavelength of the motions in a rectangular container nevertheless increases ultimately, namely when the Rayleigh number has been made high enough that the change of the wavelength, which would occur in an infinite layer, is sufficiently large to squeeze out one roll or one pair of rolls in the rectangular container – a process which does not take place gradually, but discontinuously. Afterward the rectangular container is filled with another integer number of rolls parallel to the short side of the container. The variation of the number of the rolls in two rectangular boxes as a function of the Rayleigh

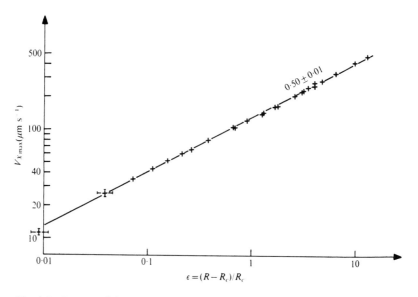

Fig. 6.5. Increase of the maximal value of the x velocity component of two-dimensional convection rolls in a rectangular container as a function of $\varepsilon = (\mathfrak{R} - \mathfrak{R}_\mathrm{c})/\mathfrak{R}_\mathrm{c}$. After Dubois and Bergé (1978).

number was shown in Figs. 9a,b of Kirchartz and Oertel (1988). The results of the velocity measurements of Dubois and Bergé discussed in the following apply without restrictions only to the case of constant wavelength. Measurements of the amplitude in experiments with convective motions with wavelengths different from λ_c but at constant λ and constant $\mathfrak{R} = 6.67\mathfrak{R}_\mathrm{c}$ were described in Dubois, Normand, and Bergé (1978).

Bergé (1975) and Dubois and Bergé (1978) established first that the velocity distribution in the parallel rolls is strictly sinusoidal in the x-direction, i.e. in the direction of the long side of the container. That observation appeared to be likely from the many visual observations of rolls in earlier convection experiments but was now put on a quantitative footing. Second, and more important, they established the form of the increase of the amplitude of the velocity as a function of the Rayleigh number. They found that in the range from $\mathfrak{R}_\mathrm{c}$ to about $10\mathfrak{R}_\mathrm{c}$ the maximum of the x velocity component increases as $\sqrt{\varepsilon}$, where ε is the reduced normalized Rayleigh number

$$\varepsilon = (\mathfrak{R} - \mathfrak{R}_\mathrm{c})/\mathfrak{R}_\mathrm{c}. \tag{6.1}$$

The measurements showing the $\sqrt{\varepsilon}$ dependence are shown in Fig. 6.5. The $\sqrt{\varepsilon}$ increase is a striking confirmation of the amplitude equation introduced by Landau (1944). The amplitude equation, which had until then lingered as an unverified assumption for supercritical convective flow, had suddenly become a certainty. In the measurements shown in Fig. 6.5 small contributions of the second and third Fourier component of the velocity distribution are involved, which do not change the validity of the $\sqrt{\varepsilon}$ dependence. The second and third Fourier components may be strongly influenced by the constant wavelength condition forced upon the flow by the geometry of the container.

In Bergé (1975) one can also find a discussion of the variation of the disturbance temperature θ as a function of the coordinate x and the Rayleigh number. Using data on the temperature distribution in a convecting fluid from Farhadieh and Tankin (1974), Bergé showed that θ varies sinusoidally in x and increases with $\Re$ as $\sqrt{\varepsilon}$. Furthermore, Bergé discussed the convective heat flux, and showed that Q_{conv} increases as ε (details will be discussed in Section 6.3). Finally we want to point out the range of $\Re$ over which these dependencies on the Rayleigh number were found. They apply to the moderately supercritical range up to $10\Re_c$ in which, in high Prandtl number fluids, regular steady rolls can be observed experimentally.

Moderately supercritical Rayleigh–Bénard convection is characterized not only by the pattern, the wavelength, and the velocity of the flow, but also by the temperature distribution in the fluid and by the heat transfer. We shall not discuss the temperature distribution in detail, other than to refer to Farhadieh and Tankin (1974) and to Chu and Goldstein (1973) for the temperature field at much higher Rayleigh numbers. We shall discuss the heat transfer by moderately supercritical convection together with the heat transfer of time-dependent and turbulent convection in Section 6.3.

6.2   Oscillatory Rayleigh–Bénard Convection

When the Rayleigh number is increased beyond the moderately supercritical range, the fluid motions become time-dependent. The expectation was that the onset of time dependence would be marked by a second spontaneous transition, analogous to the critical transition at the onset of convection. Exploratory investigations of time-dependent convection were made by Willis and Deardorff (1965, 1967a, 1970) and by Krishnamurti (1970b) in fluid layers of very large aspect ratio and with fluids of very different Prandtl numbers. Willis and Deardorff (1965), working with an air layer, observed a sudden appearance of temperature fluctuations and a gradual development into irregular fluctuations in a range of Rayleigh numbers from $\Re = 6300$ to $\Re = 10,000$. Willis and Deardorff (1967a) found that the Rayleigh numbers at which well-defined temperature fluctuations first appear increase strongly with increased Prandtl number. Power spectra of the temperature fluctuations at various Rayleigh numbers were given. The spectra had broad peaks, showing clearly that the fluctuations were not monoperiodic. Willis and Deardorff (1970) shifted the emphasis to oscillations of wavy rolls, which appear to be rather irregularly distributed over the entire fluid layer. The periods of the oscillations were determined from visual observations or from thermocouple measurements. They stated that, for air, "when R reaches about 12,000 we estimate visually that the oscillations occur essentially all the time over most of the area" (1970, p. 670). The period τ of the oscillations was found to decrease as $\mathcal{R}^{-0.4}$, fairly similar to the $\mathcal{R}^{-2/3}$ dependence suggested in a boundary layer study of Howard (1963). Willis and Deardorff determined the periods only from the most clearly sinusoidal temperature fluctuations. Actually "a fairly broad range of frequencies is present for any fixed R even when only the most sinusoidal oscillations are considered" (1970, p. 670). Willis and Deardorff also noticed some weak evidence of frequency doubling in their experiments.

In the experiments of Krishnamurti (1970b), which were also made with a container of very large aspect ratio, the value of $\mathcal{R}$ for the onset of time-dependent fluctuations depended likewise on the Prandtl number; the larger $\mathcal{P}$, the later the fluctuations appeared. The observed periods of the motions decreased approximately as $\mathcal{R}^{-2/3}$. The fluid motions were made visible by a photographic technique which shows, on one picture, a time sequence of an illuminated section through the convecting fluid layer. Three pictures made this way are shown in Fig. 6.6. On each of these photographs the abscissa corresponds to an almost 50-cm-long horizontal section through the fluid at a height $z = 0.4d$. The ordinate corresponds to the time, showing how the section through the fluid looked at consecutive times. The total time intervals over which these photographs were made are several minutes, and are stated in the caption. In the top frame, we see various cells which appear as bright patches. If the cells did not change their form or their location with time, they appear as columns, as most of them do in this frame. Note that the Rayleigh number is $28\mathcal{R}_{c\infty}$. In the middle frame, at $123\mathcal{R}_{c\infty}$, we see some cells which did not change with time, at least one cell which has moved steadily perpendicular to the line of sight, and at least two pairs of cells which varied periodically. The bottom frame shows that at $135\mathcal{R}_{c\infty}$ there were still cells which changed form or position very little, whereas other cells varied periodically. The meaning of these photographs is that, just above $30\mathcal{R}_c$, "perhaps only one cell of all the cells in the line across the tank displays oscillation during one thermal diffusion time. As R is increased, more and more cells display

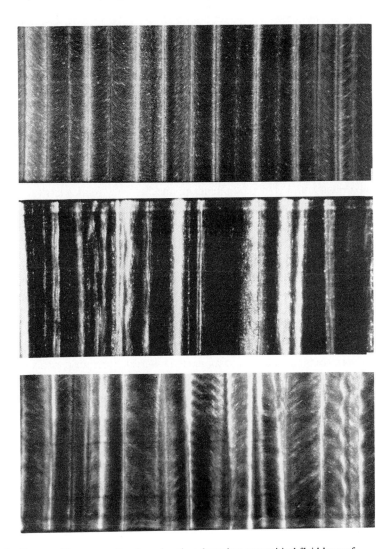

Fig. 6.6.   Photographic cross section through a time-dependent supercritical fluid layer of large aspect ratio; $\mathscr{P} = 57$. The ordinate on the photographs shows the same cross section at consecutive times. (*Top*) $28\mathscr{R}_{c\infty}$, 17-min exposure, (*middle*) $123\mathscr{R}_{c\infty}$, 6.4-min exposure, (*bottom*) $135\mathscr{R}_{c\infty}$, 15-min exposure. After Krishnamurti (1970b).

this oscillation at any instant. . . . The oscillation in any particular cell disappears after approximately one thermal diffusion time, and appears in another cell, so that on the average the same number of cells display oscillation at a given R,P'' (Krishnamurti, 1970b, p. 315). In the bottom frame we see an example of frequency doubling; compare the period of the third cell from the right with the period of the upper end of the two cells at the center of the picture. These photographs have been reproduced here because they demonstrate so clearly that the onset of time dependence is gradual and occurs locally, and extends increasingly through the fluid with increased Rayleigh number.

If we summarize the results of the convection experiments concerning the onset of time-dependent convection in large aspect ratio containers, we find that the onset of time-dependent convection apparently does not occur in the form of a discrete transition to a fluid flow with an oscillation which extends *coherently* through the entire convection chamber. Nobody has observed convection rolls oscillating synchronously over the entire area of a large aspect ratio container, or motions resembling that concept. The onset of time-dependent convection is gradual and accompanied by a broad band of frequencies. This assessment of the results of the experiments concerning the onset of time-dependent convection is not expressed this way by Willis and Deardorff, who seem to favor coherence, nor by Krishnamurti, who favors the idea of discrete transitions. The apparent fact that the onset of time-dependent convection in large aspect ratio fluid layers is not a discrete transition is, of course, of fundamental importance for the theoretical description of time-dependent convection, which is usually concerned with infinite fluid layers. The transition to time-dependent convection could in principle as well be a discrete transition, just as the transition from axisymmetric time-independent Taylor vortex flow to time-dependent wavy Taylor vortex flow is a discrete transition to a wavy flow coherent over the length of the column.

An entirely different picture of the onset of time dependence in convection emerges in containers of small aspect ratio, as was discovered by Libchaber and Maurer (1978) and Gollub and Benson (1978), and was to some extent also apparent in Ahlers and Behringer (1978). In a tiny circular cell filled with liquid helium at 4 K ($\mathscr{P} = 0.63$), Libchaber and Maurer (1978), using one local probe, observed the occurrence of very sharp frequencies in the time-dependent temperature fluctuations at Rayleigh numbers $>2\mathscr{R}_c$ if the aspect ratio was either $\Gamma = 4$ or $\Gamma = 5$. An example of such a line frequency spectrum is shown in Fig. 6.7. Since there is no way of visualizing the fluid motions in liquid helium convection experiments, we do not know to what kind of geometric oscillation these frequencies correspond, but presumably the whole pattern is oscillating. If, in the container with $\Gamma = 5$, the Rayleigh number was increased, the frequency of the oscillation doubled (within about 20%) three different times, until at around $12\mathscr{R}_c$ a continuous low-frequency noise appeared, which spread slowly over the entire spectrum as $\mathscr{R}$ was in-

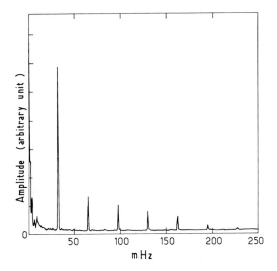

Fig. 6.7.   Line frequency spectrum of oscillatory convection in a layer of helium at 4 K with
aspect ratio $\Gamma = 5$ at $\mathfrak{R} = 5.3\mathfrak{R}_c$. After Libchaber and Maurer (1978).

creased further. In the same type of experiments in the same container but
with an aspect ratio $\Gamma = 12$, no sharp frequencies were observed (Fig. 6.8).
The noise spectrum evolved as follows: at $2\mathfrak{R}_c$ there was very low frequency
noise at zero hertz (1 Hz = 1/s). As the Rayleigh number was increased the
noise spread to higher frequencies and formed a broad peak at 0.1 Hz when
$\mathfrak{R}$ was $3.7\mathfrak{R}_c$. The broad peak shifted very slowly as $\mathfrak{R}$ was increased further
(Fig. 6.8), but a line spectrum never appeared.

There is a fundamental difference in the frequency spectrum in small containers and wide containers (Γ larger than about 10) at the onset of time dependence. The first have discrete frequency spectra; the latter have
continuous spectra with a broad peak, in agreement with what we found in
the early conventional experiments with time-dependent convection. This result has been upheld by several subsequent investigations, which were summarized by Behringer (1985). Whereas for small aspect ratio containers time-
dependent flow may be characterized as chaotic, in particular because of the
appearance of sharp frequencies which give way to broad band noise, one
wonders whether this characterization applies also to time-dependent flow in
large aspect ratio containers. This depends, of course, on how "chaos" is
defined. In time-dependent convection in large aspect ratio containers, there
does not seem to be sensitive dependence on initial conditions or random
transitions between multiple states, but only an increased sensitivity to the

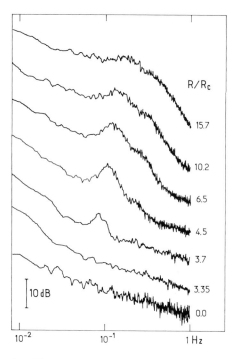

Fig. 6.8.   Broad-frequency power spectra of oscillatory convection at various Rayleigh
numbers in a layer of helium at 4 K. The aspect ratio is $\Gamma = 12$. After Libchaber
and Maurer (1978).

ever present disturbances. If, however, chaos denotes only a state of disorder
and irregularity, then this is an apt description of time-dependent convection
in large aspect ratio containers.

The transition from convection with oscillations with a line frequency
spectrum to convective motions with a turbulent noise spectrum has been
found to be quite complex. This problem has been investigated by Gollub
and Benson (1980) in experiments with rectangular containers of aspect ratio
$\Gamma = 2.42$ and $\Gamma = 3.21$, filled with water whose Prandtl number could be
varied between 9 and 2 by changing the mean temperature of the fluid. Several different flows with two or three somewhat irregular rolls along the long
side of the container were observed. A particular pattern was chosen ''by
manipulation of the boundary temperatures,'' and the time dependence then
studied at a single location. It was found that the instabilities leading to nonperiodic flow depend on the Rayleigh number, the Prandtl number, the aspect
ratio, and the type of pattern. The following sequences of instabilities were

observed: (1) from steady to periodic, to quasi-periodic with two frequencies, to phase locking of the two frequencies, to nonperiodic, i.e. turbulent; (2) from steady to periodic, to period doubling, to a second period doubling, to nonperiodic; (3) from steady to periodic, to quasi-periodic with two frequencies, to quasi-periodic with three incommensurate frequencies, to nonperiodic; (4) from steady to quasi-periodic with two frequencies, to intermittent noise, to nonperiodic. Simple rules for predicting which of these sequences will occur for a given aspect ratio, Prandtl number, and pattern have not been found. In summary, it has turned out that even in rectangular containers with very small aspect ratios the transition to turbulence occurs via many instabilities which, in view of the four parameters involved, will be very difficult to track down theoretically. It is important to note that this is so in containers of very small aspect ratio in which the freedom of motion of the fluid has been restricted radically by the lateral walls and in which, therefore, a deterministic flow appears to be much more likely than in an infinite layer.

The investigations of time-dependent convection in containers of small aspect ratio discussed above were concerned with the determination of the frequencies of the oscillations, but did not address the question of the physical mechanism responsible for the periodic motions. The latter question was studied by Dubois and Bergé (1981). They worked with a cell of small aspect ratio ($\Gamma = 2$) filled with silicone oil of 0.1 cm^2/sec viscosity, $\mathscr{P} = 130$. The pattern studied consisted of two rolls along the long side and one roll along the short side of the container. The mean velocity field of the flow was fairly asymmetric. The temperature field in the fluid was made visible with differential interferometry; the vertical velocity component was measured with laser–Doppler anemometry. Dubois and Bergé studied the sequence from stationary to monoperiodic convection, to quasi-periodic convection with two frequencies, to turbulent flow. In their small cell with a fluid of such a large Prandtl number, the onset of time-dependent convection took place only at a very high Rayleigh number, in this case at $\mathscr{R} \cong 215\mathscr{R}_{c\infty}$. The flow then remained monoperiodic until $\mathscr{R} \cong 250\mathscr{R}_{c\infty}$. The interferometric pictures showed that the oscillation originated from a periodic variation of the depth of the cold boundary layer of the fluid. The variation occurred over the entire horizontal extension of the upper part of the fluid. The period of the monoperiodic oscillations decreased with increased $\mathscr{R}$. When $\mathscr{R}$ exceeded $250\mathscr{R}_{c\infty}$, a second frequency appeared, accompanied by very strong local variations of the velocity of the motions. This periodicity originated from cold plumes emerging from the cold boundary layer and sinking along either of the short sidewalls of the container. The plumes on either side had the same frequency, but differed in their phases and their amplitudes. At first both oscillators, the boundary layer and the cold plumes, occurred simultaneously and their periods were coupled at the ratio 6/7, the cold plumes having the shorter periods. The period of the plumes decreased steadily when $\mathfrak{R}$ was increased. On the other hand, the oscillation period of the boundary layer remained virtually constant, except in the range from 270 to $290\mathfrak{R}_{c\infty}$, in which the boundary layer oscillation disappeared and the flow was monoperiodic again, but with the period of the plumes. At around $305\mathfrak{R}_{c\infty}$ convection became turbulent, as indicated by the appearance of low-frequency noise.

Although a number of questions, e.g. the existence of rising warm thermals instead of sinking cold plumes, remain unanswered, this study demonstrated clearly the pivotal role that the boundary layers and the lateral walls play in the formation of time-dependent convection with discrete frequencies in small aspect ratio containers.

If we compare what we have learned about time-dependent convection in large and small containers, we note above all the startling differences in the frequency spectra of the oscillations in both types of containers. It is obvious that the aspect ratio is an essential parameter for the theory of time-dependent convection in small containers. With small aspect ratio we have few spatial transitions; with large aspect ratio many spatial transitions can occur. The confinement of the fluid, or the few degrees of freedom, eliminate the consequences of many disturbances to which the fluid responds if it is not confined. A theoretical explanation of time-dependent convection in small containers can be concerned with only a limited number of disturbances, whereas a theoretical explanation of time-dependent convection in large aspect ratio layers must deal with the consequences of a large spectrum of disturbances. In either case the problems are formidable, because they are nonlinear and time-dependent and depend also on the Prandtl number and the aspect ratio.

6.3 Supercritical Heat Transfer

The measurement of the heat transfer by a convecting fluid layer provides information about an integral property of the moving fluid: the heat transferred through the bulk motion of the fluid – more specifically through the horizontal average of the vertical velocity. The heat transfer is usually given in nondimensional form by the Nusselt number, which is defined as

$$\mathcal{N} = (Q_{\text{cond}} + Q_{\text{conv}})/Q_{\text{cond}}, \tag{6.2}$$

where Q_{cond} is the amount of heat transferred per unit area and unit time by the thermal conductivity of the (resting) fluid alone, and Q_{conv} is the heat transferred additionally by the convective motions of the fluid. The Nusselt number is also equal to the ratio of the effective thermal conductivity of the convecting fluid to the thermal conductivity of the resting fluid. In the absence of fluid motions the Nusselt number is equal to 1.

The original measurements of heat transfer by Rayleigh–Bénard convection were made by Schmidt and Milverton (1935). These experiments served the specific purpose of determining the critical Rayleigh number, as discussed in Section 5.4. There were only four experiments and no conclusive results about supercritical conditions were reported.

These early experiments were followed by the investigations of Malkus (1954a), who was concerned with turbulent convection, which means highly supercritical conditions. Virtually all heat transfer measurements of convection cover a large range of Rayleigh numbers, because in the same apparatus very large Rayleigh numbers can easily be realized by using a fluid of small viscosity; the applied temperature differences, on the other hand, are often quite small. With the same fluid the range of Rayleigh numbers can also be increased greatly by increasing the depth of the fluid layer, making use of the d^3 dependence of the Rayleigh number. Malkus' experiments were made with water ($\mathscr{P} = 7$) and acetone layers in "quasi-equilibrium" between two brass blocks cooling down to their mean temperature. The aspect ratio of the fluid layers ranged from $\Gamma = 19.7$ for the experiments with slightly supercritical Rayleigh numbers to $\Gamma = 1.2$ for the experiments with very high Rayleigh numbers. The convective heat transfer was found to increase with increased $\mathscr{R}$, but slower than linearly. Characteristic of the heat transfer curves found by Malkus are several "transitions," by which discontinuous changes of the slope of the heat transfer curves are meant. Such slope changes were found at $\mathscr{R} = 1700 \pm 80$, at $(1.1 \pm 0.2) \times 10^4$ or $(1.8 \pm 0.1) \times 10^4$ or $(2.6 \pm 0.5) \times 10^4$, at $(5.5 \pm 0.4) \times 10^4$, at $(1.7 \pm 0.15) \times 10^5$, at $(4.25 \pm 0.2) \times 10^5$, at $(8.6 \pm 0.3) \times 10^5$, and possibly at 1.7×10^6. One can argue the reality of the observed transitions for $\mathscr{R} > \mathscr{R}_c$; it appears that within the given experimental uncertainty all supercritical data can be fitted by smooth curves.

The first slope change is, of course, genuine, marking the onset of convection. The second slope change occurs, according to Malkus (1954a, p. 185), somewhere in the range $10{,}000 < \mathscr{R} < 30{,}000$. In only three out of eight runs was the second transition at $\mathscr{R} = 18{,}000$, the others being much above or much below that value. The evidence for the slope changes above $\mathscr{R} = 55{,}000$ seems to be quite weak. Nevertheless all these transitions must be mentioned because they are the basis for a number of later studies. Malkus suggested an explanation for these transitions. He believed "that the transitions in the data seem to be those found theoretically on the assumption of a constant temperature gradient" (p. 193) in linear theory for free surfaces, and

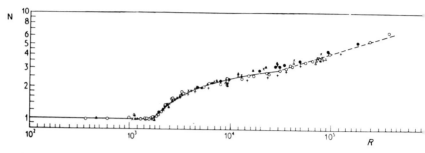

Fig. 6.9.   Heat transfer in large aspect ratio fluid layers as a function of the Rayleigh number. After Silveston (1958).

for wavelengths of one-half, one-third, one-quarter, etc. of the normal wavelength of the convective motions. We know now, e.g. from Farhadieh and Tankin (1974), that supercritical convection does not have constant vertical temperature gradients, and we know also that supercritical convection occurs with wavelengths *longer* than the critical wavelength (Section 6.1), instead of the wavelengths shorter than λ_c which the transitions are said to be associated with. It appears, therefore, that Malkus' interpretation of the transitions was not realistic. Regardless of the Rayleigh numbers at which the transitions occur, the existence of the transitions indicates an abrupt change of the velocity field, which makes them important.

These experiments were followed by the investigations of Silveston (1958), which always have been regarded as reliable. Silveston worked in steady state with fluid layers of large aspect ratio ranging from $\Gamma = 15$ to $\Gamma = 136$. He used five different fluids of very different Prandtl numbers, ranging from $\mathscr{P} = 7$ for water to $\mathscr{P} = 4400$ for a very viscous silicone oil. The top and bottom of the fluid layer were copper plates, in order to satisfy the boundary conditions in the theory with which the results are compared. As was mentioned in Section 5.4 the critical Rayleigh number was found to be $\mathscr{R}_c = 1700 \pm 51$ in this apparatus, proving thereby the correct functioning of the setup. The measurements of the Nusselt number of the fluid layer extended from $0.2\mathscr{R}_c$ to about $200\mathscr{R}_c$, as can be seen in Fig. 6.9. The data points in this figure actually extend further, to about $2000\mathscr{R}_c$, or $\mathscr{R} \cong 5 \times 10^6$. For this purpose data from the much earlier measurements of Mull and Reiher (1930) with an air layer were used by Silveston. As one can see immediately from Fig. 6.9 the heat transfer is, within the accuracy of these measurements, a steadily increasing function of the Rayleigh number. It does not appear that these data support the concept of breaks in the slope of the heat transfer curve. There is one break in Fig. 6.9 at $\mathscr{R} = 2.5 \times 10^4$, which is the artificial consequence of extending by a straight line the measurements made with this apparatus to the measurements in the completely different apparatus of Mull and Reiher, in which a different conducting medium (air) was used. It does not appear to be difficult to draw, within the scatter of the measurements, a smooth curve through Silveston's data between $\mathcal{R}_c$ and $\mathcal{R} = 4 \times 10^5$. A smooth curve $\mathcal{N}(\mathcal{R})$ means that the slope changes continuously, without breaks. The scatter of the data of Silveston can be attributed in part to the use of different fluids with different Prandtl numbers, as Silveston points out himself. We have seen in Section 6.2 that the onset of time dependence, where presumably the first supercritical slope change is, is not sharp but gradual, and the heat transfer should therefore also change gradually.

There is a consensus that Silveston's experiments were careful and that his results are accurate. One finds very little room for criticism, other than that the flow rate of the coolant was too small, producing too large a temperature difference on the lid, up to 0.8°C. We shall see that the smooth heat transfer curve measured by Silveston has been supported by later experiments.

The topic of the heat transfer transitions was picked up again by Willis and Deardorff (1967b). Their time-dependent experiments were made in a square container of 80×80 cm^2 size, with either air ($\mathcal{P} = 0.71$) or silicone oil of 5×10^{-2} cm^2/sec viscosity ($\mathcal{P} = 57$). Only five runs were made. Willis and Deardorff believe that the results of their experiments confirm the "reality and generality" of the transitions observed by Malkus. There are nevertheless discrepancies with the value of the Rayleigh number for the second transition, which Malkus places at $\mathcal{R} = 18,000$, but for which Willis and Deardorff find 8200 and 24,000. It appears that their heat transfer curves could just as well be smooth curves, especially so with the transitions at large $\mathcal{R}$. It is artificial to approximate these gentle curves with straight line segments. It is good to remember that one-third of all points which originate from averages of several measurements can be further away than one error bar from the curve if the deviations of the measurements are of stochastic origin. One point off a smooth curve does not prove a break in the slope of a curve.

In steady-state experiments with water and silicone oils in a large aspect ratio container Krishnamurti (1970a,b) found three discrete supercritical slope changes in the heat transfer curve at around $12\mathcal{R}_c \cong 2 \times 10^4$, at $35\mathcal{R}_c \cong 6.0 \times 10^4$, and at around $105\mathcal{R}_c \cong 1.79 \times 10^5$. Later Brown (1973), working with a very large container filled with air, found two transitions at $\mathcal{R} = 9.6 \times 10^3$ and $\mathcal{R} = 2.6 \times 10^4$. The data points of Brown can as well be represented by smooth curves, but one would have to know the error bars in order to make a meaningful evaluation. Chu and Goldstein (1973), working with water and aspect ratios from $\Gamma = 1.5$ to $\Gamma = 6$, believe they were

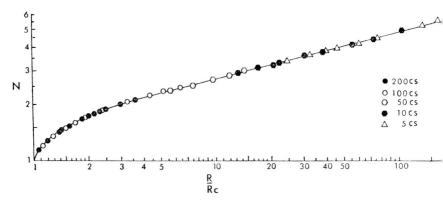

Fig. 6.10   Heat transfer through silicone oils in large aspect ratio fluid layers as a function of the Rayleigh number. After Koschmieder and Pallas (1974).

able to discern "eight discrete heat-flux transitions" (p. 157) in the range from $\mathscr{R} = 2.7 \times 10^5$ to $\mathscr{R} = 1 \times 10^8$. However, on a log $\mathscr{N}$ versus log $\mathscr{R}$ plot their data are represented very well by a straight line; the standard deviation of their data points from the line fit through their points is only 1.4%. At even higher Rayleigh numbers, between $\mathscr{R} = 1.36 \times 10^7$ and $\mathscr{R} = 3 \times 10^9$, Garon and Goldstein (1973) found four "apparent discontinuities in slope" (p. 1821) at $\mathscr{R} = 5.9 \times 10^7$, 1.3×10^8, 2.2×10^8, and 4.0×10^8, in water layers with aspect ratios 2.5 and 4.5. The authors are, however, circumspect about these slope discontinuities, saying that "their validity is open to question" (p. 1821).

The heat transfer in steady state through silicone oil layers of Prandtl numbers ranging from $\mathscr{P} = 70$ to $\mathscr{P} = 1800$ in fluid layers of aspect ratio $\Gamma = 26$ was measured by Koschmieder and Pallas (1974). Their heat transfer curve extending from $\mathscr{R}_c$ to $150\mathscr{R}_c$ is shown in Fig. 6.10. All data points on this curve are averages of several measurements taken at each point. The uncertainty of the Nusselt number was 2%, so the errors are smaller in size than the symbols marking the data points. This curve, and other curves in this paper showing segments of the heat transfer curve, are all smooth within the experimental uncertainty. Agreement with the measurements of Silveston (1958) was excellent.

The first convection experiments with helium at low temperatures (about 4 K) were made by Threlfall (1975). He determined the heat transfer in a circular container of aspect ratio $\Gamma = 2.4$; the Prandtl number was $\mathscr{P} \cong 0.7$. A very large range of Rayleigh numbers from $\mathscr{R} = 60$ to $\mathscr{R} = 2 \times 10^9$ was

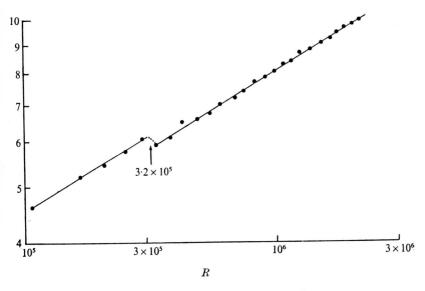

Fig. 6.11. A stepwise decrease of the heat transfer at $\Re = 3.2 \times 10^5$ without slope change in a small aspect ratio convection experiment with helium at low temperature. After Threlfall (1975).

accessible in these experiments by varying the pressure of the helium. Threlfall observed discontinuities of the heat transfer at $\Re = 9 \times 10^3$, 2.8×10^4, 5×10^4, 6.5×10^4, and 3.2×10^5, as well as four other "slight changes of gradient" which may represent further transitions. There is a difference between the discontinuities observed by Threlfall and the slope changes we have discussed so far. In the $\mathcal{N}$ versus $\Re$ plot of Threlfall (Fig. 6.11), in which his "largest discontinuity in heat flux" is presented, there is a sudden decrease of the value of $\mathcal{N}$ at $\Re = 3.2 \times 10^5$, but the slope of the curve $\mathcal{N}$ versus $\Re$ is picked up unchanged after $\Re = 3.2 \times 10^5$. There is a step in the curve, but not a slope change. It is important to keep in mind that this was an experiment with small aspect ratio. We shall find such "discontinuities" without following slope change in other experiments with small aspect ratio.

In another measurement of the heat transfer through liquid or gaseous helium Ahlers (1974) found "no evidence for discrete heat-flux transitions" in the range between $\Re_c$ and $150\Re_c$ in a circular container with aspect ratio $\Gamma = 10.5$. He states that "within a precision of 0.1% the data can be represented by a function with a continuous derivative" (p. 1187). Ahlers' measurements reflect the accuracy with which temperature differences or the corresponding heat fluxes can be measured in low-temperature physics.

These measurements are the most accurate measurements of supercritical convective heat transfer in large aspect ratio containers. Since accuracy is, after all, decisive, the concept of slope changes in supercritical convective heat transfer in large containers seems to be unrealistic. Threlfall (1975) and Ahlers (1974) disagree with regard to the slope changes in the heat transfer curves, although both employ the same technique. But Threlfall worked with a small aspect ratio container in which changes of the cell form – and there was only room for one convection cell – can have noticeable consequences for the heat transfer, whereas the aspect ratio Ahlers worked with was so large that a change common to all cells would have been required to have an impact on the heat transfer. We have seen that there is already no coherence with mildly time-dependent convection in large containers; that will also be the case when the Rayleigh numbers are much larger.

A stepwise change of the Nusselt number in a small aspect ratio container ($\Gamma = 4.16$) was found in low-temperature experiments with helium by Behringer and Ahlers (1982). They observed a large drop of the Nusselt number at approximately $3\mathcal{R}_c$, but the $\mathcal{N}(\mathcal{R})$ curve continues with its prior slope immediately after this step. There seems to be a second small but positive step in $\mathcal{N}$ at around $9\mathcal{R}_c$. Behringer and Ahlers attribute the abrupt changes in $\mathcal{N}$ to the appearance of different (nonspecified) states. For the next larger container used by them ($\Gamma = 9.46$) the Nusselt number had no detectable singularity for $\mathcal{R} \leq 24\mathcal{R}_c$.

The relation just implied between aspect ratio (or cell form) and heat transfer was confirmed in conventional convection experiments with water in a rectangular container of moderate aspect ratio by Walden et al. (1987). As is shown in their Fig. 3 different numbers of parallel rolls transfer different amounts of heat at the same Rayleigh number. At, e.g., $\mathcal{R} \cong 9\mathcal{R}_c$ 10 rolls have a Nusselt number $\mathcal{N} \cong 2.6$, whereas 6 rolls have a Nusselt number $\mathcal{N} \cong 2.4$. Note that the 6 rolls just mentioned were induced by controlled initial conditions. The procedure for the calculation of the Nusselt and Rayleigh numbers is not described in this paper, or in Kolodner et al. (1986), to which they refer for this purpose. It is therefore not possible to assess the accuracy of these measurements.

The heat transfer in helium gas ($\mathcal{P} \approx 1$) at low temperatures (5 K) was measured again by Castaing et al. (1989), in a circular container of aspect ratio $\Gamma = 1$, using different pressures and small temperature differences between 3×10^{-3} K and 1 K. These experiments cover the range of Rayleigh numbers from 5×10^3 to 10^{12}. The experimental method resembles Threlfall's (1975) work. Several domains of convective flow were found, beginning with oscillatory instability at $\mathcal{R} = 9.4 \times 10^4$, followed by a chaotic state at $\mathcal{R} = 1.5 \times 10^5$, in which the time coherence was lost, whereas the space

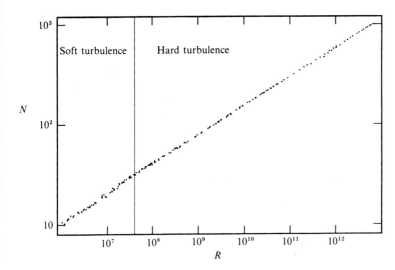

Fig. 6.12. Heat transfer by turbulent convection in helium at 5 K in a circular container with aspect ratio $\Gamma = 1$. The vertical line at $\mathfrak{R} = 4 \times 10^7$ indicates the transition from soft to hard turbulence. After Castaing et al. (1989).

coherence persisted. Then came a transition region up to 5×10^5, which was followed by a soft turbulence regime up to 4×10^7, and finally what is called hard turbulence followed up to $\mathfrak{R} = 6 \times 10^{12}$. The Nusselt number–Rayleigh number correlation found in these experiments is shown in Fig. 6.12. The concept of "hard turbulence" has been studied with a scaling analysis by Castaing et al. (1989). One of the results of this analysis is that the Nusselt number should be proportional to $\mathfrak{R}^{2/7}$, assuming that the Prandtl number is equal to 1. With $\frac{2}{7} = 0.286$ the measurements of Castaing et al. and the scaling analysis are in good agreement. One must, however, not forget that the aspect ratio $\Gamma = 1$ of the apparatus of Castaing et al. severely restricted the freedom of motion of the fluid, in this case of the helium. In a circular apparatus with aspect ratio $\Gamma = 1$ only one-half of a natural convection cell fills the container, since $\lambda_c = 2.016$. The flow in the container is also prevented from expanding its wavelength with increased Rayleigh number, as nonrestricted convective motions do. This will certainly affect the flow and the Nusselt number, at least below $\mathfrak{R} = 10^7$. It has to be shown by further experiments whether the results obtained for hard turbulence are independent of the aspect ratio.

Over the time span of more than 30 years we have accumulated many data about the heat transfer in supercritical or turbulent convection. The results of the measurements of the heat transfer remain, however, inconsistent. This is partly so because not enough attention has been paid to the fact that the heat transfer is a function not only of the Rayleigh number, but also of the Prandtl number and the aspect ratio. All too often the Rayleigh number has been raised by varying the fluid depth, which means by decreasing the aspect ratio. In order to eliminate the consequences of the dependence of the heat transfer on the Prandtl number and the aspect ratio, one has to work with one fluid only, or with fluids of high Prandtl number only, and one must work with aspect ratios greater than 10. In order to cover a large range of Rayleigh numbers under these circumstances, one must increase the temperature differences across the fluid layer to much larger values of ΔT than have been used in most experiments.

As far as the "heat transfer transitions" are concerned, a topic that has often overshadowed the heat transfer problem, one wonders whether these transitions actually exist in large aspect ratio containers. Much of the experimental evidence for the transitions in wide containers is tainted by the flawed method of drawing straight line segments through gentle curves without consideration of the errors of measurement. One wonders, furthermore, whether one actually should expect to find "discrete transitions" in the supercritical heat transfer. We have seen in Section 6.2 that the transition to time-dependent convection in large aspect ratio containers, presumably the first supercritical transition to occur, is not a discrete transition, but a gradual one. This will be all the more so if the Rayleigh numbers are higher. On the other hand, in small aspect ratio containers, where changes in one cell can have a significant effect on the heat transfer of the entire layer, heat transfer transitions, or, better, stepwise changes in the heat transfer, do apparently occur. They are, however, not accompanied by changes of the slope of the heat transfer curve.

The measured heat transfer curves have been fitted many times to theoretical or empirical formulas. The first such formula was proposed by Malkus (1954b). His investigation was based on several assumptions, among them a postulated maximal heat transfer and a "smallest scale of motion which contributes to the heat transport," in other words a minimum size for the higher modes. It follows from Malkus' calculations that the heat transfer by turbulent convection should be given by the formula

$$\mathcal{N} = k\mathcal{R}^{1/3},\qquad(6.3)$$

where k is a constant. The same relation has been suggested on other grounds by Howard (1963) and again on other grounds by Long (1976).

The first empirical formulas for the heat transfer by convection were given by Silveston (1958). He gave several formulas for the Nusselt number in different ranges of the Rayleigh number, e.g. for $1700 < \mathcal{R} < 3000$,

$$\mathcal{N} = 0.24 \mathcal{R}^{1/4} \tag{6.4}$$

and for turbulent flow with $\mathcal{R} > 3.5 \times 10^4$,

$$\mathcal{N} = 0.094 \mathcal{R}^{3/10}, \tag{6.5}$$

which follows from the slope of the dashed line in Fig. 6.9.

Instead of discussing now the heat transfer over the entire range of Rayleigh numbers, we shall split up the heat transfer curve into sections and begin with the initial slope of the heat transfer at slightly supercritical flow. Formulas for the initial slope of the heat transfer curve for different flow patterns were determined theoretically by Schlüter et al. (1965). For $\mathcal{P} > 10$ it follows that the initial slope has the value 1.43. The initial slope of the heat transfer curve was measured by Koschmieder and Pallas (1974). From the data plotted in Fig. 5.8 it follows that in these experiments the initial slope had the value $1.48 \pm 6\%$, in good agreement with the theoretically predicted slope. The formula given by Schlüter et al. is, however, valid for straight parallel rolls of infinite length, whereas the pattern with which the slope of the heat transfer was determined consisted of circular, concentric rolls in an apparatus with aspect ratio 26.6. Further measurements of the initial slope were made by Behringer and Ahlers (1982). For helium at low temperature in a circular apparatus with aspect ratio $\Gamma = 114$ they found an initial slope of 1.3 ± 0.05, but did not know the flow pattern. In their experiments with smaller aspect ratios ($\Gamma < 11$) the slopes differed significantly from the theoretical value for rolls. This can be a consequence of the large difference in the thermal conductivity of helium and the lateral stainless steel wall of the layer, which difference is likely to introduce lateral temperature gradients that cause horizontal circulations.

In the moderately supercritical range up to $10\mathcal{R}_c$ the convective heat transfer should be proportional to $(\mathcal{R} - \mathcal{R}_c)/\mathcal{R}_c = \varepsilon$, as follows from the $\sqrt{\varepsilon}$ proportionality of the vertical velocity component and the $\sqrt{\varepsilon}$ proportionality of the temperature disturbance θ. This was noted by Bergé (1975) and was demonstrated in his Fig. 17. The heat transfer in the moderately supercritical range was also found to be proportional to ε and at higher $\mathcal{R}$ to higher order terms of ε by Behringer and Ahlers (1982).

In the highly supercritical range in which infinite Rayleigh numbers are approximated asymptotically, and where the flow is certainly turbulent, the Nusselt number is proportional to $\mathcal{R}^n$, but may still depend on the Prandtl number and the aspect ratio. Various analytical and experimental determinations of the exponent n have been made and are tabulated with the Rayleigh number range and the Prandtl number, but without the aspect ratio, in Tables

1 and 2 of Goldstein et al. (1990). From these data it appears that the value of n tends either to 0.33 or to a value around 0.29. This means that more work has to be done in order to find a conclusive asymptotic relation between the Nusselt number and the Rayleigh number. It is worthwhile to note that the measured relationship between the Nusselt number and the Rayleigh number for turbulent heat transfer in very deep or quasi-infinite fluid layers heated from below is also characterized by an exponent of $\mathfrak{R}$ that has the value 0.290, according to Lewandowski and Kupski (1983).

6.4 Turbulent Convection

The term "turbulent" is not clearly defined, and we shall in the following differentiate between "weak turbulence" and "turbulence." By weak turbulence we mean irregular, nonperiodic, time-dependent flow with slow variations with time and slow motions. By turbulent we mean rapidly and randomly varying flow at high Reynolds numbers with very fast variations with time, relative to the vertical thermal relaxation time.

Since, as we shall see in Chapter 7, the time-dependent part of the convection equation is multiplied by the factor $1/\mathscr{P}$, the importance of time dependence increases very much with decreasing Prandtl number. The onset of time dependence and irregular time dependence or weak turbulence in convection is likely to occur at smaller Rayleigh numbers in fluids of small Prandtl number than in fluids of large Prandtl number. In experiments, $\mathscr{P} \to 0$ can be approached by using mercury, whose Prandtl number is 0.025, and weak turbulence seems to have been observed near $\mathfrak{R}_c$ in mercury by Rossby (1969). In experiments with helium at around 5 K Ahlers and Behringer (1978) observed the existence of nonperiodic flow with an extremely slow time scale for all temperature differences larger than ΔT_c in an apparatus with aspect ratio 114. The Prandtl number of the fluid was 2.94 or 4.40. One wonders, of course, whether in experiments in which very slow time dependences occur the time dependence is a consequence of an instability of the fluid, or whether it is the consequence of slow variations of the external parameters of the system. In order to draw conclusions from observations of slow time dependences in convection it has to be proved that the imposed conditions did not vary for time periods *longer* than the periods of the observed variations. This matter is not discussed in Ahlers and Behringer (1978). It appears also that the horizontal boundaries in this apparatus were not sufficiently parallel. From information in Behringer and Ahlers (1982), Table 1 therein, it follows that the nonuniformity of the fluid depth was (6.8

± 4.5)%. A nonuniform fluid depth causes horizontal motions in a fluid layer heated uniformly from below. These interfere with the convective motions driven by vertical instability and can be the cause of premature time dependence in the fluid layer.

The same topic, weak turbulence near the onset of convection, was studied again by Ahlers et al. (1985) in conventional experiments with water with $\mathscr{P} = 5.7$ in a circular container of aspect ratio $\Gamma = 30$. They observed persistent changes in a pattern of disordered rolls at random intervals and rates as late as 200 times the horizontal relaxation time. There are, unfortunately, problems with this apparatus, which is described in Steinberg et al. (1985). The temperature distribution on the lid of the circular layer was not axisymmetric, the fluid could not expand, and the fluid layer was not insulated laterally, although the heated bottom plate and the cooled lid extended beyond the rim of the fluid. This will cause the medium outside of the rim to convect and will distort the temperature distribution in the wall. This will certainly affect the flow inside of the wall. Convection patterns that changed slowly and persistently under slightly supercritical conditions ($\varepsilon < 0.2$) were observed as well by Heutmaker and Gollub (1987) in water. These experiments seem also to be affected by problems with the apparatus and the procedure. The initial conditions were unusual: first a highly disordered pattern was created by raising the Rayleigh number rapidly through $\varepsilon = 0$ to $\varepsilon \cong 5$, but then ε was set at 0.1, ε being $(\mathscr{R} - \mathscr{R}_c)/\mathscr{R}_c$. One wonders why ε was not increased slowly from zero to 0.1. The authors state that at $\varepsilon = 0.1$ "the horizontal inhomogeneity in ε . . . could have been as large as 15%" (p. 243). This means that the possibility existed "that some of the features of the flow near onset might be different" in an apparatus with better uniformity (p. 243). I do not believe that the experiments of Ahlers et al. (1985) and of Heutmaker and Gollub (1987) are unambiguous proof of the existence of weak turbulence near $\mathscr{R}_c$ in low Prandtl number fluids (water).

It still has to be shown that there is weak turbulence near $\mathscr{R}_c$ in low Prandtl number fluids as a consequence of a genuine instability in the fluid, when the external parameters are fixed over time and uniform in the horizontal plane. Weak turbulence near $\mathscr{R}_c$ has to be carefully separated from the consequences of stochastic external forces. It is, of course, possible to produce variations of the flow by uncontrolled changes of the external parameters. But stochastic external forces are not considered in the usual theory of Rayleigh–Bénard convection, which considers the stability of fluid layers heated absolutely uniformly from below without any time variation.

In truly turbulent convection the motions must have an average velocity field and fluctuations thereof, there must be an average temperature field with its fluctuations, and there will be the turbulent heat transfer. As far as the pattern of turbulent convection is concerned there may still be a recognizable pattern; observations of cloud cells in the atmosphere or the granulation on the sun lead one to believe that polygonal or possibly hexagonal convection cells exist at very high Rayleigh numbers in very large scale systems. But since these observations do not originate from controlled, reproducible conditions, these observations have no more than a suggestive value. In laboratory experiments turbulent convection on a uniformly heated plate should be completely random at sufficiently high $\mathcal{R}$ or should have a hexagonal pattern, because turbulence, in the absence of a preferential direction, is three-dimensional and hexagonal cells match three-dimensional flow.

Rudimentary studies of the turbulent velocity and temperature field have been made by Deardorff and Willis (1967), Garon and Goldstein (1973), and Fitzjarrald (1976). A picture of the characteristics of turbulent convective motions as a function of $\mathcal{R}$ and $\mathcal{P}$ in, for the sake of simplicity, fluid layers of infinite horizontal extent, or even the outline of such a picture, will require many more and much more sophisticated studies. The same applies for the numerical studies of convection at high $\mathcal{R}$. We shall refer to only two recent examples. Travis et al. (1990) studied two-dimensional thermal convection in rectangular containers with Rayleigh numbers up to 5×10^5. They found steady rolls with wave number $a > 2.22$ for $\Gamma = 1.41$. Deane and Sirovich (1991) and Sirovich and Deane (1991) studied chaotic or weakly turbulent convection up to $70\mathcal{R}_c$ under highly idealized conditions with aspect ratio $L/H = 2\sqrt{2}$, assuming periodic flow in the x- and y-directions. In both cases the aspect ratio seems to be much too small for the fluid motion to be representative of turbulent convection. Interferometric observations of the temperature field in turbulent convection make it clear that the horizontally averaged temperature field consists of a thin thermal boundary layer at the bottom, with a nearly isothermal section above, which fills the largest part of the layer, and another thin thermal boundary layer underneath the top plate; see, e.g., Figure 11 in Chu and Goldstein (1973). But this result will likewise have to be supplemented by additional investigations. Turbulent heat transfer has been discussed in detail in Section 6.3. One has to be careful about the relevance of some of the results discussed there for "turbulent" convection, because some of the turbulent heat transfer experiments were made with small aspect ratio fluid layers, which usually means small temperature differences, so that they have more of the characteristics of weak turbulence. They are also unduly influenced by effects of the lateral walls. If we summarize we find that, at present, we know little about genuine turbulent convection.

7

NONLINEAR THEORY OF RAYLEIGH–BÉNARD CONVECTION

7.1 Weakly Nonlinear Theory

The goal of the theoretical investigation of nonlinear convection is to find out how a steady equilibrium state is attained after the onset of convection and what the characteristics of the equilibrium state are. We want to learn what the amplitude of the motion is, what the preferred pattern is, and what the wavelength of the convective motions is; whether there is, in particular, a preferred supercritical wavelength or a continuum of possible stable supercritical wavelengths; in other words, whether the nonlinear convection problem is unique or nonunique. This is not the place to review the mathematical methods used to study nonlinear stability. Original contributions to this topic are, among many others, those of Stuart (1958, 1960) and Watson (1960). For information about the basics of nonlinear stability theory we refer to Stuart (1971) and to the extended discussion in Chapter 7 of Drazin and Reid (1981). We pursue here only the problem of the stability of nonlinear (supercritical) Rayleigh–Bénard convection as it evolved since 1950.

Supercritical convection is described by the nonlinear Navier–Stokes equation,

$$\rho\left[\frac{\partial \mathbf{v}}{\partial t} + (\mathbf{v} \cdot \nabla)\mathbf{v}\right] = -\nabla p + \rho\mathbf{g} + \mu\nabla^2\mathbf{v}, \qquad (7.1)$$

the nonlinear equation of thermal conduction in a fluid in the form

$$\frac{\partial T}{\partial t} + \mathbf{v} \cdot \nabla T = \kappa\nabla^2 T, \qquad (7.2)$$

in which heat added by viscous dissipation in the fluid is omitted, and by the equation of continuity and the equation of state of the fluid. After the variables are made nondimensional, the Boussinesq approximation is used, and ρ is eliminated, we arrive at the equations for nonlinear convection

$$\mathcal{P}^{-1}\left[\frac{\partial \mathbf{v}}{\partial t} + (\mathbf{v} \cdot \nabla)\mathbf{v}\right] = -\nabla p/\mathcal{P} + \theta \mathbf{e}_z + \nabla^2 \mathbf{v}, \tag{7.3}$$

$$\frac{\partial \theta}{\partial t} + \mathbf{v} \cdot \nabla \theta = \nabla^2 \theta + \mathcal{R}w, \tag{7.4}$$

$$\nabla \cdot \mathbf{v} = 0. \tag{7.5}$$

The analytical investigation of supercritical Rayleigh–Bénard convection began with the paper of Gor'kov (1957). He expanded the velocity components, the disturbance pressure p, and the disturbance temperature θ in equations (7.3)–(7.5) in Fourier series. He considered hexagonal and equilateral triangular cells, and showed that the amplitude of the convective motions near the critical point is proportional to the square root of $\mathcal{R}-\mathcal{R}_c$. He did not connect this result to the Landau equation (7.14), although Landau advised him on this paper.

At about the same time, finite amplitude convection was also studied by Malkus and Veronis (1958). They treated the nonlinear terms in equations (7.3)–(7.5) as perturbations of the linear convection problem. Following conventional techniques of perturbation theory they sought steady solutions of the nonlinear equations by developing the velocity, temperature, and Rayleigh number in power series of a small parameter ε; i.e. they set

$$w = \varepsilon w_0 + \varepsilon^2 w_1 + \varepsilon^3 w_2 + \cdots,$$

$$\theta = \varepsilon \theta_0 + \varepsilon^2 \theta_1 + \varepsilon^3 \theta_2 + \cdots, \tag{7.6}$$

$$\mathcal{R} = \mathcal{R}_c + \varepsilon \mathcal{R}_1 + \varepsilon^2 \mathcal{R}_2 + \cdots.$$

Malkus and Veronis equated the parameter ε with the amplitude of the motion. In order for the variables w, θ, and $\mathcal{R}$ to remain finite, the amplitude ε has to remain small, which means that the perturbation analysis is restricted to conditions just slightly above critical. If we substitute the power series (7.6) into the nonlinear convention equations, we obtain a set of linear inhomogeneous equations, because the coefficients of each power of ε must vanish individually. Using this procedure Malkus and Veronis showed that rectangular and hexagonal cells are finite amplitude solutions of the weakly nonlinear convection equations, and that the infinite degeneracy of the patterns from linear theory remains (for small ε), because the rectangular pattern permits an infinite number of different side ratios. Malkus and Veronis also discussed two-dimensional rolls, but merely as the mathematically most simple case, because they believed that rolls could not be realized in practice. In order to select the actually realized pattern from the infinite set of possible solutions, Malkus and Veronis introduced the maximum heat transport hypothesis, among other extremum principles. They found that the only stable solution is the one of maximum heat transport. They also found that square planforms are preferred to the hexagonal planform in ordinary fluids with symmetric boundary conditions, and that the preferred horizontal wave number increases with increasing Rayleigh number.

The maximum heat transport hypothesis has been abandoned over the years, although it has not been proved formally that the theorem is incorrect. On the contrary, Schlüter et al. (1965) confirmed the maximum heat transport theorem for slightly supercritical convection. On the other hand, maximum heat transport contradicts the balance theorem that was discussed in Section 2.2. The reason to give up on the maximum heat transport concept seems to be the experimental finding that supercritical convective motions have wavelengths $\lambda > \lambda_c$, as discussed in Section 6.1, rather than the wavelengths $\lambda < \lambda_c$ with which Malkus and Veronis connect maximum heat transport. Intuitively it seems to make sense that shorter wavelengths, i.e. more cells, transport more additional heat upward by the motion of the fluid, than motions with longer wavelengths, i.e. fewer cells. Busse (1967a) showed theoretically that the wave number of the solution with maximum heat transport increases will $\Re$; i.e. the corresponding wavelength decreases with $\Re$. But the actually observed supercritical wavelengths are larger than λ_c, not smaller than λ_c. The heat transport by supercritical convective rolls with different wavelengths can be computed numerically, as has been done by Lipps and Somerville (1971) and by Plows (1972). The results of Plows' computations are shown in Fig. 7.1. As expected, the calculated maximum heat transport, as expressed by the Nusselt number, is at wavelengths shorter than λ_c. But the observed supercritical wavelengths are longer than λ_c, which means that the convective motions do not follow the maximum heat transfer hypothesis.

The paper of Malkus and Veronis (1958) originally seemed to open the door to the investigation of nonlinear convection. However when we compare the results of this investigation with experimental results established over the years, we find that the maximum heat transfer hypothesis has not been confirmed, that the square pattern plays a marginal role as compared with two-dimensional rolls, and that the preferred horizontal wave number does not increase with $\Re$, but decreases. With a method similar to the one used by Malkus and Veronis, Kuo (1961) determined the mean vertical temperature field and the heat transport of supercritical convection.

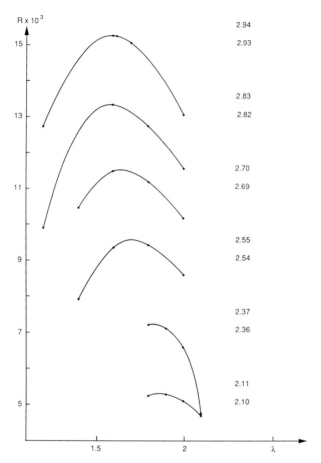

Fig. 7.1. Computed heat transport of rolls as a function of the Rayleigh number and the wavelength for infinite Prandtl number fluids. The numbers to the right of the maxima of the curves show the Nusselt number scale for the particular curve. After Table 4A of Plows (1972).

The, up to now, final formulation of the weakly nonlinear convection problem on an infinite plane with either free–free or rigid–rigid horizontal boundaries was presented by Schlüter et al. (1965). They also expanded the solutions of the nonlinear equations in power series of a small parameter ε, as in equations (7.6), and equated ε with the amplitude of the motions. Substituting the power series into the nonlinear convection equations, they obtained, just as Malkus and Veronis (1958) did, a set of inhomogeneous equations. From the existence conditions for the solutions, not from an ad hoc introduced physical principle, Schlüter et al. determined the $\mathcal{R}_i$ which appear in the power series for $\mathcal{R}$. Instead of studying, in the first approximation, the different solutions of the membrane equation for the possible patterns, they wrote the solution of the membrane equation in the general form

$$w = \sum_{-N}^{N} C_n w_n \quad (n \neq 0), \qquad w_n = \exp(i\mathbf{k}_n\mathbf{r}), \tag{7.7}$$

with

$$|k_n|^2 = a^2 \quad \text{and} \quad \sum_{-N}^{N} |C_n|^2 = 1,$$

where $\mathbf{r}$ is the two-dimensional radius vector in the horizontal plane. Solutions of the form (7.7) have a twofold infinite degeneracy, there is an infinite number of possible orientations of the vector $\mathbf{k}_n$, and there is an infinite number of possible coefficients C_n. The vertical velocity of steady motions in the first approximation is

$$w = f(z) \sum_{-N,\, n \neq 0}^{N} C_n w_n. \tag{7.8}$$

Schlüter et al. found that in the second approximation no steady solution is preferred, $\mathcal{R}_1 = 0$. Going to the third approximation they found that an infinite number of steady finite amplitude solutions exist, which incorporate parallel rolls of arbitrary direction, hexagonal, and also rectangular solutions. For the amplitude of the motion, it follows, with $\mathcal{R}_1$ being zero, that ε is proportional to the square root of $\mathcal{R} - \mathcal{R}_c$.

In order to find out which solution of the manifold of possible solutions is actually preferred, Schlüter et al. investigated the stability of these solutions with regard to steady infinitesimal disturbances of wave number a. The linear stability problem of the possible steady-state solutions of the nonlinear convection equations leads to a homogeneous set of differential equations with the growth rate s being the eigenvalue. Since the stationary solutions of the convection equations are given by power series expansions in ε, an analogous expansion was used for the disturbances $\bar{v}$ and θ, and the growth rate s. It is

$$
\begin{aligned}
s &= s_0 + \varepsilon s_1 + \varepsilon^2 s_2 + \cdots \\
\bar{v}_i &= \bar{v}_{i1} + \varepsilon \bar{v}_{i2} + \varepsilon^2 \bar{v}_{i3} + \cdots, \\
\bar{\theta} &= \bar{\theta}_1 + \varepsilon \bar{\theta}_2 + \varepsilon^2 \bar{\theta}_3 + \cdots,
\end{aligned}
\tag{7.9}
$$

with $i = 1, 2, 3$. In the second approximation s_1 was found to be zero. In the third approximation Schlüter et al. showed that all eigenvalues s_2 are real, and that only two-dimensional rolls are stable, whereas all three-dimensional solutions are unstable. This is the first fundamental result of this study, saying unambiguously that according to theory the infinite plane should be covered with *straight parallel rolls of infinite length* pointing in an arbitrary direction.

One wonders, of course, what determines, under the assumed absolutely uniform conditions, the direction of the rolls in a particular experiment. It would require the interaction of at least two local disturbances in order to establish a preferred direction. It also requires that in experiments on infinite planes with the same apparatus and exactly the same experimental procedure the orientation of the rolls is different from experiment to experiment. If the orientation of the rolls did not differ from experiment to experiment, there would be a preferred direction in the experiment, which would have to originate from a nonuniformity of the conditions, in contradiction to the assumptions. In the bounded containers we work with, such a degeneracy of the direction of the rolls has not been observed, nor is it expected. We note finally that we have invoked nonlinear motions in order to find out what the preferred planform is. One wonders, then, what the planform of the motions is in the linear state, which has to be passed before we reach the nonlinear state.

Following the investigation of the preferred pattern, Schlüter et al. discussed on half a page the range of wave numbers for which the finite amplitude rolls are stable. They investigated the stability of rolls with regard to disturbances with wave numbers a^* different from the wave number a of the rolls. They found, in the third approximation, that in the limit $|a^* - a| \to 0$ rolls are unstable for $a < a_c$, whereas for $a > a_c$ finite amplitude rolls are *stable* in a range $a_1(\varepsilon) > a > a_c$, and unstable otherwise. That means that the solution of the nonlinear convection problem is *nonunique*. For a given supercritical $\mathfrak{R}$ there exists, according to theory, a *continuum* of possible solutions with wave numbers which are larger than a_c or, what is the same, with wavelengths which are *smaller* than λ_c. That is the second fundamental result of this study. A qualitive picture of the stability range of finite amplitude rolls is shown in Fig. 7.2. Finally Schlüter et al. proved that the heat transport of rolls has an absolute maximum as compared with the other patterns, which is as much as proof of the maximum heat transport hypothesis of Malkus and Veronis (1958).

Comparing the results of Schlüter et al. with the experimentally established facts, we find qualitative agreement as far as the preferred roll pattern

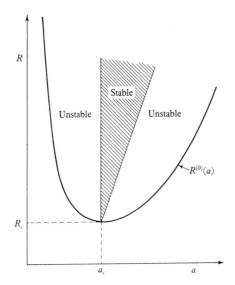

Fig. 7.2. Stability diagram for supercritical Rayleigh–Bénard convection rolls in a fluid
layer of infinite horizontal extent according to weakly nonlinear theory. $R^0(a)$
marks the marginal curve of linear theory. After Schlüter et al. (1965).

is concerned. Assuming that one can adequately approximate the infinite horizontal layer considered in theory with bounded fluid layers of large aspect
ratio, then rolls seem indeed to be the preferred planform if one accepts
straight rolls in rectangular containers and circular concentric rolls in circular
containers as the valid solution of pattern selection in these containers.
The only other solution that experiments in such containers produce is irregular three-dimensional patterns. We also note the directional degeneracy
of the roll pattern in theory. This has not been observed to occur experimentally in the bounded containers we work with. But the orientational degeneracy of theory is a consequence of the infinite plane assumption and is probably removed as soon as lateral boundaries are introduced. Apparently in
order to alleviate the problem with the orientational degeneracy of straight
rolls Busse (1978) expresses the opinion that "patches" of rolls represent the
actual pattern, patches being areas where "on a small scale of a few layer
depths convection assumes the form of nearly two-dimensional rolls" (Busse
and Clever, 1979, p. 319). Patches are, by definition, three-dimensional
and are not a solution of weakly nonlinear convection according to Schlüter
et al. (1965).

Comparing the nonuniqueness or range of stable supercritical wave numbers predicted by Schlüter et al. with the experimental facts, we find

disagreement between theory and experiment. It is, as explained in Section 6.1, a characteristic feature of supercritical convective motions in bounded layers of large aspect ratio that the wavelength of the motions is *larger* than the critical wavelength, or that the wave number is *smaller* than a_c. Just the opposite is predicted by theory. As Fig. 7.2 shows, stable rolls are predicted for $a > a_c$, but in this range we do not observe stable convective motions experimentally. On the other hand, theory predicts an absence of stable motions for all values $a < a_c$, but this is the range into which the experimentally observed wave numbers fall (Fig. 6.3). The experimental observations deal with steady motions, just as theory does. There is also, as discussed likewise in Section 6.1, no convincing evidence that supercritical convection in containers of large aspect ratio is nonunique, which means one does not observe, within the accuracy of the experiments, a continuum of wave numbers for one given steady supercritical $\Re$. Palm (1975) suggested that ''the realized wave number (and the orientation of the rolls) in an actual experiment is determined by the initial conditions'' (p. 47). There are, on the other hand, no initial conditions involved in the stability analysis of Schlüter et al. The idea that initial conditions may lead to the band of stable wave numbers that theory suggests has been tested by the experiments with the controlled initial conditions, also discussed in Section 6.1. The controlled initial condition experiments of Busse and Whitehead (1971) do not, however, produce *stable* rolls with wave numbers $a > a_c$, nor do they produce stable rolls in the range $a < a_c$, but three-dimensional patterns (patches).

Let us now recall the results of the weakly nonlinear analysis of supercritical surface-tension-driven convection (Chapter 3) made by Cloot and Lebon (1984) with the same technique which Schlüter et al. used. As shown in Fig. 3.3 Cloot and Lebon arrived, in the Rayleigh–Bénard case, at stable rolls with a continuum of supercritical wave numbers *larger* than a_c of linear theory in agreement with Schlüter et al. In the surface-tension-driven case with $400 \le \Re \le 500$ Cloot and Lebon arrived at stable hexagons with a continuum of supercritical wave numbers *smaller* than a_c of linear theory. On the other hand, for $100 \ge \Re \ge 0$ the stable wave numbers were predicted to be *larger* than a_c. Schlüter et al., as well as Cloot and Lebon, correctly predict the patterns established experimentally. On the other hand, both analyses predict stable supercritical wave numbers in ranges just opposite to the experimental findings in the Rayleigh–Bénard case. In the completely different case of surface-tension-driven Bénard convection with $\Re \le 100$ the observed wave numbers of stable hexagonal cells are clearly in a range in which they should be unstable according to Cloot and Lebon. Both analyses also predict the existence of continua of nonunique supercritical wave numbers, which is likewise not supported by the experimental findings. These contradictions between theoretical and experimental results seem to indicate that there is a principal problem with the application of the perturbation technique to supercritical convection.

Based on very general stability considerations Eckhaus (1965) predicted instability in nonlinear two-dimensional convection as well as Taylor vortex flow over wide intervals in wave numbers both larger and smaller than a_c. The interval in wave numbers is, however, smaller than the range in wave numbers given, at the same $\Re$, by the neutral curve of linear theory of convection shown in Fig. 2.1. The range of possible supercritical wave numbers, according to Eckhaus, is given by

$$a_E = a_c - \frac{1}{\sqrt{3}}(a_c - a_m), \tag{7.10}$$

where a_E is the maximal as well as minimal wave number of the Eckhaus instability at a given $\Re$, a_c is the critical wave number of linear theory, and a_m is the maximal or minimal wave number of the marginal or neutral curve of linear theory for the same $\Re$. The range of instability according to Eckhaus is indicated by the dashed line in Fig. 2.1. Note, however, that this range is valid only for just supercritical conditions. Eckhaus' results apply to two-dimensional flows and are valid only for real eigenvalues. An extended form of Eckhaus' results was given by Stuart and DiPrima (1978). The reality of the so-called Eckhaus instability has not been demonstrated in Rayleigh–Bénard convection; that would require clear proof of the nonuniqueness of the wavelength. But the Eckhaus instability is apparent in the Taylor vortex problem, where steady stable flows with $a < a_c$ as well as $a > a_c$ can be established easily by different initial conditions. The interval of nonunique wave numbers is, however, even at moderate supercritical Taylor numbers, much smaller than the interval given by (7.10). But, this is so only with the Taylor vortex problem. As Rayleigh–Bénard convection experiments tell us, some as yet not identified mechanism reduces the range of possible stable supercritical wave numbers at a particular $\Re$ to a point, or at most to a very narrow interval, with wave numbers $a < a_c$.

Although the stability diagram (Fig. 7.2) of Schlüter et al. was modified by Busse (1967a) to include a continuum of wave numbers $a < a_c$, and modified further in order to account for the influence of the Prandtl number (see Busse, 1978), there is, also in the modified versions, the range of wave numbers $a > a_c$ in which, as theory says, stable two-dimensional rolls are possible, in contradiction to the experimental observations. Since the mathematical work of

Schlüter et al. has been found impeccable, a particular element must be missing in their stability analysis that ultimately forces the fluid to select the wave numbers established by the experiments.

We note first that Busse's (1967a) paper shows that the nonlinear terms in the Navier–Stokes equations are not the cause of the appearance of the stable supercritical motions with $a > a_c$ in theory. Busse's study assumes an infinite Prandtl number, which means that the nonlinear terms in the Navier–Stokes equations drop out; see equation (7.3). As Fig. 7.3 shows there are, in theory, also stable motions with $a > a_c$ when the Prandtl number is infinite, in contradiction to the results of experiments with fluids of large $\mathscr{P}$ which approximate infinite Prandtl numbers very well. That means that part of the problem with the stable wave numbers is already in the (nonlinear) energy equation.

As far as the procedure of Schlüter et al. for the determination of the range of stable supercritical wave numbers is concerned, one wonders whether it is physically correct to put the disturbances of wave number a^*, with which they disturb the rolls of wave number a, on an equal footing with the finite amplitude rolls whose stability they study. At $\mathscr{R}_c$ and in the immediate vicinity of $\mathscr{R}_c$ there is only one finite amplitude motion in the fluid and this motion has the wave number a_c. The onset of convection at $\mathscr{R}_c$ with motions of the unique wave number a_c consumes all available energy in the fluid. If the Rayleigh number is increased quasi steadily, as is implied in Schlüter et al., there is no energy available for other disturbances to grow, except at the expense of the motion with the critical wave number. As the experiments show, the convective motions with wave number a_c are not replaced by a continuum of wave numbers different from a_c in the vicinity of $\mathscr{R}_c$. All disturbances other than the finite amplitude motion with the critical wave number are infinitesimal. It makes a difference for stability whether a roll of finite amplitude is disturbed by rolls of equal (finite) amplitude or by infinitesimal disturbances. It seems to be possible that finite amplitude rolls in a field of infinitesimal disturbances follow their own dynamics when the Rayleigh number is increased, rather than exchange stability with other infinitesimal rolls. Because of the obvious discrepancies between the range of stable supercritical wave numbers predicted by stability studies and the experimentally observed supercritical wave numbers, Getling (1991) suggests that stability considerations may not be the correct way to find the preferred wave numbers of nonlinear rolls.

One also wonders whether the class of disturbances which Schlüter et al. consider is complete. They considered disturbances with wave numbers a^* different from the wave number a of the rolls. This choice does not include the most likely disturbances in a fluid layer with random infinitesimal noise, which are three-dimensional with a continuous wave number spectrum, a continuous amplitude distribution, and also a broad-frequency spectrum. Thermal noise should be the only disturbance present in an apparatus which matches the ideal assumptions of theory, such as uniform temperatures on the top and bottom of the layer, and perfect conduction in the boundaries.

Three-dimensional disturbances in a fluid layer, such as "dislocations," "grain boundaries," and "pinches," as discussed, e.g., in Pocheau and Croquette (1984), are sometimes said to be required to permit a wave number change in bounded layers. Many of these defects seem to be the consequence of imperfections in the supposedly uniform temperature field; actually some of the defects have been introduced deliberately in this way. On the other hand, sufficiently uniform temperature fields permit a wave number change without defects in the flow field, as shown by Fig. 6.2. Although a theoretical investigation of the consequences of defects requires the consideration of the nonuniformities which are at the root of the defects, it appears nevertheless that three-dimensional disturbances ease a wave number change in a restrictive geometry. That may indicate that it is necessary to include three-dimensional, time-dependent disturbances in a comprehensive stability analysis. It appears to be natural to assume that the pattern that is stable with respect to three-dimensional time-dependent disturbances is actually the preferred pattern. Some insight into the consequences of such disturbances on convective motions can be gained from the investigation of Kelly and Pal (1978) on the effect of spatially periodic temperatures in the horizontal boundaries of the fluid and from the study of Jhaveri and Homsy (1980) on random fluctuations. But the understanding of the consequences of random three-dimensional, time-dependent fluctuations is still in its infancy.

Since the above-discussed studies of weakly nonlinear convection do not give us the correct preferred supercritical wavelength, we have to look elsewhere for an explanation of the increase in the supercritical wavelength, which is so prominent in the experiments. Such an explanation may be found in the balance theorem of Chandrasekhar (1961). As we noted in Section 2.2 the balance theorem holds under slightly supercritical steady conditions if the nonlinear equations of convection and the Boussinesq approximation are used, and three plausible assumptions about the horizontal averages of the velocity components, the pressure, and the vertical temperature gradient are made. The balance theorem states that the kinetic energy dissipated by viscosity is balanced by the internal energy released by buoyancy. Looking now at slightly supercritical conditions, it appears that more internal energy must be released than at the critical point, because the applied temperature has been increased. Viscous dissipation, on the other hand, is likely to increase rapidly after transition, because the velocity increases rapidly after the onset of convection. As the results of the experiments indicate, balance of internal energy release and viscous dissipation is apparently established by the fluid selecting a longer wavelength. Larger cells mean, under otherwise identical conditions, less dissipation. The excess dissipation in supercritical cells is apparently reduced by changing the cells to a larger size or wavelength. The reduced dissipation in *larger* supercritical cells can then balance the only moderately increased internal energy release of supercritical flow. We found similar wavelength changes before, when we discussed in Section 2.2 the different values of the critical wavelength which result from free–free, rigid–free, and rigid–rigid boundary conditions. Analogous wavelength changes occur in supercritical flow if one enhances internal energy release by rapid heating of the bottom plate. This is, as observed experimentally by Koschmieder (1969), accompanied by shorter nonsteady wavelengths. Or one can reduce the internal energy release by placing an insulating lid on top of the fluid. In the steady case the flow then selects longer wavelengths, as observed by Koschmieder (1969). In either case it seems to be the balance of internal energy release and viscous dissipation which determines what the fluid does.

Let us now compare the results of weakly nonlinear theory of convection with the goals of the theory listed in the first paragraph of this section. As far as the amplitude of the motions is concerned, weakly nonlinear theory has been a success. In Gor'kov (1957), Malkus and Veronis (1958), and Schlüter et al. (1965) the amplitude was found to be proportional to $\sqrt{(\mathcal{R} - \mathcal{R}_c)}$ or proportional to $\sqrt{\varepsilon}$, as this relationship is written in the Landau equation which we shall discuss in Section 7.3. Since the $\sqrt{\varepsilon}$ rule has been confirmed experimentally by Dubois and Bergé (1978), the amplitude problem seems to have been solved. As far as the pattern of motion in weakly nonlinear convection is concerned, weakly nonlinear theory has come up with an ambiguous answer. Rolls are said to be the preferred pattern, but that is an academic answer because this applies for an infinite layer nobody can work with. For the question of the preferred supercritical wavelength, weakly nonlinear theory has been a disappointment. It is not only that weakly nonlinear theory has come up with the result that supercritical convective motions should be nonunique, which they do not seem to be, but also that the theory has had the tendency to point to a range of nonunique supercritical wavelengths just opposite to the wavelengths found in the experiments. Why that is so is a puzzle. Weakly nonlinear theory is an analytical extension of the successful linear theory of convection into the just nonlinear range, and one should expect that this procedure should provide a solution of the problem of the preferred nonlinear wavelength.

7.2 Moderately Nonlinear Convection

The analytical methods discussed in the previous section are valid only under just supercritical conditions, where the changes brought about by the increase of the Rayleigh number are too small to be confirmed unambiguously by experiments. Rigorous verification of the theoretical results should be possible in the moderately supercritical range between, say, $1.2\Re_c$ and $10\Re_c$. The first theoretical study of supercritical convection on an infinite plane in this Rayleigh number range was made by Busse (1967a). He used the nonlinear convection equations in the case of an infinite Prandtl number. This means that the nonlinear terms in the Navier–Stokes equations vanish because they are multiplied by $1/\mathscr{P}$; see equation (7.3). The only nonlinear term in the convection equations is then the buoyancy term in the heat transfer equation. Busse solved the convection equations with a Galerkin method, assuming stationary two-dimensional solutions and expanding the variables in Fourier series. This leads to an infinite set of algebraic equations, which were solved numerically. The stability of the steady solutions of the convection equations was then studied by disturbing the solutions with nonoscillatory two-dimensional disturbances. This leads, as in the analysis of Schlüter et al., to linear homogeneous equations with the growth rate s being the eigenvalue. The stability range of the motions is shown in Fig. 7.3. According to this figure stationary two-dimensional supercritical rolls are stable in only a part of the range given by the neutral curve of linear theory. There are no stable time-independent motions above $\Re \cong 22{,}600$. Very near the critical Rayleigh number the range of stable rolls is, in the limit $|a' - a| \to 0$, compatible with the results of Schlüter et al. as expressed by Fig. 7.2. Between $\Re_c$ and $\Re \cong 22{,}000$ there are stable two-dimensional rolls with wave numbers $a < a_c$ as well as $a > a_c$. It was stated that the most stable solution has a slightly increasing wave number.

Busse's (1967a) paper has been supplemented by an investigation of the consequences of finite Prandtl numbers for the stability diagram of nonlinear convection. Busse and Clever (1979) studied the cases of water ($\mathscr{P} \cong 7$) and air ($\mathscr{P} = 0.71$) with methods similar to, but more general than those of Busse (1967a). The disturbances admitted in order to study the stability of the motions were now of infinitesimal arbitrary three-dimensional form. For $\mathscr{P} = 7$

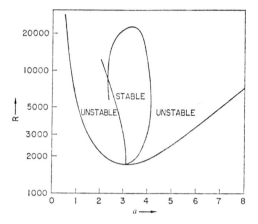

Fig. 7.3.   Stability diagram for stationary two-dimensional Rayleigh–Bénard convection in a
fluid layer of infinite horizontal extent and infinite Prandtl number, at moderately
supercritical Rayleigh numbers. After Busse (1967a).

a stability diagram fairly similar to that in Fig. 7.3, but shifted slightly to
smaller wave numbers, was obtained. For $\mathscr{P} = 0.71$ the diagram is quite different, the interval of stable wave numbers is much narrower, and most of the
stable wave numbers are at values of $a < a_c$. A composite of the stability
diagrams for various Prandtl numbers, including the two Prandtl numbers just
discussed, is shown in Fig. 7.4.

Before we discuss the meaning of Fig. 7.4 we have to consider the various
instabilities, the zigzag, cross-roll, skewed varicose, and knot instability,
which outline the boundaries of this stability diagram. These instabilities
come about by the action of three-dimensional disturbances of different wave
numbers on supercritical parallel rolls. The existence of the different instabilities was investigated in experiments with controlled initial conditions. We
discussed the controlled initial conditions before in Section 6.1. Pictures of
the cross-roll and zigzag instabilities can be found in Busse and Whitehead
(1971). The cross-roll instability of an originally parallel roll system did not
occur homogeneously throughout the layer and was indicated by disturbances
in the form of patches of rolls whose axes were perpendicular to the original
rolls. After some time practically the entire roll system turned its orientation
by 90° and increased its wavelength by about 50%. The zigzag instability occurred as a patchy transformation of a parallel roll system into zigzagging
(wavy) rolls, the amplitude of the zigzags being about one-half wavelength.
At the end, one-half of the original rolls turned 45° to the left, the other half
45° to the right of the original orientation of the rolls, except for the center

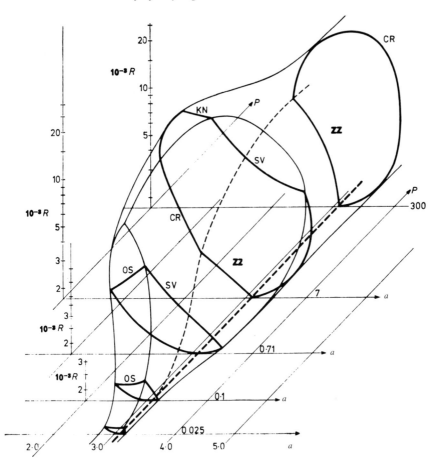

Fig. 7.4. Stability diagram for stationary, stable two-dimensional convection in fluids of different Prandtl number at moderately supercritical Rayleigh numbers. The stability boundaries are for the zigzag (ZZ), skewed varicose (SV), cross-roll (CR), knot (KN), and oscillatory (OS) instabilities. After Busse (1978).

pair of rolls, which maintained its orientation. In this case the wavelength of the motions does not seem to be affected by the transformation of the rolls. Considering these instabilities one has to keep in mind that the original systems of parallel rolls are unstable per se due to the controlled initial conditions procedure. The rolls produced by the controlled initial conditions are not the *stationary* rolls studied in theory. The skewed varicose and knot instabilities were shown in Busse and Clever (1979). Note that each of the four sets of photographs in this paper showing these two instabilities were made at wave numbers and Rayleigh numbers clearly outside the stability boundaries shown in Fig. 1 of the paper. Likewise, most of the experimental points plotted in Fig. 11 of the paper are outside the stability boundaries of theory. This is not a convincing confirmation of the theory. These discrepancies are said to be due to the poor thermal conductivity of the top and bottom glass boundaries of the fluid in their apparatus.

Any one of the four types of instabilities just discussed are rarely, if ever, seen in large aspect ratio, time-independent convection experiments without controlled initial conditions, because the wave numbers at which these instabilities occur are not encountered in quasi-steady experiments. They are sometimes referred to; e.g. Kolodner et al. (1986) discuss the cross-roll and knot instability. Their experiment was made in a rectangular container of such small aspect ratio ($\Gamma_y = 5$) that the theory of convection on an infinite plane is irrelevant. Three-dimensional motions in this apparatus are almost necessarily a consequence of the geometry of the fluid layer, and not of the cross-roll or knot instability.

Considering now the validity of the composite stability diagram in Fig. 7.4, we face problems similar to the problems found with the stability diagram of Schlüter et al., shown in Fig. 7.2. The stability diagram in Fig. 7.4 represents, however, an improvement over Fig. 7.2 since it permits stable flow with wave numbers $a < a_c$. On the other hand, in Fig. 7.4 there is, for nearly all values of $\mathcal{P}$, an extended range of stable, stationary two-dimensional flow with wave numbers $a > a_c$, in contradiction to the experimentally found absence of steady supercritical convective motions with $a > a_c$. According to Fig. 7.4 there should also be, as theory predicts, a wide range of stable, supercritical two-dimensional flow with wave numbers $a < a_c$, or wavelengths $\lambda > \lambda_c$. In experiments with large Prandtl number fluids, as summarized in Fig. 6.3, convective motions in this range of wave numbers have been observed either along curves $a(\mathcal{R},\mathcal{P})$ or in narrow intervals of wave numbers which are only a small part of the range of stable wave numbers $a < a_c$ in Fig. 7.4, so this part of the stability diagram is not in agreement with the experimental results either. For small Prandtl numbers there is the section of Fig. 7.4. for $\mathcal{P} = 0.71$ (air). This stability diagram is clearly not compatible with the measurements of the wave numbers in air by Martinet et al. (1984). According to them, stable supercritical rolls occurred up to Rayleigh numbers twice as high ($\mathcal{R}/\mathcal{R}_c \cong 7$) as the maximal $\mathcal{R}$ for which stable supercritical flow is predicted theoretically; and the first oscillatory motions occurred at Rayleigh numbers three times as high as the $\mathcal{R}$ at which they are predicted, as can be seen in Fig. 2 of Martinet et al. Summarizing,

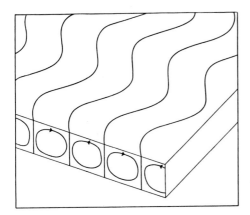

Fig. 7.5. Qualitative sketch of oscillating convection rolls. After Clever and Busse (1974).

we find that Busse's (1978) stability diagram for moderately supercritical Rayleigh–Bénard convection (Fig. 7.4) does not match reality.

The stability diagram in Fig. 7.4 does not identify the mechanism which at the end determines the selection of the wave numbers actually preferred by convection. The cross-roll, skewed varicose, and knot instabilities are separated from the actually preferred wave numbers by wide ranges of theoretically stable wave numbers, so that they have no opportunity to influence the selection mechanism. The zigzag instability is close to the actually realized wave numbers, but is above the observed wave numbers at lower $\mathcal{R} > \mathcal{R}_c$, and is at wave numbers smaller than the observed wave numbers at larger $\mathcal{R}$; that means the zigzag instability does not determine the preferred wave number either.

At higher Rayleigh numbers moderately supercritical Rayleigh–Bénard convection changes to time-dependent flow. There are only two theoretical investigations of the transition to time-dependent flow, the studies of Clever and Busse (1974, 1987). These papers are concerned with the transition to oscillatory convection for low Prandtl number fluids, $\mathcal{P} < 5$, in infinite fluid layers. The authors found that the steady solutions of moderately supercritical conditions in the form of two-dimensional rolls become unstable to an oscillatory instability and form wavy rolls, as shown in Fig. 7.5. The waves were predicted to propagate along the axis of the roll pattern. Clever and Busse (1974) believe that the agreement between experimental findings and their theoretical predictions "is nearly perfect" and refer for a comparison to the paper of Willis and Deardorff (1970). In these experiments, made with a layer of air ($\mathscr{P} = 0.71$) of large aspect ratio, a few waves not more than one and a half wavelength long appeared sporadically in a very wide field of irregular cells. One wonders whether the irregular cells or the few waves are characteristic of the velocity field. It seems that the experiments of Willis and Deardorff (1970) do not provide proof of the existence of steady, regular waves, oscillating coherently and propagating in the direction of the rolls. Actually they observed "a fairly broad range of frequencies" at a fixed $\mathscr{R}$. Broadfrequency spectra, not monoperiodic waves, appeared with the onset of time dependence in large aspect ratio experiments of Libchaber and Maurer (1978); see Fig. 6.8. Lack of coherence and absence of a clear transition to oscillatory motions has, as discussed in Section 6.2, also been typical of corresponding experiments with large Prandtl number fluids (see Krishnamurti, (1970b). We have shown pictures of those flows in Fig. 6.6. The absence of sharp frequencies in convection in moderately large containers observed by Libchaber and Maurer (1978) and in other experiments renders the theory of oscillatory convection by Clever and Busse unrealistic.

To summarize, the theoretical investigations of steady, moderately supercritical convection based on the nonlinear Navier–Stokes equations studying Boussinesq fluid layers of infinite horizontal extent has provided us with stability diagrams which predict stable flow in the form of rolls with a wide range of wave numbers that are smaller as well as larger than the critical wave number. Rolls seem to be the correct solution if due consideration is given to the presence of lateral walls in large aspect ratio containers. The predicted range of wave numbers, however, is not in agreement with the experimental results, according to which there are, for a particular $\mathscr{R}$, preferred wave numbers smaller than the critical wave number. So the principal questions concerning steady, moderately supercritical Rayleigh–Bénard convection remain: why does the fluid choose at a particular $\mathscr{R} > \mathscr{R}_c$ a particular wave number smaller than a_c, and is moderately supercritical convection unique or nonunique and why? If we have answered these basic questions, we have a solid foundation for the even more difficult investigation of convection at higher Rayleigh numbers.

7.3   The Landau Equation

The theoretical investigations of nonlinear convection discussed so far have not revealed clear information about the question of how a steady equilibrium state is attained after the onset of convection if a supercritical temperature difference is applied to the fluid layer. We shall now discuss the increase of the amplitude of the convective motions after the onset of convection. The pioneering study of this problem was made by Landau (1944), not exactly in the context of convection, but rather in connection with the question of the initiation of turbulence, which Landau perceived to be caused by a sequence of instabilities at consecutively higher Reynolds numbers. He was concerned with the increase of the amplitude of the disturbances making up the turbulent flow. So instead of writing for the amplitude of the disturbances the equation

$$A(t) = \text{const } e^{i\Omega t}, \tag{7.11}$$

where in his notation Ω is a frequency, which equation corresponds to the ansatz for the increase of the amplitude made in linear theory of convection, equation (2.13), Landau suggested describing the increase of the amplitude with time by a power series.

In the first approximation it is

$$\frac{d|A|^2}{dt} = 2s|A|^2, \tag{7.12}$$

considering complex amplitudes. In the real case the amplitude varies as e^{st}, with s being the growth rate, which is, for slightly supercritical conditions, proportional to $(\mathcal{R} - \mathcal{R}_c)/\mathcal{R}_c = \varepsilon$ according to (2.31b). The next term in the power series for the variation of the amplitude should be of third order, which term, however, does not appear because it is zero upon averaging over time for periodic disturbances. Taking the fourth order term into consideration it is

$$\overline{\frac{d|A|^2}{dt}} = 2s|A|^2 + 2a_1|A|^4, \tag{7.13}$$

the bar on the left hand side indicating averaging over time. Bifurcations to time-dependent periodic solutions had been studied earlier by Hopf (1942). Bifurcation to a time-dependent convection problem is therefore sometimes referred to as a Landau–Hopf bifurcation. If the amplitude is real it is, of course,

$$dA/dt = sA + a_1 A^3, \tag{7.14}$$

where a_1 must be < 0 in order to counteract the exponential growth of the amplitude given by the first term on the right side of (7.14). Equation (7.14) is referred to as the Landau equation or the amplitude equation for weakly nonlinear two-dimensional convection in a Boussinesq fluid. A solution of equation (7.14) was found by Davey (1962); see equation (14.10). Curves showing the variation of the amplitude of a disturbance with initial amplitude A_0 as a function of time for various values of the parameter a_1 in (7.14) can be found in Chapter 7 of Drazin and Reid (1981).

It can be seen from (7.14) that in the application of the Landau equation to the Rayleigh–Bénard problem an A^2 term cannot appear, because in equilibrium the amplitude of a two-dimensional convection roll must be the same whether the velocity is either upward or downward at the side of the roll. This must be so since a roll pattern must be invariant to a translation of the pattern by $\lambda/2$. However this is true only for a pattern of harmonic rolls. This is not so in three-dimensional flow; we have encountered the amplitude equations for hexagonal Bénard cells before [equations (3.22a)–(3.22b)]. These equations contain terms with the squares or products of the amplitudes in two perpendicular directions, and other higher order terms. Second order terms in the amplitude equation appear also when the fluid is non-Boussinesq, as was shown by Haken (1975). In bounded fluid layers, such as circular or rectangular layers, equation (7.14) cannot apply throughout the entire layer because the wavelength of the motions can increase only at the expense of one or some rolls of the layer, whose amplitude must consequently *decrease* when the Rayleigh number is increased. As we have seen the center roll of a pattern of circular concentric rolls gradually disappears as $\mathcal{R}$ is increased, and some straight rolls in a rectangular container disappear discontinuously near the ends of the container. The experimentally established increase of the wavelength of supercritical Rayleigh–Bénard convection also means that the Landau equation does not apply equally to all modes of the flow. The amplitude of the critical mode with wavelength λ_c must decrease with increased Rayleigh number in order to make room for motions with longer wavelengths, even in an unbounded layer.

Solving (7.14) for supercritical conditions ($s > 0$) gives the equilibrium amplitude A_e when $t \to \infty$. Landau showed that $A_e^2 = -s/a_1$, or

$$A_e \propto \sqrt{Re - Re_c} \, , \tag{7.15}$$

from which it follows with s from (2.31b) that in the case of convection

$$A_e \propto \sqrt{\frac{\mathcal{R} - \mathcal{R}_c}{\mathcal{R}_c}} = \sqrt{\varepsilon}. \tag{7.16}$$

A graphic illustration of equation (7.16) is the well-known pitchfork bifurcation diagram (Fig. 7.6). The validity of equation (7.15) as applied to supercritical axisymmetric Taylor vortex flow, where the Reynolds number is replaced by the Taylor number $\mathcal{T}$, was proved by Stuart (1958) in the case

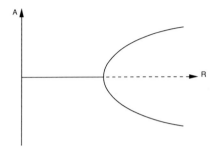

Fig. 7.6.   Bifurcation diagram for a supercritical bifurcation at $\mathfrak{R}_c$.

where the wavelength of the vortices is assumed to remain equal to the critical wavelength. Extensive work with the Landau equation applied to Taylor vortices was made by Davey (1962). The dependence of the amplitude of Rayleigh–Bénard convection on $\sqrt{\varepsilon}$ was confirmed by Chandrasekhar (1961, p. 612). Equation (7.14) in extended form was derived from the convection equations for two-dimensional flow in a Boussinesq fluid by Segel (1969) and by Newell and Whitehead (1969). The validity of equation (7.16) was verified convincingly in experiments by Bergé (1975) and by Dubois and Bergé (1978), as discussed in Section 6.1. They showed that (7.16) is valid up to about $10\mathfrak{R}_c$, far exceeding the range of validity (small ε) of the derivation of the amplitude equation from the convection equations.

7.4   Models of Supercritical Convection

Since the governing equations of fluid mechanics are well known there is, strictly speaking, no need for models in the theory of supercritical convection. The reason for the use of models, i.e. equations introduced ad hoc, is the mathematical difficulty of the nonlinear Navier–Stokes equations. It can be of advantage or may even be a necessity to simplify the difficulty of a nonlinear problem by using an abbreviated version of the Navier–Stokes equations which incorporates some term or terms that one hopes will represent the essential consequences of the nonlinear effects. Some models stay very close to the nonlinear convection equations, whereas others contain terms which do not originate from the basic equations. Models are justified by the agreement of their results with some accepted experimental finding or with the promise that such an experiment can be made and will have the predicted result. From our experience in physics, where models are sometimes necessary because the basic equations are often not known, we know that a model will prove its value only if it is able to explain several related features when it is applied in a consistent way to a variety of problems. A model does not make a contribution to hard sciences when the results of the model cannot be confirmed quantitatively by experiments corresponding to the model's assumptions.

We begin with the most famous model dealing with Rayleigh–Bénard convection, that proposed by Lorenz (1963). Following a procedure first used by Saltzman (1962b), Lorenz expanded the equations describing two-dimensional nonlinear convection on a uniformly heated infinite plane with free–free boundaries in double Fourier series. The resulting system of equations was then truncated radically, so that only three ordinary nonlinear differential equations remained. These are the so-called Lorenz equations,

$$dX/dt = -\sigma X + \sigma Y,$$
$$dY/dt = -XZ + rX - Y, \qquad (7.17)$$
$$dZ/dt = XY - bZ,$$

where $X(t)$ is proportional to the amplitude of the convective motions, $Y(t)$ is proportional to the temperature difference between the ascending and descending motions, i.e. the horizontal temperature difference across a roll, and $Z(t)$ is proportional to the deviation of the vertical temperature profile from the linear profile; t is a dimensionless time, σ stands for the Prandtl number, $r = \mathcal{R}/\mathcal{R}_c$, and $b = 4/(1 + a^2)$.

Obviously $X = Y = Z = 0$ describes the motionless state before the onset of convection, X, Y, Z being time-independent but nonzero describes supercritical convection with $X = Y = \pm\sqrt{b(r - 1)}$ for $r > 1$. At $r = \sigma(\sigma + b + 3)/(\sigma - b - 1)$ steady convective motions become unstable; for sufficiently high r steady convection is unstable if $\sigma > b + 1$. The general solutions of the Lorenz equations are obtained by numerical integration for selected values of σ, r, and b, and sets of initial conditions $X(0)$, $Y(0)$, $Z(0)$. The case made famous through Lorenz's investigation uses $\sigma = 10$, $b = \frac{8}{3}$, and $r = 28$, which is slightly above the critical value for instability ($r = 24.74$) with this choice of σ and b. The astounding result of the computations was that the solutions of (7.17) were irregularly time-dependent and did not approach either a final periodic solution or a final steady state. The extremely complex nature of this solution is demonstrated in Fig. 7.7, which shows the projection of the solution onto the X–Z plane. The apparent intersections of the curve are a consequence of the projection and do not occur in three-dimensional space. Similar results were obtained with different parameters in the unstable range. This finding of nonperiodic – chaotic – solutions resulting from these simple nonlinear equations has created extraordinary interest.

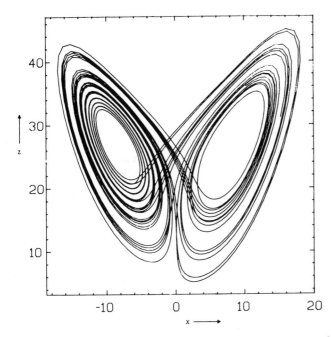

Fig. 7.7.   Sequence of solutions of the Lorenz equations with $\sigma = 10$, $b = \frac{8}{3}$, and $r = 28.0$, projected on the X–Z plane. After Sparrow (1982).

This is not the place to pursue the results of the Lorenz equations in detail; the reader is referred to, e.g., the book of Sparrow (1982) or the article of Yorke and Yorke (1981).

As far as the realities of the Rayleigh–Bénard problem are concerned, the Lorenz model does not contribute to their solution. Lorenz himself addressed this question in his (1963) paper, stating that equations (7.17) may give realistic results when the Rayleigh number is slightly supercritical, but "in view of the extreme truncation" their solution "cannot be expected to resemble" those of the complete nonlinear convection equations at larger values of the Rayleigh number, and that includes the highly supercritical Rayleigh number at which the computations leading to Fig. 7.7 were made. It is very unlikely that the convective motions on an infinite plane are still two-dimensional at $\mathcal{R} = 28\mathcal{R}_c$, as is required by the input in the solution of the Lorenz equations shown in Fig. 7.7.

The first theoretical attempt to solve the problem of the preferred supercritical wavelength with model equations was made by Segel (1962), at a time when apparently no experimental data about the supercritical cell size were available. Segel tried to find the nonlinear mechanism which, in his opinion, enabled the fluid to select a single horizontal wave number, or a single sharp peak at a certain wave number in the spectrum of the possible band of wave numbers unstable according to linear theory. He looked for the selection mechanism in the interaction of unstable modes, following the ideas outlined by Stuart (1958). Segel restricted his considerations to the interaction of two modes only, and arrived at the result that the equilibrium state slightly above $\Re_c$ will contain only one of the two original linearly unstable modes. The mode which attains finite amplitude will be the one which grows faster according to linear theory, whereas the other mode will ultimately decay to zero. Since the growth rate, as discussed in Section 2.2, is larger for modes with larger wave number (at the same $\Re > \Re_c$), Segel's result means, although it is not expressed this way, that modes with larger wave number should be preferred under supercritical conditions, which is, however, not borne out by experiments.

In a seminal paper Segel (1969) investigated the consequences of the presence of vertical lateral walls in a rectangular container on the onset and the amplitude of convection. Considering *finite* rolls parallel to the shorter side of the container with free–free horizontal boundaries and neglecting higher order terms, he derived for sufficiently small $\varepsilon = (\Re - \Re_c)/\Re_c$ an amplitude equation of the form

$$\frac{\partial A}{\partial t} = A + \delta_1(a)\frac{\partial^2 A}{\partial x^2} - 2i\delta_{12}(a)\frac{\partial^3 A}{\partial x\,\partial y^2} - \delta_2(a)\frac{\partial^4 A}{\partial y^4} - \beta A^2 A^*, \qquad (7.18a)$$

where x is the nondimensional distance perpendicular to the shorter lateral wall, as well as the direction perpendicular to the axes of the finite rolls, y the direction parallel to the shorter wall, and A^* the conjugate complex amplitude. Slow spatial changes of the amplitude in the x-direction make it possible to "fit" the solution for the infinite layer into the rectangular container via the term $\partial^2 A/\partial x^2$. The variation in the y-direction of the x-velocity of the finite rolls is accounted for by the $\partial^4 A/\partial y^4$ term. In (7.18a) δ_1, δ_{12}, δ_2, and β are constants. The nonlinear term alters the results of linear stability theory concerning the exponential growth rate of the amplitude. Equation (7.18a) is a modified version of the Landau equation.

An abbreviated two-dimensional version of this equation in the form

$$\frac{\partial A}{\partial t} = \alpha A + \beta|A^2|A + \gamma\frac{\partial^2 A}{\partial x^2} \qquad (7.18b)$$

is referred to also as the modulation equation, or nowadays most often as the Ginzburg–Landau equation, after a paper of Ginzburg and Landau (1950) dealing with superconductivity, in which the time-independent form of this equation was introduced. The Ginzburg–Landau equation appears in this or similar forms in many papers. For example, Daniels (1977) used this equation to show that a small heat transfer through the sidewalls makes the onset of convection an imperfect bifurcation in the two-dimensional, free–free case; and Brown and Stewartson (1978) used this equation in the corresponding axisymmetric case. The Ginzburg–Landau equation was derived and studied in detail by DiPrima et al. (1971) and was extended to higher order terms by Cross et al. (1983). Mathematical solutions of the Ginzburg–Landau equation are discussed in Doelman (1989).

Neglecting the nonlinear term in (7.18a) Segel (1969) derived a formula for the onset of convection in bounded finite rectangular containers,

$$\mathcal{R}_{cb} = \mathcal{R}_{c\infty}[1 + \delta_1'(a_c)L^{-2} + O(L^{-3}) + \cdot\cdot\cdot], \tag{7.19a}$$

with $\delta_1'(a_c) = \frac{8}{3}$. In the case of a circular container with free–free horizontal boundaries the corresponding formula is, according to Brown and Stewartson (1978),

$$\mathcal{R}_{cb} = \mathcal{R}_{c\infty} + 18\pi^4 L^{-2} + O(L^{-3}), \tag{7.19b}$$

where L is the radius of the layer. We referred to (7.19a) in Section 5.2. Considering the boundary condition $A = 0$ at the lateral walls Segel found that the boundary layer thickness in the x-direction is proportional to $\varepsilon^{-1/2}$, as long as the layer is wide enough that slow amplitude modulation is possible. That means that, as $\varepsilon \to 0$, the thickness of the boundary layer extends to infinity, i.e. extends throughout the entire container and thereby determines the form of the pattern in the container. On the other hand, "when the Rayleigh number is 5 to 10% above its critical value for an unbounded medium the effects of lateral walls should be confined to a boundary layer whose thickness is about a wavelength" (Segel, 1969, p. 212). This implies that the influence of the lateral boundaries decreases rapidly with ε and is, in large containers, not likely to be responsible for the increase of the supercritical wavelength at moderate values of ε. For a measurement of the amplitudes of rolls as a function of ε and the distance from the wall see Fig. 4 of Wesfreid et al. (1978). Actually it seems that the boundary layer at the wall is inversely proportional to a power of $|\varepsilon|$, because wall effects which are already present at $\mathcal{R} < \mathcal{R}_c$, or $\varepsilon < 0$, extend increasingly throughout the fluid layer as $|\varepsilon|$ approaches zero from below; see Figs. 3a,b in Koschmieder and Pallas (1974).

At the same time Newell and Whitehead (1969) published a study of the stability of finite amplitude convection on an infinite plane with free–free

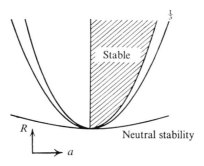

Fig. 7.8.   Stability diagram for finite bandwidth, finite amplitude convection, with the neutral stability curve of linear theory and the Eckhaus instability boundary marked by $\frac{1}{3}$ at the upper end of the curve. After Newell and Whitehead (1969).

horizontal boundaries using an amplitude equation derived independently from Segel, but of the same form as equation (7.18a). Application of (7.18a) to an infinite layer introduces a conceptional difficulty, because the $\partial^2 A/\partial x^2$ term expresses a dependence of the amplitude of the convective motions on the coordinate x, but on an infinite plane with absolutely uniform temperatures on the top and bottom of the fluid layer the coordinate x is not defined other than by the axes of a pattern of parallel rolls covering the infinite plane. The amplitude A is then periodic in x. Slow modulation of the amplitude in the x-direction, or an envelope of the amplitudes of the rolls, in the time-independent case on an infinite plane is not compatible with the uniform and isotropic input in the theory. In the time-dependent case slow modulation of the amplitude in the layer is also not compatible with the uniform input. In Segel's problem, on the other hand, the coordinate x is clearly defined by the existence of the lateral wall at $x = 0$ (and $x = L$) and from here on the amplitude must vary with x. Newell and Whitehead apply their amplitude equation to continuous packages of modes within an $O(\varepsilon)$ wide band around the neutral (critical) mode (with wave number $\pi/\sqrt{2}$). They obtain the stability diagram shown in Fig. 7.8, which is similar to the stability diagram of Schlüter et al. (1965). According to Newell and Whitehead supercritical convective motions are stable only if their wave numbers are larger than the critical wave number and a little less than the maximal possible wave number at a particular $\mathcal{R}$ given by the Eckhaus instability. All wave numbers smaller than the critical wave number are said to be unstable. The stability diagram of this investigation is therefore in contradiction to the experimental finding that stable supercritical convective motions have wave numbers *smaller* than the critical wave number, as was discussed in Chapter 6 and shown in Fig. 6.3.

Although the results of the rigorous study of Newell and Whitehead (1969) to find the stable wave numbers of nonlinear Rayleigh–Bénard convection via the amplitude equation were not encouraging, the amplitude equation or modifications thereof have nevertheless been used repeatedly in attempts to determine the preferred supercritical wave number. The first such attempt was made by Pomeau and Manneville (1979), who applied the so-called Swift–Hohenberg equation

$$\partial A/\partial t = \varepsilon A - (\nabla_2^2 + q_0^2)^2 A - A^3, \tag{7.20}$$

where q_0 is the critical wave number, to a one-dimensional stationary periodic pattern of infinite extent. The Swift–Hohenberg equation (Swift and Hohenberg, 1977) is obtained from the amplitude equation by assuming that the amplitude is of the form $A = A(x,z)\exp iqx + cc$. Pomeau and Manneville found a slight increase of the wavelength near the threshold $\mathfrak{R}_c$. This was followed by the study of Pomeau and Zaleski (1981), who used two one-dimensional model equations,

$$\frac{\partial A}{\partial t} = \varepsilon A - \left(\frac{\partial^2}{\partial x^2} + q_0^2\right)^2 A - A^3, \tag{7.21a}$$

$$\frac{\partial A}{\partial t} = \varepsilon A - \left(\frac{\partial^2}{\partial x^2} + q_0^2\right)^2 A - A\frac{\partial A}{\partial x}, \tag{7.21b}$$

in order to show "that wavelength selection is likely due to boundary effects" (p. 516). In the stationary case the solution of (7.21a) requires solving a fourth order ordinary differential equation for $A(x)$ via the envelope method, which means that $A(x)$ is assumed to be the product of a rapidly varying function that is periodic in x with period q_0 and a smooth, slowly varying function. They found near $\varepsilon = 0$ that horizontally half-infinite solutions have wave numbers in a band of $\pm\varepsilon/16q_0^3$ width around the critical wave number. The same result was found by Cross et al. (1983). For the model (7.21b) Pomeau and Zaleski found the accessible wave numbers in a band which is always shorter than the critical wave number, the band being between $-\frac{53}{48}\varepsilon q_0^{-3}$ and $-\frac{47}{48}\varepsilon q_0^{-3}$. This means that the accessible wave numbers are smaller than the critical wave number, in accordance with the experimental findings, but nonunique, for which there is no experimental support. This result was likewise confirmed by Cross et al. (1983). Pomeau and Zaleski write that their main qualitative results are "that the lateral boundary conditions control the wavelength and restrict it to a band of width $\propto \varepsilon$" (p. 524), the band therefore diverging with ε.

The temporal evolution of finite amplitude two-dimensional convection rolls after a change of $\mathcal{R}$ by an amount of the order of $O(L^{-1})$ was studied with the amplitude equation by Daniels (1984). This study matches the procedure used frequently in convection experiments in which $\mathcal{R}$ is often increased in small steps. When the Rayleigh number of the initially slightly supercritical layer is increased at $t = 0$, the fluid first adjusts its vertical temperature profile to the changed conditions. Then the amplitude of the motions increases on a time scale $O(L)$ to correspond to the increase in $\mathcal{R}$ without, however, changing the roll pattern. Then the pattern changes on an $O(L^2)$ time scale to a new roll pattern, the change being driven by influences of the sidewalls. Streamline patterns at various stages of the pattern evolution have been calculated for fluid layers of Prandtl numbers $\mathcal{P} = 10$ and $\mathcal{P} = 0.2$ in a layer with semiaspect ratio $\Gamma = 5$. Depending on the initial state, the number of rolls was found either to increase slightly or to decrease substantially. In the two examples for a decreasing number of rolls, or for an increase of the wavelength of the motions, either 7 rolls changed to 3 rolls or 7 rolls changed to 4 rolls. Such a decrease of the number of rolls, i.e. such an increase of the cell size, exceeds by far the increase of the cell size that has been observed in experiments as a consequence of a small increase of a supercritical $\mathcal{R}$.

One wonders whether the attempts to determine the preferred supercritical wave number with the help of the amplitude equation (or Ginzburg–Landau equation) and with the concept that the lateral boundaries control the selection of the wavelength can be successful. The Ginzburg–Landau equation applies just as well to the Taylor vortex problem in which the fluid column is, in practice, likewise bounded. But in this case the wavelength of supercritical flow is independent of the Taylor number if the vortices are axisymmetric, as we shall see in Section 13.2. That means that, although the same amplitude equation applies and although the fluid is bounded, the instability reacts to an increase of the control parameter in a way that is quite different from the Rayleigh–Bénard instability. This is best illustrated by Fig. 6.4, which shows the variation of the wavelength of Rayleigh–Bénard convection, of the Taylor vortices, and of surface-tension-driven hexagonal Bénard cells as a function of the control parameters. The increase of the amplitude of the motions in Rayleigh–Bénard convection and of Taylor vortices in the weakly nonlinear case is given by the Landau equation, as has been proved experimentally, and one would like to assume that this is similar in surface-tension-driven Bénard convection. The fluid layers are bounded in each case. Nevertheless the wavelength varies in each case in a clearly different way. There is a characteristic mechanism for each instability which determines the selection of the wavelength, and such a specific feature is missing in the Landau equation or the Ginzburg–Landau equation.

A comment seems also to be in order concerning the consequences of the sidewalls. In the experiments of Willis et al. (1972) with fluid layers of aspect ratio 80 the wavelength of the convective motions increased with increased $\mathcal{R}$, although the rather irregular pattern did not reflect the form of the container. The sidewalls appear, therefore, to have had little impact in this case and these experiments seem to approximate the concept of an "infinite" layer very well. Taylor vortices convey a similar message. Very long columns can be used in such experiments and the results concerning the wavelength can be safely extrapolated to columns of infinite length. Both instabilities pick, in the "infinite" cases, characteristic wavelengths, indicating that the instabilities have an *internal* ability to select a characteristic wavelength independent of the presence of sidewalls. In finite containers the quantization condition certainly has an influence on the selected wavelength within $\pm\frac{1}{2}$ wavelength. But this influence goes to zero as the size of the layer or the length of the column goes to infinity. It seems to be unlikely that the preferred wave number is determined by the lateral boundaries.

Almost the opposite approach to the above-discussed concept that wave number selection is determined by the lateral walls has been taken in a number of theoretical investigations which circumvent the (necessary) presence of lateral boundaries by introducing "soft" conditions at the sides of patterns of parallel rolls. Getling (1983) studied the temporal evolution of disturbances in the form of either one roll or a small number of rolls in the center of an initially resting fluid layer at supercritical Rayleigh numbers. Lateral boundary conditions do not appear in this formulation of the problem. This problem was studied because, according to Getling, "sufficient freedom offered to the flow seems to be important for the physical optimal wave number to be manifested" (p. 166). The initial value problem so posed was solved by numerical integration of the nonlinear convection equations, using Fourier integrals and continuous wave number spectra for the velocities and temperature. When the disturbance is introduced, the convective motions develop with time arrays of parallel rolls which have nearly the same horizontal wave number. This wave number was interpreted as the preferred wave number a_p. When $\mathcal{P}$ is sufficiently small (≤ 0.1) the preferred wave number was found to decrease with increased $\mathcal{R}$. Getling concludes that "the results presented here demonstrate the decrease in a_p with increasing R to be an intrinsic property of the convection mechanism in an infinite horizontal layer" (p. 185).

In concept a similar but less rigorous approach to wave number selection is via models which study the consequences of defects. Defects seem to enable a system to change its wave number more easily than ordered bounded systems. The concept of defects originates from experiments in which defects of the flow field appeared. However it is not at all certain whether the defects in

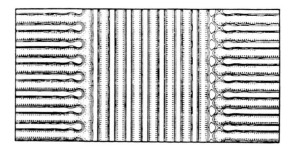

Fig. 7.9.   Grain boundaries along the short side of a rectangular container. After Tesauro and Cross (1987).

these experiments did not originate from imperfections of the apparatus, in particular from nonuniformities of the temperature field, and whether the consequences of defects can adequately be described by theories which assume uniform conditions (e.g. Siggia and Zippelius, 1981). Of the many theoretical investigations of the consequences of defects on wave number selection we can discuss only one class (out of a large number of options) of defects, the so-called grain boundaries. Grain boundaries are rolls which extend perpendicular to the side of a container into the fluid, in particular to the shorter side of a rectangular container (Fig. 7.9). In a rectangular container with grain boundaries the rolls which occupy the center of the layer and are parallel to the shorter side of the container are free to change their wave number, because the length of the grain boundaries can change easily. This arrangement is similar to the free lateral conditions investigated by Getling (1983). We note, on the other hand, that it is by all means possible to have a pattern of rolls *parallel* to the short side of a rectangular container throughout the container (Fig. 5.3), that such a pattern corresponds to the results of linear theory (Davis, 1967), and that such a pattern changes its wavelength (discontinuously) when the Rayleigh number is increased (Kirchartz and Oertel, 1988) without having to resort to the grain boundaries.

The theoretical investigation of grain boundaries was started by Manneville and Pomeau (1983). Wave number selection with grain boundaries (and other defects) was studied by Cross et al. (1986), using two model equations,

$$\dot{\psi} = \varepsilon\psi - (\nabla^2 + 1)^2\psi - \psi^3, \tag{7.22a}$$

$$\dot{\psi} = \varepsilon\psi - (\nabla^2 + 1)^2\psi + 3(\nabla\psi)^2\nabla^2\psi, \tag{7.22b}$$

where the field $\psi(x,y,t)$ represents the vertical velocity at the midplane and the critical wave number $q_c = 1$ at $\varepsilon = 0$. Equation (7.22a) is again the

Swift–Hohenberg equation. For (7.22a) Cross et al. found that the selected wave number is $q_c \pm O(\varepsilon^2)$. For (7.22b) they found wave numbers shorter than the critical wave number and decreasing with ε. But the selected wave numbers depend on the particular defect. Cross et al. made the startling statement: "There is no general wave number selection principle" (1986, p. 16). This applies, of course, only to the particular model equation (7.22b). One has, furthermore, to keep in mind that the wave numbers of the rolls in the center of the layer do not represent the *entire* flow in the rectangular container. The wave number of the flow in the entire container is given by the spatial average of the wave numbers of the rolls parallel to the short side of the container and of the wave numbers of the grain boundaries. The wave numbers of the parallel rolls in the center of the layer and the wave numbers of the grain boundaries in a unique state with stationary grain boundaries were discussed later by Tesauro and Cross (1987).

Still another approach to wave number selection via "soft" boundaries is through a so-called ramp. A ramp is a systematic nonuniformity of the temperature field at the side of a (say) parallel roll pattern. Convective motions resulting from a nonuniformity of the temperature field will be discussed in Section 8.2. Understanding the convective motions resulting from nonuniform heating is, of course, a requirement for assessing whether the idea of a ramp as a tool for a theory of wave number selection in a uniform horizontal temperature field is useful. We will therefore postpone a discussion of this topic until Section 8.2.

Other attempts to explain wave number selection in Rayleigh–Bénard convection have relied on additional physical assumptions. Into this category falls the paper of Manneville and Piquemal (1983) dealing with the consequences of curvature of convection rolls for wave number selection. Curvature induces, as was first pointed out by Cross (1983), a large-scale flow toward the center of curvature, the flow originating from the advection terms of the Navier–Stokes equations. Such a large-scale flow would mean, in the case of axisymmetric rolls, that a fluid of finite Prandtl number would pile up in the center of a layer with free–free boundaries, which is evidently not the case. Instead a radial pressure gradient forms, which returns the fluid to the outside. Manneville and Piquemal give two figures which show the vertical velocity profile of the horizontal flow resulting from slightly curved rolls, and also the vertical velocity profile in the axisymmetric case, the latter one having zero vertical average. Working with the nonlinear convection equations whose solution they expand in powers of a phase φ, which is a slowly varying function of x, y, t, Manneville and Piquemal arrive at a formula for the selected wave numbers in the axisymmetric rigid–rigid case,

$$\frac{q - q_c}{q_c} = -\frac{N'(\mathcal{P})}{R_2(\mathcal{P})}\,\varepsilon, \tag{7.23}$$

where $N'(\mathcal{P})$ and $R_2(\mathcal{P})$ are given functions of the Prandtl number, which are positive for large Prandtl number, but can change sign when $\mathcal{P}$ is small. This formula states that the selected wave numbers q are $<q_c$ for large Prandtl number, or in other words that the selected wavelengths are larger than λ_c, and that the selected wave numbers are a linear function of ε.

Now we have what we are looking for, a formula giving q as a function of the Rayleigh number and the Prandtl number. Whether this formula is the correct solution of the wave number selection problem remains to be seen. This solution is based on the assumption that the rolls are curved. It is certainly true that axisymmetric convective motions change their wave number with increased $\mathcal{R}$, and in the axisymmetric case this is actually more obvious than in the case of straight parallel rolls. There is, on the other hand, a consensus that straight parallel rolls are a valid solution of the weakly nonlinear convection problem, and it has been established experimentally that straight parallel rolls change their wave number when $\mathcal{R}$ is increased (in finite containers). So it does not seem to be that curvature is a requirement for the wave number to change. If we calculate the wave numbers from (7.23) at $\varepsilon = 1$, i.e. at $\mathcal{R} = 2\mathcal{R}_c$, with the functions $N'(\mathcal{P})$ and $R_2(\mathcal{P})$, we find that for $\mathcal{P} = 100$, i.e. for large Prandtl numbers, the wave number q should have the value $q = 3.06$, whereas the critical wave number is $q_c = 3.117$. For air ($\mathcal{P} = 0.71$) we find that q should be $q = 3.19$ at $\varepsilon = 1$. For large Prandtl numbers the wave numbers decrease according to (7.23) in agreement with the experimental results (Fig. 6.3). The experimental values of q at $\varepsilon = 1$ are at around $q = 3.00$ instead of $q = 3.06$ as predicted. The 2% difference from the theoretical value of q is well within the error of the experiments. On the other hand, the predicted wave numbers of supercritical convection in air increase with $\mathcal{R}$ according to (7.23), in contradiction to the experimental findings (Fig. 6.3).

An investigation in a similar vein was made by Buell and Catton (1986a). Using the concept of curved rolls of large radius of curvature and integrating the convection equations numerically, Buell and Catton arrive at *unique* supercritical wave numbers which are smaller than the critical wave number and decrease with increased $\mathcal{R}$ for $\mathcal{P} \geq 0.7$, whereas for $\mathcal{P} < 0.7$ the selected wave numbers are larger than the critical wave number and increase with

$\mathfrak{R}$. These results are similar to the results of Manneville and Piquemal (1983), and agree fairly well with the experimental findings for large $\mathscr{P}$, but the nearly constant supercritical wave numbers for gases ($\mathscr{P} = 0.7$) disagree with the results of both Willis et al. (1972) and Martinet et al. (1984). Buell and Catton shed some light on the puzzling question of how convection rolls can remain in the presence of the horizontal flows which result from curvature. Buell and Catton show two figures with the streamlines and isotherms of axisymmetric convection and for slightly curved rolls. According to these graphs the effects of the horizontal flows on either the streamlines or the isotherms are so small that it seems unlikely that they can be verified experimentally with current techniques.

A substantial amount of work with both analytical and numerical models has been concerned with the patterns of convection. These studies deal with complicated patterns which are not completely regular but have defects, dislocations, singularities, etc. The theoretical investigation of such patterns was begun by Cross (1982), put into rigorous form by Cross and Newell (1984), incorporated into several other models, and recently pursued again by Newell et al. (1990). Common to these models is their reliance on patterns of poor regularity as proof of the necessary existence of defects in convection; any pattern goes. The paper of Newell et al. presents a large number of irregular patterns making this point. Also common to these models is the usually tacit assumption that the fluid is heated uniformly from below and cooled just as uniformly from above, and in many cases the fluid layer is assumed to be of infinite horizontal extent, which means that there are no lateral nonuniformities either in the theoretical input. If lateral boundaries are considered, it is implied that the vertical temperature gradient along the wall is uniform. We have discussed the likelihood that irregular patterns form under these conditions in Section 5.2, where we noted that it is by all means possible to observe regular patterns under these conditions, provided that the uniform conditions assumed in theory are met in the experiment. We have also noted before that the irregular patterns observed in many experiments are most likely due to imperfections of the experiments, in particular to nonuniformities of the temperature field. An irregularity of a pattern, e.g. an off-center pattern of quasi-circular concentric rolls, is not proof of an instability or a nonlinearity, but rather an indication of an experimental deficiency. In the models studying the formation of irregular patterns it is not asked whether the observed irregularities originate from local nonuniformities of the temperature. It is also not asked whether a particular irregularity observed in an experiment is reproducible in an experiment done under nominally the same conditions by somebody else. But such a test is crucial for the reality of an effect. Since these models are used to study irregular patterns which are not really reproducible, it appears that the results of these studies are ambiguous, and we shall therefore not pursue this topic.

If we summarize what we have learned from the models of supercritical convection, we find that the models have not solved the wave number selection problem in Rayleigh–Bénard convection either. There have been some tantalizing glimpses at the possibility of a solution of the problem, but the ultimate answer does not seem to have emerged yet. The models arrive dutifully at selected wave numbers smaller than the critical wave number, responding to a by now well-known experimental fact. In many cases the predicted selected wave numbers are in an often quite narrow band, leaving thereby the question of the uniqueness of the solution unanswered. We do not seem to know why or by what mechanism the convective motions select, out of the continuum of unstable wave numbers available according to linear and weakly nonlinear theory, one particular wave number, or perhaps a narrow band of wave numbers, at a particular supercritical Rayleigh number.

8

MISCELLANEOUS TOPICS

8.1 Non-Boussinesq Convection

In the theoretical discussions of the previous chapters the Boussinesq approximation was used again and again. It is now time to consider the consequences of non-Boussinesq conditions, in other words to consider the consequences of variations of the material properties of the fluid with temperature. For most fluids the material property that changes most with temperature is the viscosity. The possibility that the variation of viscosity can have a significant impact on convection has been apparent since A. Graham's (1933) observation that the direction of circulation in hexagonal cells in gases is opposite to the direction of circulation in hexagonal cells in fluids. The existence of hexagonal cells in air does not mean that hexagonal cells are the predominant type of convective motion in air. The form of motion near the critical condition in Rayleigh–Bénard convection is, according to weakly nonlinear theory, rolls in fluids as well as in gases. In gases convection is, however, usually of poor regularity. In an irregular field of convective motions in a gas, the direction of motion in the centers of polygonal or hexagonal cells is invariably downward. As we have mentioned before, Graham suggested that the striking difference in the direction of circulation of the cells in gases and fluids may be explained with the variation of viscosity with temperature; in gases the viscosity increases with temperature, whereas in fluids the viscosity decreases with temperature.

Graham's conjecture was verified by v. Tippelskirch (1956). He used liquid sulfur as the convecting medium. The viscosity of molten sulfur decreases by about half between 119 and 153°C, and increases by three orders of magnitude from 153 to 180°C. v. Tippelskirch showed that in the temperature range of decreasing viscosity the fluid formed cells with upward velocity in the cell centers, and with increasing viscosity the same fluid at a higher mean temperature formed cells with downward velocity in the cell centers. Working at high temperatures with a smelly fluid did not, of course, produce a pattern of regular cells. There were no hexagons, but rectangular cells below 153°C and roundish cells above 153°C. Nevertheless the reversal of the direction of circulation was demonstrated convincingly. Graham's suggestion was thus confirmed, and the importance of non-Boussinesq conditions was established.

The first attempt to explain a perceived tendency for the formation of hexagonal cells in Rayleigh–Bénard convection in fluids with small viscosity variation was made by Palm (1960). Actually the presumed tendency for hexagonal cells was a common misconception at that time, when hexagonal cells were still believed to be the preferred planform of convection because the significance of surface tension effects on the formation of Bénard cells had not yet been realized. Palm's paper deals with two consequences of the temperature dependence of viscosity. First, there are the consequences that a variable viscosity may have on the onset of convection, which means on the critical Rayleigh number and the critical wavelength. Second, there are the consequences that a variable viscosity may have on pattern selection.

The onset of convection in a non-Boussinesq fluid layer of infinite horizontal extent with free–free boundaries was studied by Palm with linear theory, adding only terms to the Navier–Stokes equations which describe the variation of viscosity. In the marginal case this leads to a sixth order equation,

$$\kappa\nabla^2\{-\nu\nabla^4 w - 2\nu_z\nabla^2 w_z + \nu_{zz}(\nabla_2^2 w - w_{zz})\} + g\alpha\beta_0\nabla_2^2 w = 0, \qquad (8.1)$$

where the subscript z denotes differentiation with respect to z, and β_0 is equal to $\Delta T_{c0}/d$, where ΔT_{c0} is the critical temperature difference in the case of a constant viscosity. Equation (8.1) is a modified version of equation (2.19a). Palm assumed that the viscosity varies as

$$\nu = \nu_0 + \Delta\nu \cos \mu(T - T_0), \qquad (8.2)$$

where $\Delta\nu$ is the difference in the viscosity between the top and bottom boundaries, μ is a constant, and T_0 is the temperature at the bottom of the layer. Although this viscosity variation is a bit artificial, for a first look at the problem it is fine. In order to simplify, Palm required that $\Delta\nu/\nu_0 << 1$, and set $\mu \Delta T = \pi$. A solution to (8.1) satisfying the boundary conditions is given by

$$w = W_0 \sum_m A_{11m} \cos kx \cos ly \sin m\lambda z, \qquad (8.3)$$

with $\lambda = \pi/d$. For small viscosity variation it is sufficient to take only the first two terms in (8.3) into account, and after some algebra it follows that with variable viscosity the critical Rayleigh number is given by

$$\mathcal{R}_c = \mathcal{R}_{c0} - \pi^4 1.3(\Delta\nu/\nu_0)^2, \tag{8.4}$$

and the critical wave number is given by

$$a_c^2 = a_{c0}^2(1 - 0.048(\Delta\nu/\nu_0)^2). \tag{8.5}$$

According to (8.4) and (8.5) the critical Rayleigh number, as well as the critical wave number, decreases with increasing viscosity variation $\Delta\nu$, regardless of whether the viscosity decreases or increases with temperature.

Pattern selection or the question of the "tendency toward hexagonal cells" in finite amplitude non-Boussinesq Rayleigh–Bénard convection in a fluid layer of infinite horizontal extent with free–free boundaries was studied by Palm (1960) with the nonlinear equations of convection which contained additional terms for the variation of viscosity. Palm's attempt to prove the tendency for hexagonal cells was based on the possibility of describing hexagonal convection cells as the sum of rolls and rectangular cells, as follows from equation (2.53). From the "noise" of the subcritical fluctuations Palm chose rolls

$$A_{021}(t)\cos 2ly \sin \lambda z, \tag{8.6}$$

and rectangular cells

$$A_{111}(t)\cos kx \cos ly \sin \lambda z, \tag{8.7}$$

with the amplitudes A_{021} and A_{111} and with $k^2 + l^2 = a^2$. He investigated the growth of the amplitudes, which is described by a pair of ordinary nonlinear differential equations,

$$\dot{A}_{111} = \varepsilon A_{111} - a_1 A_{111} A_{021} - R A_{111}^3 - P A_{111} A_{021}^2, \tag{8.8}$$

$$\dot{A}_{021} = \varepsilon A_{021} - \tfrac{1}{4} a_1 A_{111}^2 - R_1 A_{021}^3 - \tfrac{1}{2} P A_{111}^2 A_{021}, \tag{8.9}$$

where the dot means differentiation with respect to time. These equations are written in this form in Segel and Stuart (1962); ε and a_1 (8.8)–(8.9) are small parameters which can be varied independently, ε representing the increase ΔR above the critical Rayleigh number and a_1 representing the variation $\Delta\nu$ of the viscosity. R, R_1, and P in (8.8)–(8.9) are given by very complicated formulas involving the material constants of the fluid, as well as l and λ.

Investigation of (8.8)–(8.9) showed that the amplitudes A_{111} and A_{021} reinforce each other and that when $t \to \infty$ the amplitude $A_{111} = 2A_{021}$. That means that the motions are hexagonal cells according to equation (2.53). The direction of motion in the cells follows from the condition $(\Delta v/v_0)A_{021} < 0$. If the viscosity increases with temperature, as in gases, the motion is descending in the cell centers; if the viscosity decreases with temperature, as in liquids, the motion is ascending in the cell centers.

Palm's (1960) paper was amended by Segel and Stuart (1962), Jenssen (1963), Segel (1965), and Palm et al. (1967). Segel and Stuart pointed out that Palm had proved that a hexagonal pattern can exist under supercritical conditions in a fluid with viscosity variation, but that he had not proved that this pattern was preferred among the other possible solutions. Segel and Stuart determined the stability of the solutions of equations (8.8)–(8.9) and found that as $t \to \infty$ "the final mode *may* be hexagonal," but for small viscosity variations, including zero, it cannot be hexagonal. On the other hand, they found two-dimensional rolls to be a stable solution for all values of viscosity variation including zero, but they discounted the likelihood of rolls. A new element of the consequences of viscosity variation was their discovery that it is theoretically possible that hexagonal patterns form at subcritical Rayleigh numbers through a subcritical bifurcation, if the viscosity variation is large enough.

The investigation of the consequences of small viscosity variation following Palm (1960) and Segel and Stuart (1962) was concluded with the paper of Palm et al. (1967), in which it was shown that for rigid–rigid, rigid–free, and free–free boundaries and small values of $\Delta v/v_0$ there is, when the Rayleigh number is increased, a sequence of the patterns, from exclusively hexagons to hexagons and rolls to exclusively rolls, as is illustrated by Fig. 8.1, which is from Segel's (1965) paper. This figure shows that the Nusselt number, or the amplitude of the motions, can in theory be unequal to zero for a range of Rayleigh numbers $\mathcal{R} < \mathcal{R}_c$. In this range hysteresis should be observed if the Rayleigh number is first increased and later decreased. At $\mathcal{R}_c$ the amplitude should increase discontinuously from zero to a finite value with the onset of convection. On the other hand, if $\mathcal{R}$ is decreased after the onset of convection, the amplitude of the motion should remain unequal to zero in the range from $\mathcal{R}_{sc} < \mathcal{R} < \mathcal{R}_c$, although the conditions are subcritical. The amplitude of the motions should decrease discontinuously to zero only at $\mathcal{R}_{sc}$, which is the minimum value of $\mathcal{R}$ with nonvanishing amplitude. At any Rayleigh number between $\mathcal{R}_{sc}$ and $\mathcal{R}_c$, the amplitude of the motions can therefore have two values depending on the initial conditions. This behavior of the amplitude is typical of a subcritical bifurcation.

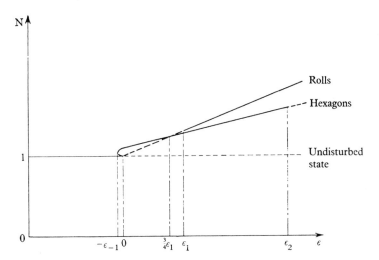

Fig. 8.1. Stability diagram for weakly nonlinear hexagons and rolls under non-Boussinesq conditions as a function of $\varepsilon = (\mathcal{R} - \mathcal{R}_c)/\mathcal{R}_c$. The ordinate is the heat flux expressed by the Nusselt number $\mathcal{N}$. Continuous lines show stable, dashed lines unstable, states. After Segel (1965).

As far as the patterns are concerned, Palm et al. found that in a 5-mm-deep layer of a silicone oil of 4.67 cm²/sec viscosity hexagonal cells should be observed between $\mathcal{R}_c$ and $1.54\mathcal{R}_c$. They were not aware that an experiment with a 5.15-mm-deep layer of silicone oil of 1 cm²/sec viscosity had already been done (Koschmieder, 1966a) and had resulted in the formation of a pattern of circular rolls. It has been argued that this roll pattern was caused by the lateral boundaries of the fluid layer instead of non-Boussinesq effects, but in view of the aspect ratio used ($\Gamma = 38.8$) this argument is not very convincing. At some point the consequences of the viscosity variation must be strong enough to overcome the weak effects of the lateral boundaries.

The consequences of the variation of all material properties with temperature, in particular of viscosity ν, thermal conductivity λ, and specific heat c_p, was studied by Busse (1967b). He assumed that the viscosity variations are small and are a linear function of temperature in the form

$$\nu = \nu_0(1 + \gamma \, \Delta T), \tag{8.10}$$

with γ being $1/\nu_0 \, \partial\nu/\partial T$. The variations of λ and c_p follow corresponding formulas. Note that the temperature distribution in the static fluid before the onset of convection differs from the linear vertical temperature profile we have always dealt with, if the thermal conductivity λ varies with temperature.

The temperature will then depend on z and z^2 if a formula for the thermal conductivity corresponding to (8.10) applies.

In order to determine the preferred pattern, Busse used the nonlinear Navier–Stokes equations and the nonlinear energy equation, both with terms for the variation of the material properties. He expanded the variables in terms of the small parameter ε, just as in Schlüter et al. (1965). That means he used

$$v = \sum_{\mu=1}^{\infty} \varepsilon^{\mu} v^{(\mu 0)} + \sum_{\mu=1}^{\infty} \sum_{\kappa=0}^{4} \varepsilon^{\mu} \gamma_{\kappa} v_{\kappa}^{(\mu 1)} + \cdots, \tag{8.11}$$

where the γ_{κ} are the coefficients for the variation of the material properties with temperature. This procedure leads again to a system of linear inhomogeneous equations belonging to the different powers of ε and γ. The possible solutions of the nonlinear equations for convection were determined from this system. The preferred pattern was determined from a stability analysis, using disturbances of the form

$$\tilde{v} = \sum_{\mu=1}^{\infty} \varepsilon^{\mu-1} \tilde{v}^{(\mu 0)} + \sum_{\mu=1}^{\infty} \sum_{\kappa=0}^{\infty} \varepsilon^{\mu-1} \gamma_{\kappa} \tilde{v}_{\kappa}^{(\mu 1)} + \cdots. \tag{8.12}$$

After a lengthy calculation it follows that hexagons may displace rolls as the preferred pattern if the viscosity is not constant. The hexagons have either descending or ascending motion in the cell centers, depending on the amplitude and on whether the asymmetry parameter representing the variation of the material properties is positive or negative. If the temperature difference across the fluid at a given value of the asymmetry parameter is increased, hexagons form first, but are replaced at slightly larger ΔT by rolls. The stability diagram is of the same kind as Segel's (1965) diagram, which we showed in Fig. 8.1. Busse found that for a given asymmetry parameter all steady solutions are unstable for small amplitudes $0 < \varepsilon \leq \varepsilon_D$. This unlikely result may be the consequence of the assumption made in the stability analysis that the wave numbers of the disturbances are equal to the critical wave number. As we have seen in Section 6.1 this is not the case; experiments show that the supercritical wave numbers increase also for just slightly supercritical conditions. The rigorous, but also much more complicated analysis of Busse leads in essence to the same results as those obtained from Palm's model.

If we summarize the theoretical results concerning the consequences of small variations of the material properties, we find that it can be expected that the critical Rayleigh number as well as the critical wave number will decrease with increased variation of the viscosity, and that the onset of convection should take place in the form of hexagonal cells which will, however, be replaced by rolls at slightly higher Rayleigh numbers.

The papers of Silveston (1958) and Somerscales and Dougherty (1970) are cited in Busse (1978) as providing proof that the transition from hexagonal cells to rolls actually occurs in non-Boussinesq convection. Silveston's visualization experiments were made with the fluid under a thick glass plate which was cooled by ambient air; this means that its temperature was probably nonuniform and the upper boundary was poorly conducting. It is doubtful whether the patterns observed under these circumstances can be called upon to prove non-Boussinesq effects, although such effects may have been present. Experiments on non-Boussinesq effects have also to prove that the experiment was made in a strictly time-independent way, so that the consequences of non-Boussinesq conditions cannot be confused with the consequences of time-dependent operation. The experiments of Somerscales and Dougherty (1970) were aimed at a verification of the non-Boussinesq transition from hexagonal cells to rolls, employing a very viscous ($10 \text{ cm}^2/\text{sec}$) silicone oil and very large critical temperature differences. Their most convincing photographs (their Fig. 8) show a pattern of cells mixed with rolls at $\mathcal{R}_c$ and a more cellular pattern at $1.07\mathcal{R}_c$. They concluded, "The cellular flow appeared to convert into a system of circular rolls at some supercritical Rayleigh number. However, this conclusion is rather uncertain" (p. 767). So it appears that this paper does not prove the non-Boussinesq transition from hexagons to rolls either.

Another experiment on the consequences of viscosity variation on pattern formation was made by Hoard et al. (1970). They worked with an apparatus with aspect ratios up to $\Gamma \approx 50$ and with silicone oil of $0.50 \text{ cm}^2/\text{sec}$ viscosity for experiments dealing with small viscosity variation and with the aromatic hydrocarbon Aroclor for experiments with very large viscosity variation. The viscosity of Aroclor decreases exponentially with temperature; the viscosity ratio $r = v_{max}/v_{min}$ across the fluid layer could be made as large as 10 at a temperature difference of 20°C. When the fluid was in contact with the glass lid of the fluid layer, they observed circular concentric rolls in a silicone oil layer 4.95 mm deep and in Aroclor 6.97 mm deep. On the other hand, in a 3.64-mm-deep layer of Aroclor, with the fluid being in contact with the lid, i.e. without surface tension effects, a pattern of regular hexagonal cells appeared at the critical temperature difference of 18.7°C. The hexagonal pattern remained to about $2\mathcal{R}_c$, when the experiment had to be discontinued. The staying power of the cells does not agree with the expected transformation of the hexagonal cells into rolls, but the Aroclor does not conform with the small viscosity variation assumed in theory. It is interesting that the hexagons in the Aroclor experiment are clearly aligned on concentric circles; see Fig. 4 of Hoard et al. That means that the fluid can very well exercise its pattern selection mechanism in spite of the presence of effects of the lateral wall. The experiments with Aroclor show that the viscosity variation can have an effect on pattern formation. Experiments with Aroclor cannot, however, be used for comparison with Palm's or Busse's theory, because the viscosity variation exceeds by far the small viscosity variation considered in theory.

The onset of convection in a fluid of variable viscosity was also studied by Stengel et al. (1982) in connection with geophysical problems. The emphasis in this theoretical and experimental study was on very large viscosity variation. Stengel et al. first solved numerically the convection equations following from the nonlinear Navier–Stokes equations and the nonlinear energy equation. They confirmed that the critical Rayleigh number as well as the critical wave number should decrease if a small viscosity variation is given by (8.2), as was assumed by Palm. For fluids with exponential viscosity variation, however, the critical Rayleigh number was found to increase to a maximum of $\mathfrak{R}_c \approx 2200$ at a viscosity ratio $r = v_{max}/v_{min} \approx 1000$. At even larger viscosity variation the critical Rayleigh number decreased again. This result was confirmed theoretically by White (1988). The critical wave number was found by Stengel et al. to be practically constant up to about the viscosity ratio 1000, but for larger viscosity ratios a_c increased noticeably. Figure 8.2 shows curves for $\mathfrak{R}_c$ and a_c as a function of the viscosity ratio.

These results were checked by Stengel et al. in an experiment with glycerol in an apparatus with aspect ratio $\Gamma = 22$. The viscosity of glycerol decreases exponentially with temperature, as is known from the literature, and was remeasured by Stengel et al. As we have discussed before, a quantitative determination of the critical Rayleigh number is usually made with the Schmidt–Milverton technique. But Stengel et al. found that a reliable determination of the critical Rayleigh number from Nusselt number measurements was not possible in their apparatus. Instead they used as a criterion for the onset of convection the appearance of a pattern on a shadowgraph screen. The critical Rayleigh numbers so observed increased with increased viscosity ratio in the range from $r \approx 7$ to $r \approx 700$. The increase of the critical Rayleigh number seems to be unambiguous; the absolute values of the Rayleigh numbers, however, seem to be uncertain. The planform of the motions in this apparatus was discussed in Oliver and Booker (1983). When they worked with silicone oil and small viscosity variation, they observed a pattern of circular concentric rolls. When the viscosity variation in glycerol was large, they observed hexagons in one part of the layer and bent rolls in the other part.

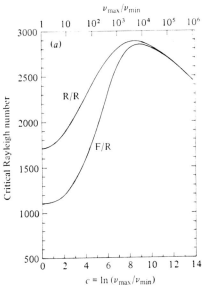

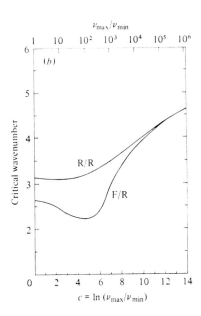

Fig. 8.2.   Critical Rayleigh number and critical wave number in glycerol as a function of the viscosity ratio for rigid–rigid and rigid–free boundaries. After Stengel et al. (1982).

Whether this is proof of the appearance of a hexagonal pattern and its subsequent conversion into rolls seems to be uncertain.

Square cells in convection in fluids with temperature-dependent viscosity were investigated with a weakly nonlinear theory by Jenkins (1987). He found that near the critical Rayleigh number rolls are preferred for lower values of the viscosity ratio, and squares are preferred for higher values of r. Jenkins compared his results with the outcome of the experiments of White (1988) in layers of golden syrup. White found hexagonal cells at the onset of convection with a viscosity ratio $r = 50$. This extends to higher values of r the observation of Hoard et al. (1970), made at $r = 10$. It would be interesting to know whether these hexagonal cells transform into rolls when $\mathscr{R}$ is increased, as the theory for small viscosity variation predicts. The other experiments of White concerning the stability of rolls, square cells, and other types of cells in golden syrup are misleading. These patterns were forced upon the fluid with the controlled initial conditions technique by which any pattern can be produced, regardless of whether the fluid layer is Boussinesq or not. Indeed, as is shown in Fig. 10 of White (1988), a stable pattern of rolls, a new square planform, hexagonal, and triangular cells can all be produced in this way under about the same conditions. "Stable" means in this context that the patterns "remain unchanged for at least 24 h," i.e. 3.5 times d^2/κ, but the real relaxation time is L^2/κ, which is 250 times d^2/κ. We note that the new square planform does not match the solution of the membrane equation for square cells [equations (2.49)–(2.50)]. These new square cells are an artifact of the controlled initial conditions. The experiments with controlled initial conditions say nothing about which pattern actually evolves from infinitesimal disturbances in a fluid layer with uniform temperatures on the top and bottom of the layer, regardless of whether the viscosity is temperature dependent or not.

Non-Boussinesq effects resulting from variations of material properties other than the viscosity were studied experimentally by Walden and Ahlers (1981), who investigated the consequences of the temperature dependence of the thermal expansion coefficient α. They found hysteresis in the heat flux near $\mathcal{R}_c$, but could not observe the pattern of the flow. The dependence of the Rayleigh number on the asymmetry parameter P could also not be verified due to the large systematic errors in the fluid properties. Non-Boussinesq effects in water near the density maximum of water at 4°C were studied by Dubois, Bergé, and Wesfreid (1978). They too found hysteresis effects in the velocity of the motions near the critical Rayleigh number.

Looking back at the studies of the consequences of non-Boussinesq conditions, in particular of the variation of viscosity with temperature, we see a not entirely clear picture. The decrease of the critical Rayleigh number and of the critical wave number which was predicted from linear theory by Palm (1960) and was confirmed theoretically by Stengel et al. (1982) has not been verified experimentally so far. However in view of the moderate accuracy with which we can currently determine $\mathcal{R}_c$ and a_c it will be quite difficult to verify the changes in the critical values for small viscosity variation. The possibility that the variation of viscosity may have an influence on pattern formation and may cause the formation of hexagonal cells at the onset of Rayleigh–Bénard convection, which would change to rolls at slightly higher Rayleigh number, has caught the attention of many. But there is, so far, no firm verification of this effect. So whether the variation of viscosity has this remarkable influence on pattern formation, in particular with small viscosity variation, has still to be proved.

8.2   Convection with Nonuniform Heating

In the previous chapters we referred many times to the requirement in the theory of convection that the temperatures on the top and bottom boundaries of the fluid be absolutely uniform. This requirement is matched to various degrees of accuracy in the experiments. Let us now look at the consequences of systematic nonuniformities of the temperatures on the horizontal boundaries on Rayleigh–Bénard convection. Any nonuniformity of the temperature on a horizontal boundary produces a horizontal pressure gradient in the fluid, because the hydrostatic pressure under a warm fluid column is smaller than the hydrostatic pressure under a cold fluid column if both columns are of the same length. And that means that any ever so small temperature difference in a horizontal boundary will make the fluid move from high pressure to low pressure, independent of the magnitude of the vertical temperature gradient, even with stable stratification. Nonuniformities of the temperatures in the horizontal boundaries therefore ought to have a profound impact on the form and the onset of convection caused by vertical instability. Nonuniformity of the temperature necessarily introduces an orientation into the fluid layer which is likely to eliminate the degeneracy of the patterns that we found on a uniformly heated plane.

The investigation of convection with nonuniform heating began with an experiment of Koschmieder (1966b). In this experiment the bottom plate of a circular apparatus of aspect ratio $\Gamma = 18$ had an axisymmetric radial temperature gradient, whereas the top plate of the silicone oil layer was cooled uniformly. The result of these experiments was of a surprising simplicity. The fluid motions in the unstable layer turned out to be a superposition of an overall circulation, caused by the radial temperature gradient, upon the circular rolls caused by vertical instability. The overall circulation consists of one long loop rising over the warm outside of the bottom and sinking over the cold center of the bottom. This circulation is driven by the pressure differences between the fluid over the warm part of the bottom and the fluid over the cold part of the bottom. We shall refer to this circulation as the density circulation. If vertical instability is also present in a nonuniformly heated fluid layer, each second convection roll will turn in a direction opposite to the motion of the density circulation; each second roll is therefore smaller than its neighbor which turns with the density circulation. That can easily be seen in Fig. 8.3, as well as in other photographs in Koschmieder (1966b), in which it is also shown that, as must be, the same superposition leading to alternately larger and smaller rolls occurs when the direction of the radial temperature gradient in the bottom plate is reversed. The superposition of the two motions holds also for moderately supercritical conditions. Since the radial temperature gradient in this experiment was only in the bottom plate and the temperature of the lid was uniform, the vertical temperature difference in the fluid in Fig. 8.3 was a function of r. If the fluid was critical at the center of the layer,

Fig. 8.3. Convection rolls of alternately larger and smaller section on a circular plate with a radial temperature gradient in the bottom. The fluid is silicone oil of 1 cm²/sec viscosity. The temperature at the center of the bottom is 23°C, at the rim of the bottom 37.5°C, and on the glass lid 31°C. The vertical temperature difference in the fluid at the rim is $\cong 2\Delta T_c$. Visualization with aluminum powder. After Koschmieder (1966b).

as in Fig. 8.3, the fluid at the rim must have been supercritical, and was in this case at $\Delta T/\Delta T_c \cong 2$.

If the radial temperature gradient was too large, the rolls changed the orientation of their axes from the azimuthal to the radial direction (Fig. 8.4). Vertical instability in Fig. 8.4 extends only to the end of the bright rolls near the rim; further inward the vertical temperature gradient is subcritical, but motions driven by the radial temperature gradient continue. These motions depend on the azimuth angle. The nodal lines of sin $n\phi$ are clearly visible in this picture.

The two essential results of these experiments were the observation that the motions on a nonuniformly heated plate are a superposition of the rolls caused by vertical instability upon the radial circulation caused by the radial temperature gradient, and that the circular rolls become unstable if the radial temperature gradient is too large and rearrange themselves into rolls pointing into the radial direction.

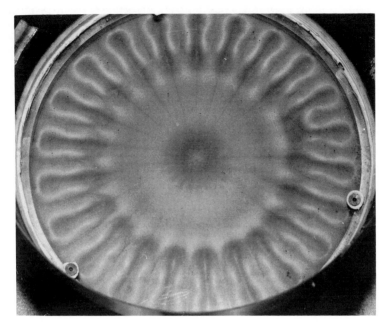

Fig. 8.4. Longitudinal rolls in a silicone oil layer on a circular plate with a radial temperature gradient in the bottom, a warm lateral wall, and under a uniformly cooled glass lid. After Koschmieder (1966b).

The experiments with nonuniform heating also shed some light on the question of pattern selection in a uniformly heated circular container. It appears natural to expect that there is a smooth transition from the solution obtained in the presence of a finite radial temperature gradient to the solution obtained without a radial temperature gradient, as the radial gradient goes to zero. According to the experiments just discussed the pattern should change from circular concentric rolls with alternating larger and smaller size in the presence of a finite radial temperature gradient to circular concentric rolls of equal sections (Fig. 5.1) in the case of a vanishing radial gradient. The case of small or vanishing radial temperature gradients has been treated theoretically by Brown and Stewartson (1978), and axisymmetric rolls have been found. Rolls which are perpendicular to the circular lateral wall have been observed only when the radial temperature gradients were comparatively large. Such rolls are characteristic of the disordered patterns discussed in Section 5.2.

The only other two experimental investigations of Rayleigh–Bénard convection with nonuniform temperatures in the horizontal boundaries are from

Fig. 8.5.   Differential interferogram of the convective motions in a silicone oil layer in a rectangular container with the same horizontal temperature gradient in the top and bottom boundaries. $(\partial T/\partial x)/(\partial T/\partial z)_c = 0.28$. The Rayleigh number at the right side of the container is $2.04\mathcal{R}_c$. After Srulijes (1979).

Berkovsky and Fertman (1970) and Srulijes (1979), both made with rectangular containers. Of particular interest are Srulijes' experiments in which the top and bottom boundaries of the container had the same linear horizontal temperature gradient. The vertical temperature difference was then the same in the entire fluid layer, but the mean temperature varied steadily in the horizontal direction. Since the vertical temperature gradient was uniform throughout the layer, the onset of convection occurred simultaneously in the entire layer. The pattern was a sequence of alternately smaller and larger cells; corresponding rolls had the same amplitude (disregarding lateral effects). The motions are shown in Fig. 8.5. They can be interpreted again as a superposition of a density circulation, caused by the horizontal temperature gradient, on rolls caused by vertical instability. Since the top and bottom of the apparatus of Srulijes were made from copper, visualization had to be accomplished from the side with differential interferometry, which provides a picture of lines of equal density differences. The density field is equivalent to the temperature field, which in turn is determined by the velocity field.

Until now we have discussed the consequences of temperature gradients which depend on one horizontal coordinate only. Considering temperature distributions which depend on both horizontal coordinates introduces the potential for a large variety of patterns. It does not appear to be productive to pursue all these possibilities. Suffice it to say that any desired pattern can be forced this way: regular periodic patterns such as fields of triangular, square, or hexagonal cells, or combinations thereof, or irregular patterns as one's own initials.

The theoretical investigation of Rayleigh–Bénard convection with nonuniform temperatures in the horizontal boundaries was started by Müller (1966). He studied the steady two-dimensional case with a constant temperature gradient $\partial T/\partial x$ in the bottom and a uniform temperature on top of the fluid, the fluid being on an infinite plane. He assumed that $|\partial T/\partial x| \ll |\partial T/\partial z|$. The solution of the equation of thermal conduction is then

$$T(x,y) = T_0 + \beta\left(1 + \frac{\varepsilon x}{d}\right)(z - d), \qquad (8.13)$$

where T_0 is the temperature at the upper boundary, β the local vertical temperature gradient, and ε the horizontal temperature gradient. Müller used the linear Navier–Stokes equations, the continuity equation, the equation of state, and the energy equation in the form

$$u\frac{\partial T}{\partial x} + w\frac{\partial T}{\partial z} = \kappa\nabla^2 T, \qquad (8.14)$$

which now incorporates the term $u\,\partial T/\partial x$, which has to be considered because $\partial T/\partial x$ is finite. This leads, following the usual procedures, to the nondimensional equation

$$\nabla^6 u - \mathcal{R}_c\left\{(1 + \varepsilon x)\frac{\partial^2}{\partial x^2} - \varepsilon(z - 1)\frac{\partial^2}{\partial z\,\partial x}\right\}u = 0, \qquad (8.15)$$

which, in the case $\varepsilon = 0$, corresponds to equation (2.19a) for the marginal state under uniform conditions.

Equation (8.15) is solved with

$$u_1(x, z) = \sum_{n=1}^{N}(g_n(x) + a_n)\cos n\pi z. \qquad (8.16)$$

A system of ordinary coupled differential equations determines the $g_n(x)$. The velocity u_1 consists of two parts, one valid for $x < 0$, where the vertical temperature is below critical, and a second part for $x > 0$, where the fluid is just above critical. A similar situation with a stepwise increase of ΔT_z just at the critical temperature difference was studied earlier by Zierep (1961). The complete solution of (8.15) consists of $u_1(x,z)$ and an additional function $u_2(z)$, which originates from the equation

$$\frac{d^4 u_2}{dz^4} = -\varepsilon\mathcal{R}_c. \qquad (8.17)$$

This is a special solution of equation (8.15) describing a shear flow. The velocity u_2 represents the density circulation in the fluid.

Solutions of (8.16) in combination with the solution of (8.17) were determined numerically for small values of ε and two ratios of the amplitude of the convective motions caused by vertical instability, and the amplitude of the motion caused by the horizontal temperature gradient. It followed that the velocity field is composed of alternately smaller and larger cells. In the subcritical region $x < 0$ there is a semiopen cell which extends to $x \to -\infty$. This is in complete qualitative agreement with the features of Koschmieder's (1966b) experiment.

The consequences of a horizontal temperature gradient on the onset of convection were studied by Unny and Niessen (1969), who assumed a quiescent state of the fluid before the onset of instability. The onset of convection and the orientation of the convection rolls were studied by Weber (1973, 1978). He assumed that the horizontal temperature gradient is small, that the Boussinesq approximation is valid, that the horizontal boundaries are perfectly conducting, and that the necessarily present lateral boundaries do not affect the motions. In order to determine the density circulation, a special solution of the Navier–Stokes equations, the continuity equation, and the energy equation were studied, assuming that $\partial/\partial t = v = w = 0$ and that the boundaries are free. It follows that

$$U(z) = \frac{1}{2}\beta\mathcal{R}\left(\frac{z}{4} - \frac{z^3}{3}\right), \tag{8.18}$$

and

$$T(z) = \frac{1}{24}\beta^2\mathcal{R}\left(\frac{9z}{80} - \frac{z^3}{2} + \frac{z^5}{5}\right) - z, \tag{8.19}$$

where z is counted from the middle of the layer, and β is the dimensionless horizontal temperature gradient. With $\beta = 0$ the usual linear vertical temperature profile reemerges. The velocity U can be approximated very well by

$$U(z) \cong \frac{1}{24}\beta\mathcal{R}\ sin\ \pi z. \tag{8.20}$$

For sufficiently large values of $\mathcal{R}$ and also of β the flow described by (8.18) or (8.20) may become unstable. The stability of the flow with regard to infinitesimal disturbances was studied, neglecting the nonlinear terms. The disturbances were normal modes of the type considered in (2.13) and (2.14). The disturbance equations which follow were solved with power series in β for u, v, w, θ, $\mathcal{R}$, assuming that the horizontal temperature gradient β is small.

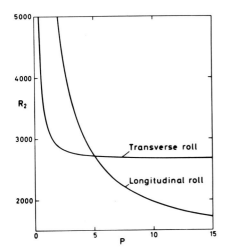

Fig. 8.6. $\mathcal{R}_2 = (\mathcal{R} - \mathcal{R}_c)/\beta^2$ as a function of the Prandtl number for transverse and longitudinal rolls. β is the dimensionless horizontal temperature gradient. After Weber (1973). By permission of Pergamon Press, Inc.

The zeroth order system in β is the traditional Rayleigh–Bénard problem. In the first approximation in β the Rayleigh number $\mathcal{R}_1 = 0$, and also all $\mathcal{R}_{2n+1} = 0$, because reversing the sign of β means only a reversal of the direction of the density circulation, which cannot affect the stability of the flow. Solving for $\mathcal{R}_2$ Weber found that the growth rate $s_2 = 0$; i.e. the onset of convection is not oscillatory. $\mathcal{R}_2$ was found to be positive, which means that the onset of convection in the presence of a horizontal temperature gradient will take place at a Rayleigh number *larger* than the critical Rayleigh number on a uniformly heated plane. A horizontal temperature is stabilizing for Rayleigh–Bénard convection. The reason for this is the density circulation, which brings warm fluid up and cold fluid down, reducing thereby the vertical temperature gradient. $\mathcal{R}_2$ depends also on the wave numbers k_x and k_y and on the Prandtl number. The minimum of $\mathcal{R}_2$, meaning the critical condition, is different for transverse rolls (rolls whose axes are perpendicular to the direction of the density circulation) and longitudinal rolls (rolls whose axes are parallel to the direction of the density circulation). $\mathcal{R}_2$ as a function of the Prandtl number is shown in Fig. 8.6. For Prandtl numbers >5.1 longitudinal rolls are preferred, whereas for $\mathcal{P} < 5.1$ transverse rolls are preferred. For infinite Prandtl number the critical Rayleigh number is given by

$$\mathcal{R}_c = \mathcal{R}_{c0}\left(1 + \frac{3}{16}\pi^2\beta^2\right) + O(\beta^4), \tag{8.21}$$

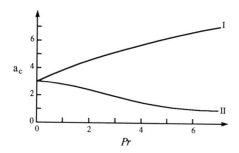

Fig. 8.7.   Critical wave numbers for longitudinal rolls (I) and transverse rolls (II) as a function of the Prandtl number for rigid–rigid boundaries and Grasshof number 1000. The Grasshof number is equal to $\mathcal{R}/\mathcal{P}$. After Weber (1978).

which means that the increase of the critical Rayleigh number will be quite small if β is small.

These investigations were supplemented by Weber (1978) with studies covering a wider range of conditions, including rigid–rigid boundaries. Of the results obtained we shall discuss only those dealing with the critical wavelength. As Fig. 8.7 shows, the critical wavelength of longitudinal and transverse rolls is a fairly strong function of the Prandtl number, and they vary in opposite direction. Neither the critical Rayleigh number nor the critical wavelength has been measured with sufficient accuracy to make a meaningful comparison with the theoretical findings. The orientation of the rolls, on the other hand, is easily observed. Srulijes (1979) reported that rolls in silicone oil, $\mathcal{P} \cong 1800$, were transverse as in Fig. 8.3. If $(\partial T/\partial x)/(\partial T/\partial z) \geq 0.29$, the rolls were longitudinal. The fact that the rolls in the very high Prandtl number silicone oil first were transverse does not agree with the expectations (see Fig. 8.6). Srulijes also made experiments with nitrogen, which has a Prandtl number $\mathcal{P} = 0.71$. Since in the experiments with nitrogen the nonuniformity of the temperature was only in the lid, Weber's theory is not strictly applicable. We note, nevertheless, that Srulijes found transverse rolls, which changed with increasing horizontal gradient to longitudinal rolls.

Results similar to those of Weber were found for rigid–rigid boundaries by Bhattacharyya and Nadoor (1976). Sweet et al. (1977) showed theoretically that oscillatory solutions are likely to occur in low Prandtl number fluids even with small horizontal temperature gradients. Walton (1983) made an analysis of the effects of the necessarily present end walls on the onset and form of convection in a nonuniformly heated layer of fluid. The end walls are of much greater importance in the case of nonuniform heating than in the case of uniform heating, because they have such a strong effect on the basic circulation. At, e.g., the warm end of the container the basic circulation is forced upward, a transverse roll will necessarily be formed, and the formation will occur at subcritical values of $\mathcal{R}$. The cold end wall will be the location of downward motion, and consequently only odd numbers of rolls can form between both ends.

The supercritical wave numbers in convection with nonuniform temperatures in the horizontal boundaries have been investigated by Buell and Catton (1986b). The analysis of Buell and Catton was influenced by an earlier paper of Kramer and Riecke (1985). As we remember from Section 7.1 wave number selection under supercritical conditions is a cumbersome problem for theory which still awaits its final solution in the case of uniform temperatures on the horizontal boundaries. The question raised by the paper of Buell and Catton is whether the added nonuniformity of the temperatures in the horizontal boundaries makes the problem easier. This may indeed be so because a horizontal temperature gradient eliminates some of the degeneracy which plagues the problem in the uniform case; translational and rotational invariance do not exist in the nonuniform case. Buell and Catton determined the wave numbers of transverse rolls between two plates of horizontally nonuniform temperature. These rolls must be, as we have seen, of alternately smaller and larger size. Using the nonlinear equations and developing the field variables in powers of the small horizontal temperature gradient, they arrived at a solvability condition for their system of equations from which the possible wave numbers follow. The results of their numerical calculations are shown in Fig. 8.8. As can be seen, "the selected wave number" was found to *decrease* as a function of the Rayleigh number for fluids with $\mathcal{P} > 0.8$. On the other hand, if $\mathcal{P} \leq 0.7$ the wave numbers increase with $\mathcal{R}$.

Experimental data for the wave numbers of supercritical convective motions in the presence of horizontal temperature gradients are not available. However the trend of the wave numbers in Fig. 8.8 is the same as the trend of the wave numbers of convective motions on uniformly heated plates in fluids of Prandtl numbers $\mathcal{P} \geq 0.7$. As a matter of fact, the wave number at, say, $\mathcal{R} = 5\mathcal{R}_c$ in Fig. 8.8 is, in the case of infinite Prandtl number, compatible within the experimental error with the wave number $a = 2.5$ found by Koschmieder and Pallas (1974) at $5\mathcal{R}_c$ in silicone oil with $\mathcal{P} \cong 950$ on a uniformly heated plate. At $5\mathcal{R}_c$ the wave number of the motions on a nonuniformly heated plate with a small horizontal temperature gradient must be practically equal to the wave number of the motions on a uniformly heated plate, because the amplitude of the vertical motions is then much larger than the amplitude of the density circulation caused by a small horizontal temperature gradient. That is evident in Fig. 8.5, where the Rayleigh number is only

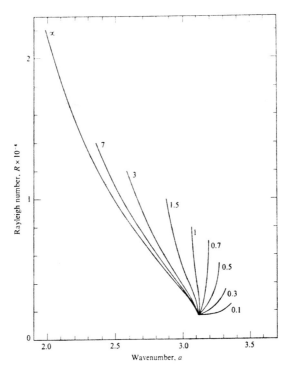

Fig. 8.8.   Theoretically predicted supercritical wave numbers on a nonuniformly heated plate
as a function of the Rayleigh number for different Prandtl numbers (given on the
curves). After Buell and Catton (1986b).

$\approx 2\mathcal{R}_c$, but the alternating cells are of about the same size, although the horizontal gradient in this figure is much larger, namely $(\partial T/\partial x)/(\partial T/\partial z) = 0.28$, than the gradient considered by Buell and Catton. Nevertheless the rolls in Fig. 8.5 are nearly of the same size, because the rolls are supercritical. What Buell and Catton actually found in Fig. 8.8 are the wave numbers of convective motions on a uniformly heated plate, but they arrived at that result studying the motions on a nonuniformly heated plate. We note that the increasing wave numbers for $\mathcal{P} = 0.7$ (gases) in Fig. 8.8 are opposite to the decreasing wave numbers in air found experimentally on a uniformly heated plate by Martinet et al. (1984)

Finally we consider the question of whether the "selected wave numbers" with nonuniform heating are unique as suggested by Buell and Catton. A pattern consisting of a periodic sequence of alternately smaller and larger rolls is described by Fourier analysis as the sum of, say, a sine function with a basic wave number plus a sine function of the same wave number but shifted in phase and the sines of all the higher harmonics thereof. It does not appear that a sequence of alternately smaller and larger rolls, so characteristic of convection with nonuniform heating, can be described with one wave number; actually one can see with the naked eye that the rolls on a nonuniformly heated plate are of alternately different size or wavelength. Only harmonic rolls of the same size can be described with *one* wave number.

That is how far we have come with convection on a nonuniformly heated plate; we have just scratched the surface of the problem, experimentally as well as theoretically. As far as theory is concerned, our efforts are hampered by the lack of an adequate description of supercritical convection on uniformly heated plates. The theory of convection on a nonuniformly heated plate requires an understanding of supercritical flow because the horizontal temperature gradient must, in order not to deal only with the trivial density circulation, include the just critical condition and then necessarily extends into the supercritical range.

The basic features which have become apparent in the case of nonuniform heating are also present in convection in a tilted container with uniform temperatures on the top and bottom, as well as in a container with uniform temperatures on nonparallel boundaries. We cannot pursue these topics.

8.3 Convection with Rotation

Chapter III in Chandrasekhar's (1961) book is devoted to a discussion of the consequences of rotation on the instability of a layer of fluid heated from below. Convection with rotation is an obvious topic for an extension of the theory of convection; it seems to be primarily of interest in connection with astrophysical and geophysical problems. The theoretical investigation of Rayleigh–Bénard convection in a rotating fluid layer of infinite horizontal extent heated uniformly from below was started by Chandrasekhar (1953a) and Chandrasekhar and Elbert (1955). Chandrasekhar used the Navier–Stokes equations incorporating as an additional force the Coriolis force,

$$\rho\frac{d\mathbf{v}}{dt} = -\nabla p + \rho g \mathbf{e}_z - 2\rho \mathbf{v} \times \mathbf{\Omega} + \mu\nabla^2\mathbf{v}, \qquad (8.22)$$

and also the energy equation, the continuity equation, and the equation of state of the fluid, as we used them in Chapter 2, equations (2.1) and (2.3)–(2.4). The symbol $\mathbf{\Omega}$ in equation (8.22) stands for the rotation rate. In order to study with these equations the stability of a rotating uniformly heated fluid layer Chandrasekhar assumed "an initial state in which a steady adverse temperature gradient β is maintained and there are no motions" (1961, p. 87). In modern terminology we say that he assumed that the fluid is, before becoming unstable, in rigid rotation. For the sake of simplicity he assumed also that the axis of rotation coincides with the vertical. After linearization and the elimination of the pressure and the velocity components u and v, and after the introduction of normal mode disturbances for the vertical velocity component w and for the temperature θ, which are of exactly the same form as equations (2.13)–(2.14), he arrived, in the marginal case, i.e. when the growth rate $s = 0$, at the dimensionless equation

$$(D^2 - a^2)^3 w + \mathcal{T} D^2 w = -a^2 \mathcal{R} w, \tag{8.23}$$

with the dimensionless operator $D = d/dz$, the wave number a, the Rayleigh number $\mathcal{R}$ as usual, and the Taylor number

$$\mathcal{T} = \frac{4\Omega^2 d^4}{v^2}, \tag{8.24}$$

which is a nondimensional measure of the rotation rate and uses as the unit of length the depth d of the fluid layer, not a radial gap width as in the Taylor vortex problem. The boundary conditions for rigid–rigid or free–free boundaries are the same as in Section 2.2.

Assuming, just as in the nonrotating case, that the vertical variation $W(z)$ of the velocity component w is of the form

$$W(z) = W_0 \sin n\pi z \tag{8.25}$$

(with the amplitude W_0 and an integer n), the lowest eigenvalues of equation (8.23) are, in the free–free case and with $n = 1$,

$$\mathcal{R}_c = \frac{1}{a^2} \{ (\pi^2 + a^2)^3 + \pi^2 \mathcal{T} \}. \tag{8.26}$$

The critical Rayleigh number $\mathcal{R}_c$ is an increasing function of the Taylor number, but in the case $\mathcal{T} = 0$ the critical Rayleigh number is, of course, the same as in equation (2.22). As before, at $\mathcal{R}_c$ the onset of convection should occur spontaneously in the entire rotating fluid layer if the fluid is heated uniformly from below with mathematical perfection.

The critical wave numbers in the free–free case are given by

$$a_c^2 = a_{c0}^2 \left(1 + \frac{\mathcal{T}}{\pi^2 + a_c^2} \right), \tag{8.27}$$

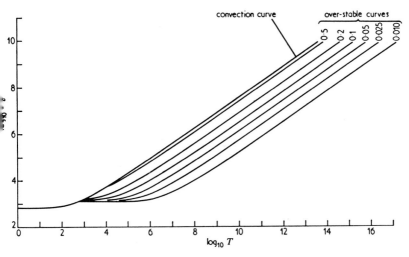

Fig. 8.9. The critical Rayleigh number as a function of the Taylor number and the Prandtl
number for Rayleigh–Bénard convection in a rotating fluid layer of infinite extent,
with free–free horizontal boundaries. The curve labeled ''convection curve'' determines the onset of convection as a steady motion; the overstable curves are for
the Prandtl numbers indicated at the end of the curves. After Chandrasekhar and
Elbert (1955). By permission of Oxford University Press.

which means that the critical wave numbers increase with $\mathcal{T}$ and can become
quite large, or that the cells can become very narrow. Chandrasekhar (1953a)
gives tables of the critical Rayleigh numbers and the critical wave numbers
for different values of $\mathcal{T}$ and three combinations of the horizontal boundary
conditions. The critical Rayleigh numbers as a function of the Taylor number
are shown in Fig. 8.9. As Chandrasekhar showed, in the asymptotic case of
infinite Taylor number, the critical Rayleigh number increases as $\mathcal{T}^{2/3}$ in the
free–free case, the critical wave number increases as $\mathcal{T}^{1/6}$, and the critical
temperature gradient decreases as $d^{-4/3}$ at constant Ω.

Note the inhibiting or stabilizing effect that rotation has on the onset of
convection, which is apparent from the increase of $\mathcal{R}_c$ with $\mathcal{T}$. The increase
of $\mathcal{R}_c$ is explained qualitatively with the argument that disturbances in the
fluid will not move up or down as easily as without rotation, because of the
presence of the Coriolis force. The path of a rising parcel of fluid will now be
curved. Its path will be longer, hence the parcel will experience more dissipation, and in order to overcome the increased dissipation a larger temperature difference has to be applied to commence convection. It follows from
(8.24) that in an inviscid fluid the Taylor number is infinite, and consequently
the critical Rayleigh number in a rotating inviscid fluid is infinite. In other

words, an inviscid fluid with rotation is stable for all practical vertical temperature gradients. This is a consequence of the Taylor–Proudman theorem, which states that in a rotating ideal fluid steady slow motions are two-dimensional, $v = v(x,y)$.

So far we have considered the marginal case in which the growth rate s of the disturbances is zero. If the growth rate is retained, the vertical velocity is described by a nondimensional equation of the form

$$(D^2 - a^2 - \sigma \mathcal{P})\{(D^2 - a^2 - \sigma)^2(D^2 - a^2) + \mathcal{T}D^2\}w$$
$$= -a^2 \mathcal{R}(D^2 - a^2 - \sigma)w \qquad (8.28)$$

with the nondimensional growth rate $\sigma = sd^2/\nu$. If $\sigma = 0$ we obtain of course equation (8.23). It must be remembered now that s and consequently σ can be complex, which means that the disturbances do not necessarily have to increase monotonically, but that the increase can vary periodically or also that the amplitude may not increase at all, but is a periodic function of time, which would mean that σ is imaginary. Chandrasekhar showed that a complex eigenvalue equation follows from (8.28) if (8.25) is used. The real and imaginary parts of the eigenvalue equation must vanish separately. From this it follows that complex solutions in the free–free case for the lowest mode $n = 1$ are possible only if

$$\mathcal{T} \geq \pi^4 \frac{(1 + \mathcal{P})}{(1 - \mathcal{P})}\left(\frac{a^2}{\pi^2} + 1\right)^3, \qquad (8.29)$$

from which it obviously follows that complex solutions are possible only if $\mathcal{P} < 1$. Closer analysis showed that the maximal Prandtl number at which overstability can occur is $\mathcal{P} = 0.676$.

Overstability means that the onset of convection takes place as an oscillatory motion. The oscillations are characterized by their frequencies, which were determined at the onset of convection for mercury, $\mathcal{P} = 0.025$, for either free–free or rigid–rigid boundaries for different Taylor numbers, and are listed in Tables XI and XII of Chandrasekhar's (1961) book as the ratio s/Ω. Since at the onset of convection the real part of the complex growth rate s is zero or very small, the ratio s/Ω gives the frequency of the overstable motions in terms of the rotation rate Ω. For not too high Taylor numbers the frequency of the overstable oscillations is of the same order as the rotation rate of the fluid layer. Overstability is a novel feature that distinguishes convection with rotation from convection on a nonrotating plate on which overstability cannot occur, as was shown by Pellew and Southwell (1940) and as mentioned in Section 2.2. The critical Rayleigh numbers for the onset of convection in the form of overstability as determined by Chandrasekhar and Elbert (1955) are also shown in Fig. 8.9 for different values of the Prandtl number. Weiss (1964) showed that overstability is followed at higher Rayleigh numbers by a transition from overstable to nonoscillatory convective modes. This transition occurs at Rayleigh numbers which depend on the Prandtl number and the Taylor number. Finite amplitude axisymmetric overstable convection has been analyzed by Daniels et al. (1984).

Chandrasekhar (1961) cites a number of papers which confirm, as he believed, the dependence of the critical Rayleigh number on the Taylor number. The experiments to which Chandrasekhar refers measured the critical Rayleigh number as a function of the Taylor number in layers of water from 2 to 17 cm deep in a cylindrical container of 30 cm diameter. Via the dependence of $\mathcal{R}$ on d^3, Rayleigh numbers in the range from 10^5 to 10^8 were reached, although the temperature differences were only a very few degrees. Also, via the dependence of $\mathcal{T}$ on d^4, Taylor numbers in the range from 10^5 to 10^{10} were reached, although the rotation rates were smaller than 1 rev/sec. Steady, uniform, linear vertical temperature gradients, as assumed in theory, were not established in these experiments. Since in the asymptotic case of large Taylor numbers the critical vertical temperature differences decrease as $d^{-1/3}$ the critical vertical temperature difference becomes smaller as the depth is made large, making it increasingly difficult to establish uniform temperature gradients. Heating rates, not temperature differences, are given in these experiments; cooling from above was usually accomplished by evaporation at the free surface. It appears that, because of these and other experimental problems, the reported agreement of the measured critical Rayleigh numbers with Chandrasekhar's theory is fortuitous.

Chandrasekhar refers also to experiments made in order to verify overstability. These experiments were made with mercury layers ($\mathcal{P} = 0.025$) because mercury is the only conventional fluid with a Prandtl number below 0.676. The fluid was from 4 to 8 cm deep in a container of 14 cm diameter; the Taylor numbers ranged from 10^8 to 10^{12}. There was no steady state, the upper surface was air, and there were capillary waves on the surface. A feature as intricate as overstability cannot be verified in this way. It appears that, in order to verify convincingly the existence of overstability in rotating convection, it is necessary to measure the frequency of the overstable oscillations in the steady state. The experiments listed by Chandrasekhar are therefore only of historical interest. The occurrence of overstability in convection with rotation has still to be confirmed experimentally.

The patterns of convection in the presence of rotation depend on both horizontal coordinates and the Taylor number. There is, again, degeneracy; an infinite number of patterns is theoretically possible at the same critical $\mathcal{R}$. The patterns can be rolls, square cells, rectangular cells of all side ratios, and hexagonal cells. The velocity fields of the various patterns were calculated and illustrated beautifully by Veronis (1959). Rolls were dismissed by Veronis with the argument that they "cannot occur in an experimental investigation," which serves to show that as late as 1959 rolls were not considered to be realistic. On the other hand, Küppers and Lortz (1969) showed that with rotation and under slightly supercritical conditions "all three-dimensional convective flows are unstable." This leaves rolls as the only alternative. But straight parallel rolls of infinite length are hard to imagine on a rotating plane. We shall therefore focus attention on a rotating pattern of circular concentric rolls which was studied by Müller (1965). The axisymmetric ring cells of Zierep (1958) seem to be a natural solution in the case of convection with rotation, which will, in its most simple experimental realization, amount to rotation about the central axis of a circular container and therefore be axisymmetric. Müller showed that the vertical velocity component w is, in the case of convection with rotation, determined by the equation

$$\nabla^2 \left\{ \nabla^6 - \mathcal{R}\nabla_2^2 + \mathcal{T} \frac{\partial^2}{\partial z^2} \right\} w = 0, \tag{8.30}$$

which can be solved, in the free–free case, by

$$w = W_0 I_0(ar)\sin n\pi z, \tag{8.31}$$

with the Bessel function I_0, the wave number a, and the integer n, just as with the ring cells. Examples of streamlines of rotating ring cells, projected on the bottom of the layer, are shown in Fig. 8.10. The cell boundaries, given by the maxima and minima of the vertical velocity component, are a sequence of concentric circles. The radial velocity components are on logarithmic spirals, which for large values of Ω can wind many times around the center.

An experiment on convection with rotation at moderate Taylor numbers was made by Koschmieder (1967b). In this experiment a 1-cm-deep layer of silicone oil of 1 cm^2/sec viscosity on a uniformly heated copper plate of 20 cm diameter and under a uniformly cooled glass lid was rotated around its axis while the vertical temperature gradient was slowly increased. With rotation rates <1 rad/sec ($\mathcal{T} \approx 4$), patterns of circular concentric rolls were observed which were indistinguishable from circular concentric rolls on a nonrotating circular plate (Koschmieder, 1966a). There is, as must be, a smooth transition from convection with $\Omega = 0$ to convection with $\Omega \neq 0$.

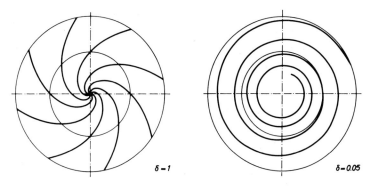

$\delta - 1$ $\delta - 0.05$

Fig. 8.10. Streamlines and cell boundaries of rotating ring cells projected on the bottom of the layer. The parameter δ is equal to $(\pi^2 + a^2)/\sqrt{\mathcal{T}}$. After Müller (1965).

The flow in the rotating ring cells with $\Omega < 1$ rad/sec must correspond to the motions discussed by Müller (1965). At $\Omega = 1.5$ rad/sec ($\mathcal{T} \approx 8$), however, the pattern formed clearly as a sequence of circular rolls with alternating larger and smaller sections. At higher rotation rates ($\mathcal{T} \approx 20$) this was only more so. We show an extreme example of alternating larger and smaller rotating circular rolls in Fig. 8.11, in which the amplitudes of the small rolls are practically zero, so that one sees only a sequence of rolls all turning in the same direction, namely outward on top, except for the small outermost counter-roll, which turns inward. The sequence of alternately larger and smaller rolls can be explained as the consequence of a superposition of an overall circulation on a pattern of equal rolls of alternating circulation caused by vertical instability. Rolls caused by vertical instability which turn with the overall circulation are strong; rolls caused by vertical instability which turn against the overall circulation are weak. The cause of the overall circulation is the centrifugal force, which moves cold, heavy fluid on top of the layer outward and light, warm fluid on the bottom inward. Any rotating fluid layer with a vertical temperature gradient experiences such an overall centrifugal circulation, regardless of whether the vertical temperature gradient is negative (unstable) or positive (stable). This observation invalidates Chandrasekhar's assumption that in a rotating fluid layer "there are no motions" prior to the onset of instability. This is already noticeable at Taylor numbers as low as 10. The centrifugal circulation makes the problem of convection with rotation very difficult. Since the centrifugal circulation necessarily increases in intensity with increased r, the critical Rayleigh number will be a function of r; in other words, there will be only a local critical Rayleigh number.

Fig. 8.11. Convective motions in a uniformly heated layer of silicone oil, 1 cm deep and 20 cm in diameter, rotated around the axis of the plate. All rolls turn outward on top except for the single, very narrow counter roll inside of the outermost roll. $\mathcal{T} \approx 20$. After Koschmieder (1967b).

Another experiment on convection in a thin, uniformly heated rotating fluid layer was made by Rossby (1969). His observations concerning the form of the convective motions are misleading because his photographs showing convection with rotation were made after the rotation of the apparatus was stopped. However his observation that at larger Ω the motions in the fluid transform into radially oriented rolls is correct and significant. If the centrifugal circulation becomes too strong, the shear of the centrifugal circulation becomes dominant and the rolls align in the radial direction, similar to the behavior of rolls in convection with nonuniform heating on a resting plate. The principal objective of Rossby's study was the exploration of the stability properties of the rotating fluid. He measured the critical Rayleigh number as a function of the Taylor number and found, as shown in Fig. 11 of his paper, that in the range from $\mathcal{T} \approx 10^5$ to $\mathcal{T} \approx 10^8$ the critical Rayleigh number is below the value calculated by Chandrasekhar (1961); at $\mathcal{T} = 10^8$ "the measured critical Rayleigh number is about one-third the expected value" (p. 323). Although Rossby is reluctant to give an explanation, he wrote that it cannot be "entirely ruled out" that the centrifugal force is the cause of the discrepancy between theory and experiment. Rossby's measurements have been interpreted by Homsy and Hudson (1971). They noticed that Rossby's $\mathcal{N}(\mathcal{R})$ relations were characteristic of an imperfect bifurcation, and they re-calculated the critical Rayleigh numbers. The recalculated Rayleigh numbers are shown in Fig. 5 of Homsy and Hudson and are in good agreement with Chandrasekhar's marginal stability curve of linear theory of convection with rotation.

We shall now continue with the discussion of the very few experimental studies of convection with rotation and return to theory later. A series of convection experiments in a rectangular container with aspect ratios $\Gamma_x = 10$ and $\Gamma_y = 4$, rotating about the center of the container, was made by Bühler and Oertel (1982). Although the rectangular configuration of the container is odd, the effects of the centrifugal force on the convective motions at Taylor numbers from 200 to 500 were obvious (Fig. 8.12). On the other hand, Bühler and Oertel pointed out a way to reduce the consequences of the centrifugal force by using a gas, instead of a fluid, as the convecting medium. Since the magnitude of the centrifugal force per unit volume depends on the density of the medium, and since the density of gases is of the order of a thousandth of the density of normal fluids, the effects of the centrifugal force on convection with rotation should be much smaller in gases. And indeed they are, as shown by Bühler and Oertel, Fig. 13 therein. Nevertheless centrifugal effects can be noticed even in this figure. Bühler and Oertel also showed that at around $\mathcal{T} = 2 \times 10^3$ the onset of convection in nitrogen occurred at $\mathcal{R} = 10,620$ if one defines the critical condition by the Rayleigh number at which cellular convective motions extend in the radial direction throughout the entire layer. The flow was then time-dependent.

Lucas et al. (1983) and Pfotenhauer et al. (1987) made convection experiments with a sample of rotating liquid helium at 2.6 K. The Prandtl number is then $\mathcal{P} = 0.49$. They detected subcritical convection below the theoretical $\mathcal{R}_c$ by a small increase of the heat flux above $\mathcal{N} = 1$. There is a long discussion in Pfotenhauer et al. about the origin of the subcritical motions which produced the increased heat flux. Overstability and finite amplitude effects were ruled out. Centrifugal effects were checked experimentally by varying the aspect ratio of the container. It was found that the subcritical instability occurred in small containers ($\Gamma \approx 2$) at lower values of $\mathcal{R}$ than in larger

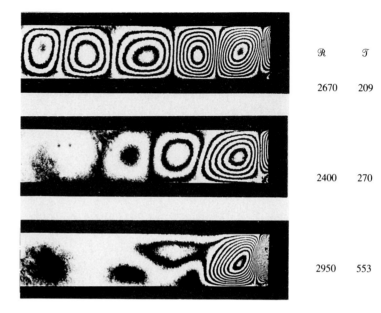

	$\mathscr{R}$	$\mathscr{T}$
	2670	209
	2400	270
	2950	553

Fig. 8.12. Convective motions in a uniformly heated layer of silicone oil in a rectangular container rotating around its center. Visualization with interferograms. After Bühler and Oertel (1982).

containers ($\Gamma \approx 8$). One should, however, expect the opposite from centrifugal effects, which should increase with Γ and cause rather more subcritical motions than less. This argument holds, however, only if the lateral stainless-steel wall of the container does not affect the fluid. If the lateral wall affects the stability of the fluid, its influence will increase with smaller values of Γ. Having also ruled out centrifugal effects, Pfotenhauer et al. saw an explanation of the subcritical motions in their experiments in asymmetric modes, which occur according to a theoretical investigation of Buell and Catton (1983b) in containers with aspect ratios $\Gamma \leq 2$.

The only other experiment on convection with rotation is that of Boubnov and Golitsyn (1986), who used deep water layers with a free surface, which was cooled by evaporation. The "cold film" at the surface caused all the motions in the water; the motions in the fluid do not, therefore, correspond to the conditions usually studied in the context of Rayleigh–Bénard convection.

The theoretical analysis of convection with rotation moves at present on two different tracks. The more common track simply ignores the disturbing consequences of centrifugal effects; the other track takes centrifugal effects into consideration. We shall first discuss the studies that are, in essence, extensions of Chandrasekhar's (1953a) model, which considers only the Coriolis force. Küppers and Lortz (1969) investigated the stability of the patterns in convection with rotation with a weakly nonlinear analysis in the same way as Schlüter et al. (1965) did in the nonrotating case. Küppers and Lortz considered a fluid layer of infinite horizontal extent, of infinite Prandtl number, and with free–free boundaries. We note that a stratified fluid layer on an infinite rotating plate experiences at its outside infinite centrifugal forces which invalidate the assumptions of Chandrasekhar's model. The stability analysis of Küppers and Lortz showed that stable, small amplitude, two-dimensional parallel rolls can exist in a rotating fluid layer under the assumptions made. However that is so only if the Taylor number is smaller than a critical Taylor number $\mathcal{T}^* = 2285$, above which no stable steady convective flow can exist for slightly supercritical Rayleigh numbers. All flows are then time-dependent, which Küppers and Lortz equate with thermal turbulence. Küppers (1970) extended this analysis to the case of finite Prandtl number and rigid–rigid boundaries. He found that $\mathcal{T}^*$ decreases as $\mathcal{P}$ decreases; e.g. for $\mathcal{P} = 0.7$ (gases) he found $\mathcal{T}^* = 480$, whereas for infinite Prandtl number $\mathcal{T}$ increased to about 3000.

The weakly nonlinear investigation of Küppers and Lortz (1969) was extended to moderately supercritical Rayleigh numbers by Clever and Busse (1979). Their method of analysis follows that of Busse (1967a) and Clever and Busse (1974) in the nonrotating case. They determined stability regions for two-dimensional rolls on an infinite rotating plane as a function of the wave number for given values of the Prandtl number and the Taylor number, the latter being below the critical Taylor number $\mathcal{T}^*$ of Küppers and Lortz. The stability regions resemble those we showed earlier in Fig. 7.4. These stability regions imply that there are, at a given supercritical Rayleigh number and below $\mathcal{T}^*$, nonunique continua of stable wave numbers, ranging from values below the critical wave number to values above the critical wave number. It seems unlikely that the results of this study can be verified experimentally, because in experiments with two-dimensional rolls of necessarily finite length centrifugal effects will immediately become apparent and invalidate the assumptions made in this theory, as is apparent in the experiments of Bühler and Oertel (1982). Furthermore we note that the predictions of Clever and Busse (1979) do not match reality even in the simple special case $\Omega = 0$, $\mathcal{R} > \mathcal{R}_c$, as discussed in connection with Fig. 7.4.

The latest theoretical paper working with Chandrasekhar's model of convection with rotation is that of Galdi and Straughan (1985). They applied nonlinear energy stability theory and found a nonlinear stability boundary that is, for Taylor numbers between 10^5 and 10^8, very close to the experimental results of Rossby (1969). A much simpler explanation of Rossby's results was given by Homsy and Hudson (1971), as mentioned before.

Homsy and Hudson (1969, 1971) made the first analyses of axisymmetric convection with centrifugal effects. Considering the Coriolis force *and* the centrifugal force, the Navier–Stokes equation becomes

$$\rho \frac{d\mathbf{v}}{dt} = -\nabla p - 2\rho\Omega(\mathbf{e}_z \times \mathbf{v}) - \rho\Omega^2\mathbf{r} + \mu\nabla^2\mathbf{v}. \tag{8.32}$$

It is obvious that Chandrasekhar's assumption of rigid rotation before the onset of convection can be valid only if the ratio of the acceleration of gravity g divided by the centrifugal acceleration $\Omega^2 r$ is much larger than 1. Actually of more interest is the question of what is the smallest acceleration ratio at which the centrifugal circulation in an unstably stratified rotating fluid layer can be neglected. Koschmieder's (1967b) experiment showed that centrifugal effects were clearly noticeable at the onset of convection at Taylor numbers of around 8. The acceleration ratio in this case was $A \approx 50$. That means that the centrifugal acceleration has to be considered in convection with rotation if the centrifugal acceleration is only 2% of the acceleration of gravity.

Homsy and Hudson (1971) showed how centrifugal convection causes a slight destabilizing effect near the outer wall of a rotating cylinder. They considered a basic laminar axisymmetric state brought about by the centrifugal force. The basic state was described by the velocity Q and the temperature Θ, both of which depend on the vertical and radial positions. They wrote the velocity and temperature following from gravitational instability as the sum of the basic state and small disturbances v and θ, and linearized and used a modified Boussinesq approximation. In the modified Boussinesq approximation all material properties are constant, except for the density, which varies with temperature in the direction of gravity *and* the centrifugal force. In the marginal case the dimensional perturbation equations are then

$$\nabla \cdot \mathbf{v} = 0, \tag{8.33}$$

$$\rho\mathbf{Q} \cdot \nabla\mathbf{v} + \rho\mathbf{v} \cdot \nabla\mathbf{Q} = -\nabla p - 2\rho\Omega(\mathbf{e}_z \times \mathbf{v}) - \alpha\rho\Omega^2\theta\mathbf{r}$$
$$+ \alpha\rho g\theta\mathbf{e}_z + \mu\nabla^2\mathbf{v}, \tag{8.34}$$

$$\mathbf{Q} \cdot \nabla\theta + \mathbf{v}\cdot\nabla\Theta = \kappa\nabla^2\theta. \tag{8.35}$$

These equations are too complicated to be solved analytically without simplifying assumptions. Homsy and Hudson (1971) studied the asymptotic case of large Taylor numbers, and Torrest and Hudson (1974) solved the simplified asymptotic equations with a Galerkin method for aspect ratios from 1 to 10, assuming that the disturbances are axisymmetric. They determined the critical Rayleigh numbers and the disturbance velocity profiles. The disturbances are large numbers of axisymmetric ring cells of small wavelength, so that with aspect ratio 1 about 10 ring cells fit into the container. They determined the maxima of the vertical velocity of the disturbances, which are plotted in Figs. 5 and 6 of their paper. These figures have a striking qualitative resemblance to the interferogram of convection with rotation of Bühler and Oertel (1982), which we have shown in the bottom frame of Fig. 8.12.

The case when in a bounded circular layer the motions caused by vertical instability and the circulation caused by the centrifugal force are of similar magnitude was studied by Daniels (1980). Assuming axisymmetric motions and studying flow with Rayleigh numbers smaller as well as larger than $\mathfrak{R}_c(\mathfrak{T})$, Daniels derived an equation for the amplitude A (s,t), where s is the nondimensional distance from the center of the layer measured in units of the radius L of the layer, and t is a nondimensional time. The amplitude equation is then

$$\frac{\partial A}{\partial t} = \frac{\partial^2 A}{\partial s^2} + \delta A + s^2 \gamma A - \frac{1}{s} A |A|^2 \operatorname{sgn} \xi, \qquad (8.36)$$

containing a new term $s^2 \gamma A$, with γ being either positive or negative, depending on the values of the other parameters of the problem. Here δ, γ, and ξ are constants which depend in a complicated way on the four independent parameters of the problem, namely the Rayleigh number, the Taylor number, the Prandtl number, and the centrifugal parameter $\mathfrak{G} = \kappa \nu / d^3 g$, which appears because the centrifugal circulation not only is a function of the Taylor number but also depends on the depth d of the fluid layer, the material properties of the fluid, and the value of the gravitational acceleration g.

Daniels found that, if the centrifugal term is sufficiently large ($\mathfrak{T} >> 1$), convection cells appear first at the outer edge of the container as the Rayleigh number is increased. This leads, if $\mathfrak{T}$ is large enough, "to a decrease of the Rayleigh number at which cellular convection sets in" (1980, p. 79) in contrast to the increase of the critical Rayleigh number predicted by Chandrasekhar's model, but in agreement with the results of Homsy and Hudson (1971). Daniels continued: "The amplitude of the motions increases with radius and initially is restricted to the region outside the critical radius. As the Rayleigh number increases, this radius decreased to zero and for [sufficiently large $\mathfrak{R}$] cellular convection occurs throughout the container, initially

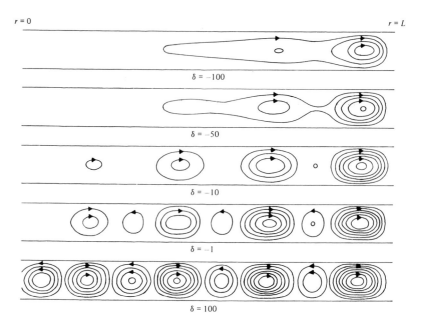

$r = 0$ $r = L$

$\delta = -100$

$\delta = -50$

$\delta = -10$

$\delta = -1$

$\delta = 100$

Fig. 8.13. Transition to finite amplitude convection in axisymmetric convection in a rotating fluid layer as the Rayleigh number, here symbolized by the parameter δ, is increased. It is $\mathcal{P} = 1$, $\mathcal{T} = 100$, and $\mathcal{G}L^2 = 0.1$. After Daniels (1980).

increasing in strength with radius, but ultimately [for very large $\mathcal{R}$] attaining a uniform distribution'' (pp. 79–80). He found that ''the flow pattern in the radial cross-section of the cylinder predicted by [this] theory is a superposition of the centrifugal circulation and the motion due to the vertical instability'' (p. 83). A sequence of flow fields with increasing Rayleigh number was computed by Daniels and is shown in Fig. 8.13. In these flow fields the vertical lateral wall is omitted, since the solution ceases to be valid at the lateral wall because of the boundary condition. Daniels found that the main features of his calculations were remarkably similar to the experimental results of Koschmieder (1967b) (compare Fig. 8.13 and Fig. 8.11). His investigation goes a long way toward explaining convection with rotation.

Summarizing, we find that the investigation of convection with rotation is in disarray. The first order of business for this topic is, of course, the investigation of convection with rotation at moderate Taylor numbers, say between 1 and 100. It seems that only in gases in this lower range of Taylor numbers will Chandrasekhar's model match reality. Crucial for the determination of the reality of the model considering the Coriolis force only will be an experimental determination of the wavelength of the motions, which must be shorter than the critical wavelength of convection without rotation in order to confirm Chandrasekhar's predictions. The theoretical investigation of convection with rotation is obviously more simple in the asymptotic case of large Taylor numbers, but it appears that with large Taylor numbers the consequences of the centrifugal force cannot be neglected in the analysis of the problem, and hence Chandrasekhar's model may not apply. In the situation which we normally deal with in convection experiments, namely shallow fluid layers with aspect ratios larger than, say, 10, the model of convection with rotation considering the Coriolis force only does not apply even at quite low rotation rates because of the consequences of the centrifugal force. If the centrifugal force is involved in convection with rotation, one cannot work with fluid layers of infinite horizontal extent, and there will be no critical Rayleigh number for the entire layer, but only a local critical Rayleigh number, which means that the analysis of the problem will become much more difficult.

8.4 Convection in and on Spheres

Following earlier studies of Wasiutynski (1946) and Jeffreys and Bland (1951) the problem of convection in spheres was formulated by Chandrasekhar (1952). Convection in spherical shells was formulated by Chandrasekhar (1953b). The results of these studies are summarized in Chapter VI of Chandrasekhar's (1961) book. The motivation for these studies was astrophysical and geophysical problems. Chandrasekhar's studies are the first application of the theory of Rayleigh–Bénard convection to a practical problem.

The procedure in Chandrasekhar (1952) follows closely the approach taken in linear theory of Rayleigh–Bénard convection on a plane, of course with appropriate modification for the spherical geometry. The continuity equation and the linearized Navier–Stokes equations are used, with the constant acceleration of gravity replaced by the gravitational acceleration within a sphere. This means that the gravitational field is described by

$$g(r) = -\tfrac{4}{3}\pi\rho G, \tag{8.37}$$

where G is the gravitational constant and ρ the density of the material in the homogeneous sphere. Chandrasekhar also used the equation of state in its usual form (2.4) and the Boussinesq approximation. The energy equation, however, had to be modified because heating in this case is not from below but results from internal heat sources, which were, for the sake of simplicity, assumed to be uniformly distributed. So the equation is

$$\kappa\nabla^2 T = -\varepsilon, \tag{8.38}$$

where ε is the temperature increase by internal heating in the absence of thermal conduction. From (8.38) follows the radial temperature distribution in the static case.

Following procedures similar to those in the plane case, and switching to spherical coordinates by using the relation

$$\sum x_i v_i = r v_r, \tag{8.39}$$

Chandrasekhar arrived at an equation for the radial velocity component v_r

$$\left(\frac{\partial}{\partial t} - \kappa\nabla^2\right)\left(\frac{\partial}{\partial t} - \nu\nabla^2\right)\nabla^2(rv_r) = -2\beta\gamma L^2(rv_r), \tag{8.40}$$

where $\gamma = \alpha(4/3)\pi G\rho$ and $\beta = \varepsilon/6\kappa$. Equation (8.40) corresponds completely to equation (2.10) in the plane case.

The operator L^2 on the right hand side of (8.40) is given by

$$L^2 = \frac{\partial}{\partial r}\left(r^2\frac{\partial}{\partial r}\right) - r^2\nabla^2 = -\frac{1}{\sin\vartheta}\frac{\partial}{\partial\vartheta}\sin\vartheta\frac{\partial}{\partial\vartheta} - \frac{1}{\sin^2\vartheta}\frac{\partial^2}{\partial\varphi^2}, \tag{8.41}$$

where ϑ is the polar angle and φ the azimuth angle. From the form of the operator L^2 it follows immediately that the general solution of (8.40) in the marginal (or neutral) case must be expressible as the product of a function $W(r)$ and a series of spherical functions $S_l(\vartheta,\varphi)$, because

$$L^2 S_l(\vartheta,\varphi) = l(l + 1)S_l(\vartheta,\varphi), \tag{8.42}$$

which corresponds to the membrane equation in the plane case. It follows immediately from (8.40) with (8.41) that, for $v_r = W(r)S_l(\vartheta,\varphi)$, in the marginal case the equation for $W(r)$ is

$$\left[\frac{d^2}{dr^2} + \frac{2}{r}\frac{d}{dr} - \frac{l(l+1)}{r^2}\right]^3 W = -l(l + 1)C_l W, \tag{8.43}$$

where the eigenvalues $C_l = 4\pi\alpha\rho G\varepsilon R^6/9\kappa^2\nu$ are different for different values of l. Equation (8.43) was solved by Chandrasekhar with sums of Bessel func-

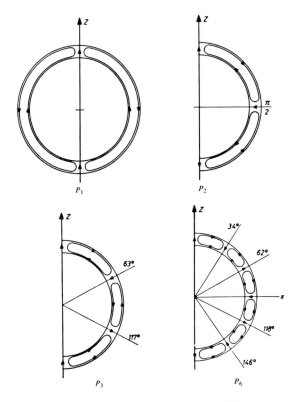

Fig. 8.14. Several possible convective motions on a uniformly heated, nonrotating gravitating sphere according to linear theory. After Koschmieder (1959).

tions of order $l + \frac{1}{2}$ for either a free or rigid boundary at the surface of the sphere. He determined the eigenvalues of the first 15 modes, which were found to increase continuously with l. Similar results were obtained for the convective motions in spherical shells (Chandrasekhar, 1953b).

The form of the convective motions on a sphere follows immediately from the nodal surfaces of the spherical functions, as was noted by Koschmieder (1959). In connection with atmospheric problems he studied the convective motions on a sphere with a modified version of Chandrasekhar's (1953a) theory. Some of the possible axisymmetric convective motions on a sphere are shown in Fig. 8.14, and the most simple convective motion in a sphere is shown in Fig. 8.15. These motions are circular rolls around the axis of the sphere. The reality of these solutions, at least of the solutions in spherical shells, is supported by the observations of circular concentric rolls in circular containers in the plane case, which are in essence the motions on a sphere

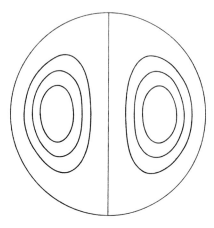

Fig. 8.15. The most simple convective motion in a fluid sphere heated uniformly from within. After Chandrasekhar (1961). By permission of Oxford University Press.

projected on the equatorial plane. Linear theory also permits nonaxisymmetric flows whose azimuthal boundaries are determined by the nodal planes of, say, sin $m\varphi$. We note that the integer m must be smaller than or equal to l, because the index m in the spherical function Y_l^m has to be $|m| \leq l$. The infinite degeneracy of the pattern, so characteristic of linear theory of convection on an infinite plane, disappears in the finite, but nonbounded fluid layers on spheres. In the plane case nonaxisymmetric motions corresponding to the nonaxisymmetric motions on a sphere are the generalized ring cells of Zierep (1959). We noted in Section 2.3 that the generalized ring cells seem to be unrealistic, because in theory the azimuthal wavelength of the motions soon becomes much larger than the radial wavelength when r is increased. Experimental observations, however, tell us that the fluid realizes motions whose azimuthal wavelength is nearly the same as the radial wavelength. This will, in all likelihood also apply to the spherical case. We mention finally that the ring cells in the plane case are the asymptotic case of axisymmetric convection cells on a sphere, because the nodal surfaces of spherical functions and Bessel functions are the same in the limit of infinite numbers of cells, i.e. in thin layers on a large sphere.

Chandrasekhar (1957) has also investigated the convective motions in a rotating sphere, considering the Coriolis force only. This leads in the axisymmetric, linear, stationary case to the equation

$$\nabla_5^6 U + \mathcal{T}\frac{\partial^2 U}{\partial z^2} = C\frac{\partial^2}{\partial \mu^2}[(1-\mu^2)U],\tag{8.44}$$

where ∇_5^2 is a five-dimensional Laplace operator, $\mathcal{T}$ the Taylor number, and U a scalar function from which by simple differentiation the radial and latitudinal velocity components can be determined. C is the eigenvalue parameter and $\mu = \cos \vartheta$. The general solution of (8.44) is a sum of products of Bessel functions and Gegenbauer polynomials. Special solutions, as in the spherical case without rotation, cannot be split off so one cannot find the form of the motion without solving the eigenvalue problem. Although the importance of (8.44) for some geophysical problems is apparent, it appears that this interesting equation has not been studied in detail.

Over the years the investigation of convective motions in spheres, in spherical shells, and on spheres has been carried much further. There are so many studies that they cannot be discussed here; the reader must refer to reviews of the astrophysical applications by Gough (1977), the applications to solar physics by Gilman (1986), the geophysical applications by Machetel and Yuen (1988), and the applications to planetary circulations by Koschmieder (1978). Much of this work is numerical and exceeds in some cases the Rayleigh number range about which we have secure experimental data in the simple plane case. Experiments with convection in spheres or spherical shells cannot be made in the laboratory, because the gravitational field of the earth interferes with the supposedly radial gravitational field on the sphere. On the other hand, the problem can be modeled easily with experiments on plane circular plates where the radial gravitational field on a sphere is replaced by the gravitational field of the earth, which is perpendicular to the bottom of the fluid. This arrangement should, for most purposes, be a good model of convection in and on spheres. The results of the convection experiments on circular plates have been described in Chapter 5. The existence of circular concentric rolls on a uniformly heated sphere, as shown in Fig. 8.14, has been made quite likely by experiments in plane circular fluid layers (Fig. 5.1).

A very unusual experiment investigating convection in a real rotating spherical shell was made by Hart et al. (1986). In this experiment the gravitational force of the earth was eliminated in the weightlessness of the space shuttle. A radial gravitational acceleration on a sphere was simulated by the polarizing forces of an alternating electric field acting on a dielectric fluid. The fluid covering the sphere was heated from the interior of the (hemi)-sphere. Experiments were made with Rayleigh numbers ranging from 4.7×10^3 to 3.8×10^5 and with Taylor numbers ranging from 1×10^3 to 6×10^5. Irregular motions similar to banana-shaped rolls whose axes were oriented perpendicular to the equator were observed. These motions are not similar to the, in the mean, axisymmetric circulation of the atmosphere of

Fig. 8.16. The convective motions in the atmosphere of Jupiter as seen by *Voyager 1* on Jan. 17, 1979. Easily visible are the bands of clouds covering the planet. Jupiter's moon Io is seen in front; Io's shadow is the black oval at the upper right. Courtesy NASA.

Jupiter (Fig. 8.16), which these experiments try to model. Control experiments at lower Rayleigh and Taylor numbers with, e.g., the well-known cases of uniform heating with or without rotation or with stable stratification and nonuniform heating, where certain results must be obtained, have apparently not been made. Therefore one cannot assess the proper functioning of the complicated apparatus, and must consider the results of these experiments inconclusive.

Extraordinary progress has, however, been made over the past decade in the observation of convection on the rotating planets. It seems to be fitting to end this discussion of convection with Fig. 8.16, in which the largest convective motion observed in the solar system is shown in spectacular detail, made visible, fortuitously, by the clouds covering the planet. We note that the motions of the atmosphere on Jupiter, Saturn, or Neptune are driven partially if not primarily by heat released from the interior of these planets, which means that these atmospheres are heated predominantly from below. Convection caused by heating from below is inseparable from the atmospheric circulations on Jupiter, Saturn, and Neptune. It appears that, after all, Bénard's comments on the relevance of his observations for the "atmospheric circulation" may be true. We have made progress over the years, but we still have a long way to go to understand even principal problems which are apparent in the pictures of convection on the planets.

REVIEW

Ninety years after Bénard discovered the hexagonal convection cells and introduced, by the excellence of his experiments, a lasting problem to science, we look back and consider how the problem has fared over the years. The most important event in the development of the understanding of Bénard convection undoubtedly occurred when Rayleigh published his theory of convection driven by heating from below. Small wonder that this study is so important in the history of Bénard convection, because Rayleigh was an exceptional scientist who made basic contributions to a number of problems. Strangely enough, Rayleigh's work resulted in the long-lasting misconception that the hexagonal Bénard cells are an example, and the most characteristic example, of buoyancy-driven convection. Whereas the development of linear theory of buoyancy-driven convection made slow but steady progress, the experimental work necessary to clear up the misconception about the hexagonal cells was not forthcoming until the end of the 1950s. Only then was it understood that the hexagonal Bénard cells are surface-tension-driven, not buoyancy-driven. Understanding the driving mechanism of the Bénard cells was a fundamental step forward, but it does not explain the selection of the cell form. In the 30 years since Block's and Pearson's papers we have made little headway in the explanation of the selection of the hexagonal cell pattern. We have also learned very little about nonlinear surface-tension-driven convection. Progress has been slow.

At the end of the 1950s, when linear theory of buoyancy-driven convection was practically complete and experimentally confirmed, the concept of Rayleigh–Bénard convection as a separate problem emerged. Then the investigation of the nonlinear aspects of buoyancy-driven convection began. Nonlinear theory of convection focused first on the question of the preferred pattern, the hexagon versus roll problem. I wonder whether this problem debated for so long is actually a genuine problem; hexagons play no role in time-independent buoyancy-driven convection with Boussinesq fluids. It seems to be that the configuration of the lateral walls, rather than nonlinear effects, determines the form of the patterns. Of the other nonlinear properties of buoyancy-driven convection the wavelength, or the size of the cells, seems to be the most characteristic. We have adequate experimental data about the variation of the wavelength, but we do not have an adequate theoretical explanation for the variation of the wavelength. Going a little further into the nonlinear regime we encounter time dependence. We do not seem to understand very well the transition to time dependence, which does not appear to be a discrete transition (in fluid layers of large aspect ratio). All work beyond the transition to time dependence is exploratory and lacks a firm theoretical foundation. If we summarize the results of 30 years of effort on nonlinear buoyancy-driven convection we find that we have not progressed with theory much further than the weakly nonlinear case, and that basic theoretical problems concerning moderately supercritical convection have still to be solved. Much remains to be done in the future.

Part II

TAYLOR VORTEX FLOW

The first summary of the early work on the Taylor vortex instability can be found in Chapter VII of Chandrasekhar's (1961) book. The 33 references there deal with linear theory of the instability and end with an apparently incidental paper by Schultz-Grunow and Hein (1956), which turned out to be an essential stepping stone for the experimental breakthrough to nonlinear Taylor vortex flow. Since the 1960s the investigation of the nonlinear aspects of the Taylor vortex problem have dominated the field. Linear and nonlinear aspects of the Taylor vortex problem have been reviewed by DiPrima and Swinney (1981) and by Kataoka (1986). The first book devoted entirely to the Taylor vortex instability was written recently by Chossat and Iooss (1992); it contains a rigorous formal presentation of the theory of Taylor vortex flow. The number of publications dealing with Taylor vortex flow is about 300. We shall refer to only about half of these publications, putting emphasis again on regular journal articles and omitting again most publications in the form of contributions to conference proceedings, letters, short communications, etc., because in most cases the results reported in such publications have been duplicated in later journal articles and because usually a greater effort is required to make a lasting contribution. It would only be distracting to mention each and every one of these papers. We must focus attention on the essential results of theory and experiment.

The investigation of the Taylor vortex instability has progressed remarkably differently from the way the Bénard problem has been pursued. We recall that for about 40 years there was confusion about the origin of the hexagonal Bénard cells, and that until now we have not come to an agreement about the preferred pattern in Rayleigh–Bénard convection. The studies of the Taylor vortex instability, on the other hand, are marked by close agreement between theoretical and experimental results and by a near unanimity of opinion about the essential experimental facts. This seems to reflect G. I. Taylor's famous paper in which this accord between theory and experiment is exemplified. The agreement on the essential experimental facts probably originates from the ease with which unambiguous experiments with a vertical fluid column between a rotating inner and a concentric outer cylinder can be made, whereas in the Bénard and Rayleigh–Bénard convection problems it is difficult to make unambiguous experiments because it is difficult to make uniform temperatures on the bottom and in particular on the top of the fluid layer and it is difficult to increase the temperature difference substantially. Also it has turned out that Taylor vortex flow is much less sensitive to small imperfections of the apparatus as compared with the extreme sensitivity of convection experiments to any imperfection. This is probably so because in convection the initial state is a fluid at rest, for which even a very small imperfection is significant, whereas the initial state of the Taylor vortex instability is an azimuthal flow, for which small imperfections are less significant. The ease with which the rotation rate of the inner cylinder of a Taylor vortex apparatus can be controlled and increased has also resulted in the extension of our present experimental knowledge about Taylor vortices over the entire spectrum of Taylor numbers, from the just critical to the highly turbulent. It appears that the basic experimental facts about Taylor vortices are known over the entire range of Taylor numbers and that, at present, the experimental knowledge of Taylor vortex flow is far ahead of the theoretical explanation.

It is practical to focus attention on a centerpiece of the Taylor vortex problem, as we did in the case of convection. The centerpiece of the Taylor vortex problem is, in my opinion, the instability of an infinitely long fluid column between a rotating inner and a resting outer cylinder. In his original investigation Taylor (1923) went one step further and studied the instability when both cylinders rotate. I believe, however, that the problems facing us today concerning this instability, in particular the nonlinear problems, are outlined more clearly in the simpler case with a resting outer cylinder. The rule that science proceeds from the simple to the complex applies, of course, also to this case. Each additional variable makes the theoretical explanation of a nonlinear problem one order of magnitude more difficult. The centerpiece, Taylor vortex flow between a rotating inner and a resting outer cylinder, has four easily recognizable segments. These are linear and nonlinear axisymmetric Taylor vortex flow, nonlinear wavy Taylor vortex flow, irregular or as it is now called chaotic Taylor vortex flow, and finally turbulent Taylor vortex flow. These segments of the Taylor vortex problem can be easily recognized with the naked eye and are demonstrated in Figs. I.1–I.4, which will serve as a guide throughout the following.

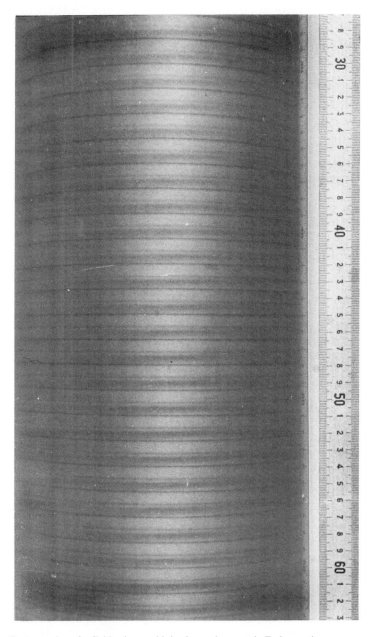

Fig. I.1.   Center section of a fluid column with laminar axisymmetric Taylor vortices
between a rotating inner and resting outer glass cylinder at $\mathcal{T} = 1.16\mathcal{T}_c$.
Visualization with aluminum powder. Heavy dark lines are the sinks, location of
radial inward motion; thin dark lines are the sources, location of radial outward
motion. $\eta = 0.896$, $\Gamma = 122$. The fluid was silicone oil of 0.10 cm^2/sec viscosity.

Fig. I.2.   Center section of a fluid column with wavy Taylor vortices at $\mathcal{T} = 8.49\mathcal{T}_c$ in water. $\eta = 0.896$, $\Gamma = 122$. Visualization with aluminum powder. The axial wavelength is $\lambda = 2.60$; there are six azimuthal waves.

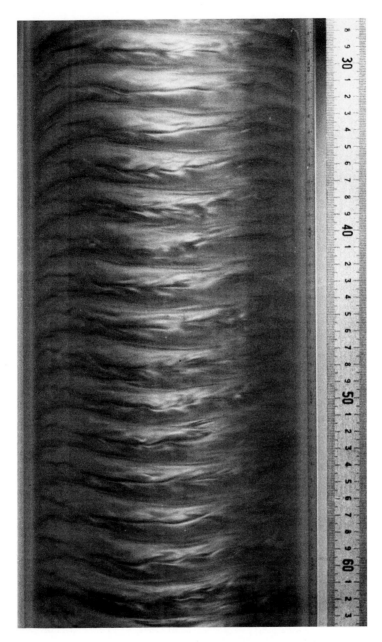

Fig. I.3. Center section of a fluid column with so-called chaotic flow in water at $\mathcal{T} = 120\mathcal{T}_c$. $\eta = 0.896$, $\Gamma = 122$. The axial wavelength is $\lambda = 3.395$; there are two azimuthal waves with transient disturbances.

Fig. I.4.   Center section of a fluid column with turbulent Taylor vortices in the steady state after a sudden start of the inner cylinder. $\mathcal{T} = 1625\mathcal{T}_c$; $\lambda = 2.396$. The fluid is water. After Koschmieder (1979).

9

CIRCULAR COUETTE FLOW

The shear flow in the fluid between a long resting inner and a long concentric rotating outer cylinder was studied by Couette (1890) in order to measure the dynamic viscosity μ of the fluid, and in order to learn whether the equations of Navier (as he put it) are exact and general. Couette wondered about the exactness of Navier's equations because of the existence of the two flow regimes which Darcy and Reynolds had observed, the regimes that we now call laminar and turbulent, the latter of which did not seem to follow Navier's equations. Couette stated that in order to measure the dynamic viscosity one had to work in the laminar regime. He pursued two experimental approaches, measuring the moment exerted by the rotation of an outer cylinder on a resting inner cylinder and measuring the flow rate through tubes, a method that had been introduced earlier by Poiseuille. We shall focus on the determination of the viscosity through the measurement of the moment exerted by the rotation of the outer cylinder on the inner cylinder.

The torque exerted by the rotation of one cylinder and transmitted by the fluid to the other concentric cylinder is given by the equation

$$M = A\tau(r)r, \tag{9.1}$$

where A is the surface area of the cylinder and $\tau(r)$ is the shear stress, given at the inner cylinder by

$$\tau_i = \mu\left\{\left(\frac{dv}{dr}\right)_{r=r_i} - \frac{v_i}{r_i}\right\}. \tag{9.2}$$

With the velocity distribution for Couette flow [equation (10.3)], it follows that the torque exerted by circular Couette flow on the inner cylinder with radius r_1 is given by

$$M = -4\pi\mu\left\{\frac{r_1^2 r_2^2 h}{(r_2^2 - r_1^2)}\right\}\Omega, \tag{9.3}$$

with the radius r_2 and rotation rate Ω of the outer cylinder and the length h of the inner cylinder on which the torque is measured.

The inner cylinder of the apparatus with which Couette determined the moment M was suspended with a piano wire. The twist of the wire can be measured easily and is proportional to the torque on the cylinder. The length of the suspended cylinder was 7.68 cm, the gap width was 0.2465 cm, and the gap was filled with water. Above and below the suspended cylinder were two fixed guard cylinders in order to keep the suspended cylinder centered and to reduce end effects. Both guard cylinders were about half as long as the suspended cylinder, which means that the entire length of the inner cylinder was about 15 cm, and that the aspect ratio, i.e. the ratio of the length of the fluid column to the gap width, was about 60. The experimental arrangement with the suspended cylinder and the guard cylinders has changed little with time; it is, in essence, still used in modern experiments (e.g. Donnelly, 1958; Debler et al., 1969).

From Couette's formula (9.3) it follows that the ratio of the torque to the rotation rate Ω is constant for a given apparatus. For low rotation rates ≤ 1 rev/sec Couette's measurements were in agreement with this prediction, and these rotation rates defined the first (laminar) flow regime. For rotation rates faster than 1 rev/sec the torque increased faster than the rotation rate, which means that a second (turbulent) regime existed. From the torque measurements in the laminar regime Couette determined the viscosity of distilled water, arriving at the value $\mu = 0.01255$ g/cm $\times$ sec at 16.7°C. This value is 15% larger than the present-day value of the viscosity of water, $\mu = 1.091 \times 10^{-2}$ g/cm $\times$ sec at this temperature. Couette himself noted the difference between his result and the value of μ obtained by Poiseuille, which was $\mu = 0.01096$ g/cm $\times$ sec at 16.7°C. The viscosity measurement was the principal result obtained with the first regime of flow; there was no determination of the velocity field of the flow between the cylinders, which was determined much later by Taylor (1936b). The second flow regime did not serve for any quantitative measurements, but for the qualitative argument that there was, in his experiment, a regime which was not described by Navier's equations, a regime similar to what we now call turbulence. The existence of such a second regime in stable flow between a resting inner and rotating outer cylinder was confirmed by Wendt (1933) and Taylor (1936a). This regime is believed to be caused by effects due to the presence of the horizontal ends of the fluid column between the vertical cylinders.

Although Couette's experiments served a mundane purpose, the determination of the viscosity of fluids, and although the viscosity he measured was not particularly accurate, his experiments were quite successful in that they introduced the concept of what we now call Couette flow, the shear flow between a moving and a resting plane or circular boundary. That turned out to be a lasting contribution to fluid dynamics.

The stability of Couette flow at very high speed of the outer cylinder was investigated anew by Schultz-Grunow (1959). Using the linear approximation of the Navier–Stokes equations he found analytically that the flow should be stable for sufficiently small gap width, e.g. up to a Reynolds number of the outer cylinder $R_o = 1670$ for the radius ratio $r_i/r_o = 0.9$. He verified his calculations with experiments with long cylinders with fluid columns of aspect ratios up to 100 and radius ratios between 0.95 and 0.84. He made the fluid motion visible with suspended particles and observed stable axisymmetric flows up to $R_o = 5 \times 10^5$ with a radius ratio 0.92. He attributed the earlier observations of instability of Couette flow at very high Reynolds numbers of the outer cylinder to imperfections of the apparatus used in these experiments. In order to make his point Schultz-Grunow studied the consequences of a small eccentricity of the cylinders or small deviations of the inner cylinder from the circular form, e.g. an elliptic cylinder. In both cases he observed the onset of instability at Reynolds numbers an order of magnitude smaller than the highest Reynolds numbers at which he had observed stable Couette flow with the circular concentric cylinders he had used before.

Experiments with two concentric cylinders similar to those of Couette were made by Mallock (1896). According to the title of this paper these were experiments on fluid viscosity, but Mallock wrote that "the object of the experiments was chiefly to examine the limits between which the motion of the fluid in the annulus was stable, and the manner in which the stability broke down. For obtaining the actual value of the coefficient of viscosity, other methods, such as flow through capillary tubes, would be more suitable" (p. 41). The apparatus he used was comparatively short; the depth of the fluid was in most experiments only 9.3 times as long as the gap width. There were some delicate arrangements at the bottom of the fluid to eliminate or reduce the end effects. A one-page critique of Mallock's apparatus can be found in Taylor's (1923) paper. Mallock's apparatus had one essential novel feature: he could rotate not only the outer cylinder but also the inner cylinder, an aspect which made his study important for others to follow. Mallock wrote, "Only a few experiments were made with this arrangement, as the motion of the fluid was eddying and unstable, even at very low speeds" (p. 45). That is about all that was written about this problem. The somewhat uncertain results of this paper were that (a) instability occurred easily when the inner cylinder was rotated, and (b) instability also occurred when the outer cylinder was rotated too fast, as Couette had already observed.

10

RAYLEIGH'S STABILITY CRITERION

In the same year in which Rayleigh published his celebrated theory of convection he published another paper (Rayleigh, 1916b), which contained a basic contribution to the dynamics of rotating fluids. The motivation for this study was a meteorological problem, the dynamics of cyclones and anti-cyclones, which are large-scale weather systems that circulate either counter-clockwise or clockwise around a center. Considering an inviscid fluid rotating around a center and writing Euler's equation in cylindrical coordinates, it follows immediately for the azimuthal velocity component v that

$$\frac{d}{dt}(rv) = 0, \tag{10.1}$$

from which, of course, it follows that for any future time $r_0 v_0 = r v$, where r is the distance from the axis. This statement expresses conservation of angular momentum.

Considering the stability of an imposed radial distribution of the azimuthal velocity $v(r)$ in a rotating fluid layer in equilibrium, Rayleigh advanced an argument which is similar to the following. Assume that the fluid is in equilibrium; this means that at any location the centrifugal force is balanced by the radial pressure gradient, i.e. $\partial p/\partial r = \rho V^2/r$. Suppose that an (axisymmetric) ring of fluid of unit mass is displaced outward in this layer from an initial position r_1 to a position r_2. The rotating fluid layer is stable if the displaced ring experiences a force which tends to return the ring to its original position. Going outward in the rotating fluid, the centrifugal force per unit mass varies with the velocity of the fluid in the ring, and the velocity varies in accordance with conservation of angular momentum. The centrifugal force experienced by the ring is then compared with the centrifugal force at r_2 of the fluid moving with the imposed radial distribution of the velocity around the center. It is $v^2/r = r\Omega^2 = k^2/r^3$, with the circulation $k = rv$. If the fluid in the displaced ring moves too slowly so that its centrifugal force at r_2 is smaller than the inward directed pressure gradient of the imposed azimuthal velocity field, the displaced ring will be pushed back to its original position; the distribution $v(r)$ of the velocity around the axis is *stable*. In formulas, the fluid is stable if $F_2 = v_2^2/r_2 = k_2^2/r_2^3 > F_1 = v_1^2/r_1 = k_1^2/r^3$. If, however, the fluid ring is displaced outward into fluid which moves with an azimuthal velocity smaller than that in the displaced ring, then the fluid is *unstable* because the centrifugal force acting on the displaced ring exceeds the inward-directed pressure gradient of the fluid at r_2, and the ring will continue to move away from its initial position. In order to be stable it is necessary for the radial distribution of the azimuthal velocity to be $k_2^2/r_2^3 > k_1^2/r_1^3$ or $dk^2/dr > 0$. This is expressed as the stability criterion

$$\frac{d(rv)^2}{dr} = \frac{d}{dr}(r^2\Omega)^2 > 0. \tag{10.2}$$

In Rayleigh's words, "A circulation always increasing outwards . . . ensures stability" (1916b, p. 151). On the other hand, if $(r^2\Omega)^2$ should decrease in the radial direction, the fluid is unstable in the inviscid case, for axisymmetric flow. Rigorous proof that the stability criterion (10.2) is necessary and sufficient for stability was given by Synge (1933) and can be found in Chandrasekhar (1961).

If we now consider Couette flow of a viscous fluid between two rotating cylinders, with the radius r_1 of the inner and r_2 of the outer cylinder, the distribution of the azimuthal velocity component around the axis of the cylinders is, in the most general form,

$$V(r) = Ar + B/r, \tag{10.3}$$

as follows from the Navier–Stokes equation for the azimuthal velocity component, which reduces in the steady state to

$$\nu(\nabla^2 v - v/r^2) = 0, \tag{10.4}$$

if the radial velocity component $u = 0$, the vertical velocity component $w = 0$, and $v = v(r)$. The constants A and B in (10.3) are determined by the angular velocities with which the inner and outer cylinders rotate. Applying (10.3) in Rayleigh's stability criterion (10.2) it follows, if both cylinders rotate in the same direction, that the stability criterion can be written as

$$\Omega_2/\Omega_1 > (r_1/r_2)^2 \quad \text{or} \quad (rv)_2 > (rv)_1. \tag{10.5}$$

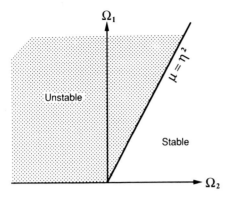

Fig. 10.1. Stability diagram illustrating Rayleigh's criterion for rotating flow of an inviscid fluid. Ω_1 is the rotation rate of the fluid at the inner cylinder with radius r_1, and Ω_2 the rotation rate of the outer cylinder with radius r_2.

This relation is usually illustrated with the stability diagram in Fig. 10.1. The stability of a fluid layer between two rotating cylinders can also be studied with the energy method. The results of these investigations have been discussed in detail in Joseph's (1976) book.

11

TAYLOR'S WORK

11.1 Linear Theory of the Instability

Taylor's* (1923) paper "Stability of a Viscous Liquid Contained between Two Rotating Cylinders" focused on Rayleigh's stability criterion in the search for an example of an instability for which a mathematical representation could be made and in which instability could actually be observed, "so that a detailed comparison can be made between the results of analysis and those of experiment" (p. 290). Taylor had noted that the experiments of Couette (1890) and Mallock (1896) were not particularly suited to confirm the correctness of Rayleigh's stability criterion equation (10.5). Taylor then built a "rough apparatus" from which he learned "that the criterion $\Omega_2 r_2^2 > \Omega_1 r_1^2$ is approximately satisfied in a viscous fluid, but that Rayleigh's result is not true when the two cylinders are rotating in opposite directions. The experiment also indicated that the type of disturbance which is formed when instability occurs is symmetrical" (1923, p. 293). This preliminary experiment is very briefly discussed in Taylor (1921). He then embarked on a systematic study which was published two years later (Taylor, 1923). This paper is unique among the nearly 500 papers discussed in this book, in that it is the only one in which the basic theoretical and experimental results are combined in the same paper.

Following Taylor's presentation we shall discuss theory first. We consider Couette flow with a velocity field $V(r) = Ar + B/r$ between two rotating cylinders. With Ω_1 being the rotation rate of the inner cylinder of radius r_1 and Ω_2 the rotation rate of the outer cylinder with radius r_2 it follows for A and B that

* Sir Geoffrey Ingram Taylor, 1886–1975, professor at Cambridge University.

$$A = \frac{r_1^2 \Omega_1 - r_2^2 \Omega_2}{r_1^2 - r_2^2} = -\Omega_1 \frac{\eta^2 - \mu}{1 - \eta^2}, \tag{11.1}$$

$$B = \Omega_1 \frac{r_1^2 (1 - \mu)}{1 - \eta^2}, \tag{11.2}$$

where $\mu = \Omega_2/\Omega_1$ and the so-called radius ratio $\eta = r_1/r_2$. We disturb the velocity field of the Couette flow with small axisymmetric disturbances u, v, and w, which are functions of r, z, and t only, where r is the radial distance from the common axis of the cylinders and z the coordinate parallel to the axis of the cylinders; u is the radial velocity component, v the azimuthal velocity component, and w the vertical velocity component. The disturbances u, v, and w are small compared with $V(r)$, so small that their squares and products can be neglected. Since Couette flow can be established with virtual perfection, Couette flow is well suited to the study of the consequences of infinitesimal disturbances on its velocity field. The linearized Navier–Stokes equations in cylindrical coordinates are then

$$\frac{\partial u}{\partial t} - \frac{2Vv}{r} = -\frac{\partial}{\partial r}\left(\frac{p'}{\rho}\right) + v\left(\nabla^2 u - \frac{u}{r^2}\right), \tag{11.3}$$

$$\frac{\partial v}{\partial t} + \left(\frac{dV}{dr} + \frac{V}{r}\right)u = v\left(\nabla^2 v - \frac{v}{r^2}\right), \tag{11.4}$$

$$\frac{\partial w}{\partial t} = -\frac{\partial}{\partial z}\left(\frac{p'}{\rho}\right) + v\nabla^2 w, \tag{11.5}$$

with the disturbance pressure p', the kinematic viscosity v, and

$$\nabla^2 = \frac{1}{r}\frac{\partial}{\partial r}\left(r\frac{\partial}{\partial r}\right) + \frac{\partial^2}{\partial z^2}. \tag{11.6}$$

The continuity equation in the axisymmetric case is

$$\frac{1}{r}\frac{\partial}{\partial r}(ur) + \frac{\partial w}{\partial z} = 0. \tag{11.7}$$

The boundary conditions are

$$u = v = w = 0 \quad \text{at} \quad r = r_1 \text{ and } r = r_2. \tag{11.8}$$

Boundary conditions at the horizontal ends of the fluid column between the cylinders are not considered, which means that we are dealing with a fluid column of infinite length.

A solution of equations (11.3)–(11.5) and (11.7) in the form of the axi-symmetric disturbances,

$$
\begin{aligned}
u &= u(r)\cos kz\ e^{st},\\
v &= v(r)\cos kz\ e^{st}, \qquad\qquad (11.9)\\
w &= w(r)\sin kz\ e^{st},
\end{aligned}
$$

is tried, where s is the growth rate of the disturbances and k the dimensional axial wave number. The origin of the postulated infinitesimal axisymmetric disturbances with a specific wave number k, which are supposed to be present over the entire infinite length of the fluid column, must be fluctuations of the velocity distribution of Couette flow. A disturbance with a specific wave number k can only be one component of the Fourier spectrum of the overall disturbances. It appears that the naturally occurring small disturbances of Couette flow should be three-dimensional but the axisymmetry of Couette flow makes axisymmetric disturbances predominant. The reality of three-dimensional disturbances in Couette flow is confirmed by the formation of wavy Taylor vortices, which will be discussed later. Assuming disturbances of the form (11.9) anticipates the existence of a final fluid flow periodic in z; in other words, the disturbances (11.9) anticipate the Taylor vortex solution. One either has to have exceptional intuition to make this assumption, or has to have prior knowledge of what the solution is likely to be, as Taylor had from his preliminary experiments. The disturbances (11.9) determine the final form of the motion of the fluid; there is not, in this case, the problem with the degeneracy of the pattern which we encountered in Rayleigh–Bénard convection, but in the latter case we admitted a larger class of disturbances which depend for their form on the two coordinates x and y. Equations (11.9) leave, of course, the value of the wave number k undetermined, which will be found only later as part of the solution of the eigenvalue problem posed by equations (11.3)–(11.5) and (11.7) and the boundary conditions.

From equations (11.3)–(11.5) and (11.7) Taylor eliminated the pressure, reducing thereby the number of equations by one, and then proceeded to a solution of the equations, noting that any continuous function $f(r)$ which vanishes at r_1 and r_2 can be developed in a Bessel–Fourier series. In order to reduce the numerical difficulties originating from the sum of the Bessel functions he made the so-called narrow gap approximation, which means that he assumed that the gap $d = r_2 - r_1$ between the cylinders is much smaller than

r_1 or that $\eta \to 1$. Using the narrow gap approximation he could use the asymptotic expressions for the Bessel functions; in other words, the Bessel functions were replaced by trigonometric functions. After laborious calculations he arrived at an equation for the critical condition for instability as a function of the rotation rates of the cylinders, the radii of the cylinders, and the viscosity of the fluid.

We shall not follow Taylor's approach now; instead we shall outline the linear theory of the Taylor vortex instability as it evolved in the many years after Taylor's original paper. A first modification of the solution of equations (11.3)–(11.5) and (11.7) was introduced by Jeffreys (1928). He reduced these equations to a single sixth order differential equation for the azimuthal disturbance velocity component v. He did this in order to establish an analogy between the Taylor vortex problem and the Rayleigh–Bénard convection problem. This analogy applies to the results of linear theory, but is not valid, as we shall see later, for nonlinear problems. Jeffreys (1928) wrote: "Prof. G. I. Taylor and Major A. R. Low both have suggested to me that there should be an analogy between the condition in a layer of fluid heated from below and in a liquid between two coaxial cylinders rotating at different rates" (p. 202). Jeffreys showed that in the marginal case ($s = 0$), and in the narrow gap, which permits us to ignore $1/r$ as compared with $\partial/\partial r$, and when distances are measured in units of the gap width d, the azimuthal velocity component v of axisymmetric disturbances between two rotating concentric cylinders is described by the equation

$$\left(\frac{\partial^2}{\partial r^2} - a^2\right)^3 v = \frac{4Aa^2d^4}{\nu^2}\left(A + \frac{B}{r^2}\right)v, \tag{11.10}$$

where a is the nondimensional wave number $a = kd$. Equation (11.10) is similar to equation (2.19b), which describes the Rayleigh–Bénard instability, except for the term $A + B/r^2$ for the angular velocity on the right hand side of (11.10). Since with a narrow gap $A + B/r^2 \cong \Omega_1$, and since then $A \cong -\Omega_1$ if $\mu \to 1$, it follows that under these conditions the equation describing Taylor vortex flow is

$$\left(\frac{\partial^2}{\partial r^2} - a^2\right)^3 v = -\frac{4\Omega_1^2 d^4}{\nu^2}a^2v. \tag{11.11}$$

The boundary conditions are then

$$v = 0, \quad (D^2 - a^2)v = 0, \quad D(D^2 - a^2)v = 0 \quad \text{at} \quad r = r_1, r_2, \tag{11.12}$$

where D is the nondimensional operator $\partial/\partial r$. Equation (11.11) is formally the same as equation (2.19b) for the vertical velocity component of the disturbances of a uniformly heated fluid layer. To the Rayleigh number $\mathcal{R}$ in (2.19b) corresponds in (11.11) the nondimensional number

$$\mathcal{T} = \frac{4\Omega^2 d^4}{\nu^2}. \tag{11.13}$$

This number is now referred to as the Taylor number. The designation "Taylor number" was apparently first used by Chandrasekhar (1953a). The Taylor number in the form (11.13) is a nondimensional measure of the rotation rate of a fluid layer rotating with angular velocity Ω; it is proportional to the square of the Reynolds number $R = \nu r_1/\nu$ of the flow. In many papers the Reynolds number, or different versions of the Reynolds number, are used as the control parameter instead of the Taylor number. The length d in (11.13) is not necessarily a radial gap width, but can as well be the depth of a plane fluid layer, as in the case of a rotating uniformly heated fluid layer (Chandrasekhar, 1953a).

The formal similarity of (11.11) and (2.19b) does not, of course, mean that both problems have the same solution, because the boundary conditions of both problems are different. Nevertheless, as Jeffreys noted, the value of the critical Rayleigh number for rigid–rigid boundaries, $\mathcal{R}_c = 1708$, and the value of the critical parameter calculated by Taylor for the conditions η, $\mu \to 1$ are very nearly the same, $\mathcal{T}_c = 1706$, when the critical condition is expressed by the Taylor number.

The onset of the Taylor vortex instability in a narrow gap was studied again with a sixth order differential equation by Meksyn (1946a,b). A more general solution covering the entire range of μ was found by Chandrasekhar (1954) and is described in Chandrasekhar's (1961) book. In the narrow gap case and when the marginal state is stationary Chandrasekhar solved the nondimensional equations

$$(D^2 - a^2)^2 u = (1 + \alpha\zeta)v, \tag{11.14a}$$

$$(D^2 - a^2)v = -\mathcal{T}a^2 u, \tag{11.14b}$$

with the operator $D = d/d\zeta$, the coordinate $\zeta = (r - r_1)/d$, the wave number $a = kd$, $\alpha = -(1 - \mu)$, and the Taylor number $\mathcal{T} = -4A\Omega_1 d^4/\nu^2$. The boundary conditions are

$$u = Du = v = 0 \quad \text{for} \quad \zeta = 0 \text{ and } 1. \tag{11.15}$$

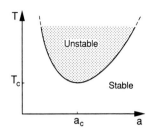

Fig. 11.1. Linear stability diagram for a viscous fluid layer between two rotating cylinders
with a given ratio μ of the rotation rates of the outer and inner cylinders,
according to equation (11.17). $\mathcal{T}$ is the Taylor number, and a is the wave number
of the flow.

Equations (11.14a)–(11.14b) with the boundary conditions (11.15) are
solved with

$$v = \sum_{m=1}^{\infty} C_m \sin m\pi\zeta. \qquad (11.16)$$

From the secular equation it follows that the Taylor number at which instability occurs in the narrow gap case according to linear theory is in the first
approximation given by the formula

$$\mathcal{T} = \frac{2}{1 + \mu} \frac{(\pi^2 + a^2)^3}{a^2\{1 - 16a\pi^2 \cosh^2(a/2)/[(\pi^2 + a^2)^2(\sinh a + a)]\}}, \qquad (11.17)$$

which is, except for the factor $2/(1 + \mu)$ the same as formula (2.29) for the
marginal curve of Rayleigh–Bénard convection, another example of the analogy between both instabilities. The stability diagram according to (11.17) for
a particular value of μ is shown in Fig. 11.1; there is for each value of μ a
stability diagram similar to Fig. 11.1. The neutral or marginal curve in Fig.
11.1 is obviously of the same type as the curve in Fig. 2.1. There is a critical
Taylor number $\mathcal{T}_c$ at the minimum of the curve or, in a given apparatus with
a given fluid, a minimal rotation rate at which the instability in the form of
toroidal axisymmetric vortices will set in spontaneously in the entire infinite
fluid column if the rotation rate is increased quasi steadily to the critical
value. For $\mathcal{T} < \mathcal{T}_c$ all axisymmetric disturbances are dampened and the flow
remains independent of z, whereas for $\mathcal{T} > \mathcal{T}_c$ all axisymmetric disturbances
within the wave number interval given by the neutral curve grow exponen-
tially with time and the flow becomes dependent on z. At $\mathcal{T}_c$ there is, in

mathematical terms, a (supercritical) bifurcation from Couette flow to Taylor vortex flow. According to experience the onset of instability occurs with an amplitude increasing monotonically with time. The mathematical proof of the "principle of the exchange of stability," which requires that the imaginary part of the growth rate s is proved to be zero, is difficult. Davis (1969b) has shown that the growth rates are real for $\eta > 0.84$ and $0.71 < \mu \leq 1$. A complicated proof that the onset of instability of Couette flow is nonoscillatory was given by Yih (1972) for $\mu > 0$ and all η.

At the critical Taylor number the vortices are characterized by the critical wave number a_c or, in other words, a critical size because the wavelength of the vortices is given by $\lambda = 2\pi/a$. The wavelength is a nondimensional measure of the vertical extension Δz of a *pair* of Taylor vortices; it is $\lambda = \Delta z/d$. The wavelength is the most characteristic feature of the Taylor vortices; it can be seen with the naked eye on the photographs of Taylor vortices shown in this book. The critical vortex size, or the critical wave number, is *unique*. However, for supercritical Taylor number $\mathcal{T} > \mathcal{T}_c$ there is, according to linear theory, a *continuum of nonunique solutions* with all the wave numbers possible according to Fig. 11.1 or equation (11.17). The reality of the nonunique supercritical solutions is illustrated in Fig. 13.6.

The critical Taylor number is a function of μ; it is, in a first approximation,

$$\mathcal{T}_c = \frac{3430}{1 + \mu}, \qquad a_c = 3.12, \qquad 0 < \mu < 1. \qquad (11.18)$$

The critical conditions for different values of Ω_1 and Ω_2 are plotted in Fig. 11.4 together with the results of Taylor's experiments. For the case $\mu \to 1$ it follows from (11.18) that the critical Taylor number is $\mathcal{T}_c(\mu \to 1) = 1715$, and is, in a first approximation, the same as the critical Rayleigh number for convection with rigid–rigid boundaries, as Jeffreys noted when he compared his results with Taylor's calculations. The critical wave number in the rigid–rigid convection case is $a_c = 3.117$, i.e. for all practical purposes the same as for the Taylor vortices. From $a_c = 3.12$ it follows that the critical wavelength of a pair of vortices is $\lambda_c = 2.014$; this means that the individual Taylor vortices at the onset of the instability are, in the range from $\mu = 0$ to $\mu = 1$, practically as long in the z-direction as they are wide in the radical direction.

Tables of the critical Taylor numbers and wave numbers as a function of μ are in Chandrasekhar's (1961) book. From these tables it follows that, in the narrow gap case, the critical Taylor number in the case $\mu = 1$ has, in the third approximation, the value 1708, being therefore exactly the same as the crit-

Table 11.1. *The critical parameters for the onset of Taylor vortex flow with a resting outer cylinder as a function of the radius ratio* η

η	$\mathcal{T}_c$	a_c
0.975	1723.89	3.1268
0.950	1754.76	3.1276
0.925	1787.93	3.1282
0.900	1823.37	3.1288
0.875	1861.48	3.1295
0.850	1902.40	3.1302
0.750	2102.17	3.1355
0.650	2383.96	3.1425
0.500	3099.57	3.1631

ical Rayleigh number in the rigid–rigid case. From these tables it also follows that the critical Taylor number increases substantially as μ is decreased from $\mu = 1$ (co-rotating cylinders) to negative values of μ (counter-rotating cylinders). It also follows from these tables that the critical wave number increases as μ is decreased. The critical wave number changes very little from $a_c = 3.12$ in the interval from $\mu = 1$ to $\mu = 0$, but increases significantly for $\mu < 0$, up to $a_c = 8.14$ at $\mu = -3$. That means that for counter-rotating cylinders the vertical extension of the vortices can be much smaller than the gap width. At $\mu = -3$ the critical wavelength is $\lambda_c = 0.77$ as compared with the wavelength $\lambda_c \cong 2$ at $\mu = 0$ (resting outer cylinder). Calculated values of the wavelength as a function of μ together with the wavelengths measured by Taylor (1923) are shown in Fig. 11.5.

So far we have dealt with the narrow gap case only. The stability problem in a wide gap was first studied by Chandrasekhar (1958) for the case in which the radius ratio $\eta = 0.5$, and then for any radius ratio by Kirchgässner (1961). The critical Taylor numbers and wave numbers for a large number of radius ratios were determined by, e.g., Sparrow et al. (1964) for the case in which both cylinders rotate and by Roberts (1965) for the case $\mu = 0$. Some of his results are listed in Table 11.1. The Taylor number as a function of the radius ratio η used in this table is given by

$$\mathcal{T} = \frac{2\eta^2}{1 - \eta^2} \frac{\Omega^2 d^4}{\nu^2}. \tag{11.19}$$

A list of the corresponding critical Reynolds numbers which are often used in practice was compiled from different sources by DiPrima and Swinney (1981).

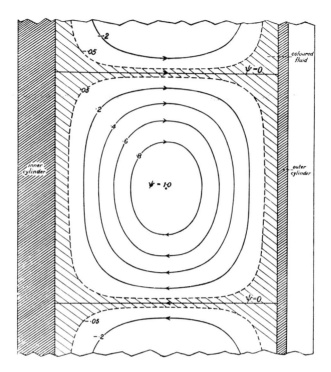

Fig. 11.2. Streamlines in the Taylor vortices at the onset of the instability in a narrow gap according to Taylor's calculations, $\mu \geq 0$. The shaded areas in the fluid mark the regions into which the dye released at the inner cylinder spread in the experiments (see also Fig. 11.3). After Taylor (1923).

The other basic feature determined by linear theory of the Taylor vortex instability is the form of the fluid flow after the onset of the instability. The streamlines of the motion in a vertical cross section through the fluid column as calculated by Taylor (1923) are shown in Fig. 11.2. In the axial plane the flow consists of a series of vortices with opposite circulations in adjacent vortices. Superposed on these motions is the overall azimuthal Couette flow $V(r)$, so that the fluid particles actually move on spiral paths in the azimuthal direction within the tori which outline individual vortices. This motion is apparent in Figs. 11.3 and I.1. If the cylinders move in opposite directions the instability does not extend through the entire gap between both cylinders, because the fluid is, according to Rayleigh's criterion, unstable only in the interior near the inner cylinder. The fluid motion then consists of comparatively small vortices adjacent to the inner cylinder and one or more weaker circulations in the outer part of the gap, which have the same vertical extension as the vortices at the inner cylinder. Streamlines for such cases were determined by Taylor (1923) and can also be found in Chandrasekhar (1961) and Harris and Reid (1964). We mentioned before that the wavelength of Taylor vortices decreases noticeably for negative values of μ (counter-rotating cylinders). It appears, however, that the definition of the wavelength as the ratio of the vertical extension of two vortices divided by the gap width is not suitable if the vortices do not extend in the radial direction through the entire gap. The appropriate length to normalize the size of the vortices is in this case the radial extension d' of the Taylor vortices at the inner cylinder. Normalized in this way, the size of the vortices is slightly smaller than the critical wavelength of the vortices that fill the entire gap. With the conditions for the onset of instability and the form of the motion after the onset we have covered the essential results of linear theory.

11.2 Taylor's Experiments

The apparatus which G. I. Taylor built in order to verify the results of his theory conforms with the assumptions on which his theory is based. The second sentence of the section "Design of Apparatus" in his (1923) paper reads: "In the first place the cylinders were made as long as possible so as to eliminate end effects." The cylinders were 90 cm long, the outer radius was 4.035 cm, and the radii of the inner cylinders ranged from 3.80 to 3.00 cm. In most experiments the thickness of the layer of liquid was less than 1 cm, which means that the fluid layer satisfied the narrow gap approximation. In modern terms, the aspect ratio Γ of the fluid column, defined as the ratio of the length of the column divided by the gap width d, was of the order of 100 or more, and went as high as $\Gamma \cong 380$ in a series of experiments with $d = 0.235$ cm. The radius ratio $\eta = r_1/r_2$ ranged from $\eta = 0.74$ to $\eta = 0.94$. Since he had "considerable trouble" obtaining an accurately bored and polished glass tube 90 cm long, only the 20-cm-long center section of the outer cylinder of his apparatus was made from glass; the lower and upper parts of the outer cylinder were made from metal. He noted, "The inside bore of the outer cylinder did not vary as much as $\frac{1}{10}$ mm in its whole length" (1923, p. 330), uniformity of the gap width being, of course, an elementary requisite of an adequate experiment. The inner cylinder was made of a large number of turned and bored sections of paraffin wax on a central steel shaft. Both cylinders were rotated by an electric motor.

The fluid used in his experiments was water, which is not a particularly viscous fluid. But this choice was probably necessitated by the visualization

Fig. 11.3. Taylor vortices in water at the onset of the instability. Motions made visible by dye spreading from the inner cylinder. Radius ratio $\eta = 0.805$; aspect ratio $\Gamma = 81$. After Taylor (1923).

technique. Following the example of Reynolds he used a "colouring matter" (dye) to visualize the flow. The dye was introduced into the layer through six small holes in the wall of the inner cylinder, and its dispersion was observed through the outer glass cylinder. If the fluid layer was stable, the dye would stay at the inner cylinder. If the fluid layer was unstable, "the layer of coloured fluid suddenly gathered itself into a series of equidistant films whose planes were perpendicular to the axis of rotation" (1923, p. 332). That marked the appearance of the toroidal Taylor vortices (see Fig. 11.3) and means the onset of instability. The cylinder speeds at which the instability occurred were reproducible with an accuracy of about 1 or 2%. From such observations followed the stability diagram shown in Fig. 11.4, which demonstrates a spectacular agreement of the data and the theory. Shown also as a dashed line in Fig. 11.4 is Rayleigh's stability criterion marked by the line

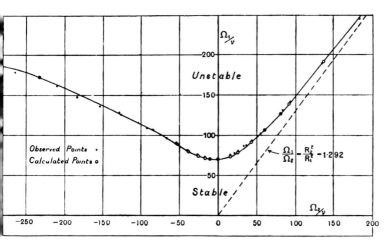

Fig. 11.4. Calculated and observed conditions for instability in water between two rotating
cylinders. $\eta = 0.88$; $\Gamma = 185$. Also marked is the line separating unstable and
stable conditions according to Rayleigh's criterion. After Taylor (1923).

$\mu = \eta^2$. We note the difference between Rayleigh's stability criterion for an
inviscid fluid (Fig. 10.1) and the stability diagram of a viscous fluid (water)
(Fig. 11.4). There is a clear difference in the onset of instability in the case
of a resting outer cylinder ($\Omega_2 = 0$). According to Rayleigh's criterion any
arbitrarily small rotation of the inner cylinder ($\Omega_1 > 0$) means instability if
$\Omega_2 = 0$, whereas actually according to Fig. 11.4 a finite minimal rotation rate
is required for the onset of instability in a viscous fluid. That means, in practical terms, that in an apparatus with a resting outer cylinder and a narrow
gap filled with a viscous oil of, say, 1 cm²/sec viscosity the inner cylinder can
be rotated with rotation rates of the order of 1 rev/sec without instability occurring. If the cylinders rotate in opposite directions, an even higher rotation
rate of the inner cylinder is required to cause instability. On the other hand,
if both cylinders rotate in the same direction, Rayleigh's criterion is approached asymptotically for large rotation rates.

Besides establishing the stability diagram Taylor's experiments confirmed
the streamlines predicted theoretically (compare Figs. 11.2 and 11.3). The
size of the vortices at the onset of instability followed directly from the observation of the streamlines. The vortex sizes which Taylor had calculated as
a function of the ratio μ of the rotation rates of both cylinders and the observed vortex sizes, shown in Fig. 11.5, are obviously in close agreement.
The size of the vortices decreases noticeably with decreasing μ, but we must
remember that with counter-rotating cylinders the vortices do not extend

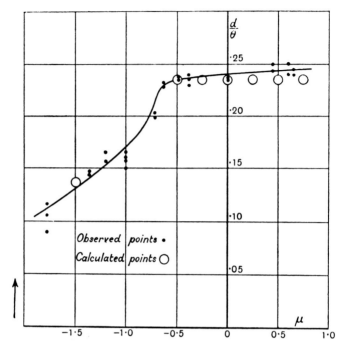

Fig. 11.5. Calculated and observed sizes of individual Taylor vortices at the onset of instability in a gap of 0.235 cm width as a function of μ. The size is given in centimeters and is that of one vortex; i.e. d/θ is equal to one-half dimensional wavelength. After Taylor (1923).

throughout the entire gap. It would be important to learn why the fluid selects a particular value of the critical wavelength, which is $\lambda_c \cong 2$ for $0 \leq \mu < 1$. In Rayleigh–Bénard convection the critical wavelength in the rigid–rigid case is $\lambda_c = 2.016$. This value of λ_c is selected as a consequence of the balance theorem, which we discussed in Section 2.2. Stuart (1958) suggested that a similar principle holds for the Taylor vortices. He suggested that there is a balance of the rate of transfer of kinetic energy from the mean flow to the disturbances and the rate of viscous dissipation of the disturbances. The consequences of this suggestion have, however, not been pursued.

Experiments made at the onset of instability almost automatically extend into the supercritical, nonlinear range. Taylor described his observations in this range at the end of his paper. He noted that a moderate increase in the speed "merely increased the vigour of the circulation in the vortices without altering appreciably their spacing or position" (1923, p. 342). However Taylor noted that the vortices would "break up," i.e. lose their axisymmetry,

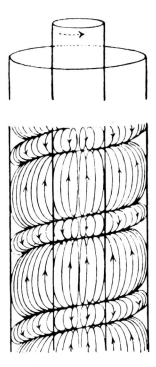

Fig. 11.6. Spiral flow between a rotating inner and resting outer cylinder when the steady motion before instability is not strictly limited to two dimensions. After Taylor (1923).

in the case of counter-rotating cylinders ($\mu \cong -1$) at slightly supercritical speed. His visualization technique was not suited to observing nonaxisymmetric flow in detail. But as seen from the side, "it looked as though each vortex was pulsating so that its cross-section varied periodically" (p. 342). In other words, he saw wavy Taylor vortices, which, as we know now, appear at slightly supercritical conditions in a narrow gap.

If we summarize the results of Taylor's experiments we find that all essential features of linear theory of the Taylor vortex instability were verified experimentally by Taylor himself. As so often with outstanding papers, nothing really new was added to the results for a long time, for 40 years actually.

One result of Taylor's experiments remains to be discussed, the spiral form of the instability, which is illustrated in Fig. 11.6. There is also a photograph of this type of flow in Taylor's paper. Spiral instability was observed "in many cases" and was very noticeable when $\mu < -1$, when the two helical vortices became so different in size that one of them almost disappeared altogether. The formation of the spiral instability was always connected with a circulation in the axial planes during the steady motion before instability appeared. The question is, what caused this circulation?

A possible reason seems to be a radial temperature difference in the fluid, which should have caused a thermal circulation rising along one cylinder and sinking along the other. Superposition of such a circulation upon the vortices will amplify the vortices turning in the direction of the overall circulation, and will weaken the vortices which turn against the overall circulation, so that a sequence of alternately larger and smaller vortices should be formed. This is analogous to the alternately larger and smaller convection rolls that form on a nonuniformly heated plate (Fig. 8.3). This argument may explain the different sizes of the vortices in spiral flow, but does not explain the origin of the thermal circulation and does not explain the spiral form.

The only explanation that I can find for a thermal circulation in the gap between the cylinders is that there was possibly a temperature difference between the water filling the gap of the apparatus and the apparatus itself. Taylor remarked (p. 331) that the water in the gap had been boiled in order to expel air. If the water, after it was filled in, was not in equilibrium with the cylinders, an axisymmetric thermal circulation should have formed. This circulation should have been intense, because the cylinders were long, and strong cooling should have occurred at the outer cylinder, which was metallic and a good thermal conductor over much of its length. The significant differences in the vortex sizes shown in Fig. 11.6 could have been caused only by a strong thermal circulation, if there was one.

Three experiments have been made to check the consequences of radial temperature differences on Taylor vortex flow. Snyder (1965) studied mainly the critical conditions for the onset of spiral flow caused by a radial temperature gradient. He provided no data about the ratio of the sizes of the small and large vortices as a function of either the Reynolds number representing the temperature gradient or the Taylor number. There is also no picture of spiral flow. The experiments of Sorour and Coney (1979) are inconclusive because the heated outer cylinder of their apparatus was made from Perspex, which is a very poor thermal conductor, so that the heat flux through the outer cylinder and the fluid layer was limited. Furthermore the inner cylinder of this apparatus was not cooled, so its temperature had to increase gradually by heat transferred from the outer cylinder, which means that the radial temperature gradient must have decreased steadily. Ball and Farouk (1989) worked with a large gap and larger temperature differences. Their results are qualitative; they seem to confirm the existence of spiral vortices, but also produced several other forms of flow. A definitive experiment concerning Taylor vortex flow with a radial temperature gradient remains to be done.

Linear stability analyses of the onset of Taylor vortex flow in the presence of a radial temperature gradient which considered buoyancy effects and used the Boussinesq approximation were made by Roesner (1978), Kolesov (1980, 1984), and Ali and Weidman (1990) for fluid columns of infinite length, and by Ball and Farouk (1988) for short fluid columns. In these studies either axisymmetric disturbances or helical disturbances were investigated, and the condition for instability determined. Streamlines of spiral flow for different radial temperature gradients with the outer cylinder being at rest can be found in Ali and Weidman. So far no solution has been found which would have the two vortices of clearly different size shown in Fig. 11.6.

Taylor's (1923) experiments with axisymmetric vortices were repeated soon afterward by Lewis (1928) with a carefully constructed apparatus. These experiments broke no new ground, although they differed from Taylor's experiments in one important aspect, the visualization, for which Lewis used aluminum powder suspended in the fluid instead of the dye technique of Taylor. Using aluminum powder for visualization opened, as we shall see, the door to the experimental investigation of supercritical Taylor vortices. Although Lewis mentions that he noticed "pulsating" flows, in other words wavy, periodic motions, he did not realize the existence of the wavy vortices. It is a puzzle how he could have missed the opportunities that were at hand. A long hiatus followed until Schultz-Grunow and Hein (1956) returned to the Taylor vortex experiments with the visualization technique using suspended particles. Using suspended particles for flow visualization was not really novel; we remember that Bénard (1900) used aluminum powder (among other methods) to make his convection cells visible (Fig. 1.2). Schultz-Grunow and Hein showed pictures of Taylor vortices over an extremely large range of Reynolds numbers. Wavy Taylor vortices and turbulent Taylor vortices can be seen clearly on their photographs. But these observations were not pursued; the authors were content with the demonstration of their visualization technique and left the realization of the different types of flow to others.

12

TORQUE MEASUREMENTS

The first study following Taylor's (1923) experiments that provided a new perspective and exceeded by far the critical condition was made by Wendt (1933). He measured the torque exerted by the rotation of either the inner or outer cylinder on the other cylinder. The fluid column in his experiments had aspect ratios ranging from $\Gamma = 42$ to $\Gamma = 8.5$ and had radius ratios ranging from $\eta = 0.935$ to $\eta = 0.68$. The fluid was water or water–glycerol mixtures. The torque exerted by the rotation of one cylinder on the other cylinder is given by Couette's formula (9.3). If, after the onset of instability, the radial velocity distribution differs from the velocity distribution of Couette flow, the transfer of momentum between both cylinders changes, and the moment M changes accordingly. The values of M measured in the unstable range are then compared with the moment M_l resulting from laminar circular Couette flow. The onset of instability is apparent from the break of the curve M versus Ω, when M/M_l begins to increase steadily above $M/M_l = 1$, and the onset of instability can be determined this way, as we shall see later. The results of Wendt's torque measurements for various values of the rotation rate of the inner and outer cylinders were summarized by Coles (1965) in a figure which we reproduce here as Fig. 12.1.

As can be seen in Fig. 12.1 Wendt's experiments extended far into the supercritical range, where, as we know now, the fluid motions are actually turbulent. Wendt did not visualize the flow. He also measured the torque when the fluid was stable, i.e. when either only the outer cylinder rotated or the outer cylinder rotated more than η^2 times faster than the inner cylinder. Under stable conditions the torque should always be equal to M_l. Actually the finding that $M = M_l$ under stable conditions is a necessary test of the proper functioning of the apparatus. But Wendt observed an increase of the torque above M_l when the outer cylinder rotated very fast, as is shown in Fig. 12.1. That

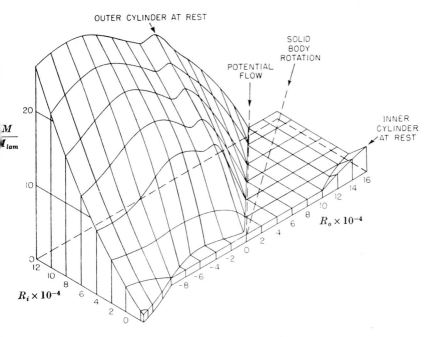

Fig. 12.1. Results of the torque measurements of Wendt as a function of the Reynolds number of the outer cylinder $R_o = \omega_o r_o^2/\nu$ and the Reynolds number of the inner cylinder $R_i = \omega_i r_i^2/\nu$. After Coles (1965).

means that another instability occurred under these conditions, in agreement with Couette's corresponding observations. It is believed that this instability is caused by the presence of the horizontal boundaries at the ends of the fluid column. Figure 12.1 provides, as Coles (1965) wrote, "a satisfactory overall view of the transition problem," except for minor detail.

Wendt also measured the velocity distribution in the interior of the fluid, as well as the pressure. The measurements of the velocity distributions are informative but probably only qualitative, because the upper surface of the fluid column was free and hence curved, which would distort the velocity field. In general the azimuthal velocity field was characterized by a strong increase of the velocity in the outward direction when the fluid was stable and rather flat velocity profiles with boundary layers along the cylinder walls when the fluid was unstable. There are similar measurements of the velocity and pressure distributions in Taylor (1936b) and Pai (1943). Modern measurements of velocity components of Taylor vortex flow with laser–Doppler anemometry were made in short fluid columns under slightly supercritical conditions by Pfister and Rehberg (1981) and Heinrichs et al. (1986). Measurements of transient velocity fields after sudden starts of the inner cylinder with ultrasound echography were described by Takeda et al. (1990).

The measurements of the torque exerted by the fluid on either the inner or the outer cylinder were repeated with a long apparatus and various radius ratios by Taylor (1936a). The end effects of the fluid column should be much smaller in his apparatus than in Wendt's setup. But there are no significant differences in the results of both studies. Taylor also observed the instability of the fluid in the stable case at very large rotation rates of the outer cylinder, which originally had been found by Couette.

After Schultz-Grunow and Hein (1956) demonstrated the possibility of visualizing Taylor vortex flow with suspended particles, the torque measurements were combined with observations of the flow patterns by Debler et al. (1969). The primary result of this work was the observation of a second break in the curve torque versus rotation rate when the flow changes from laminar axisymmetric to laminar wavy Taylor vortex flow. The second break reduces the increase of the torque brought about by the appearance of the axisymmetric vortices. This change is, of course, caused by the change in the radial velocity field of the motions from axisymmetric to wavy flow. Eagles (1974) showed theoretically that with increased azimuthal wave number the increase of the torque above the torque of Couette flow decreases. The second break in the torque curve observed by Debler et al. occurred only in experiments with narrow gaps, because wavy Taylor vortex flow is prominent only in narrow gaps whereas in large gaps the vortices remain axisymmetric to fairly high Taylor numbers. The second break in the torque curve caused by the formation of wavy vortices was also observed by Cole (1976). A comparison of the moments measured by Wendt, Taylor, Donnelly (1958), and Debler et al. at different values of the Reynolds number and different radius ratios can be found in Table 4 of Debler et al. There is, in general, qualitative agreement of the measurements; the differences are in the 5% range and can probably be accounted for by the systematic uncertainties resulting from the errors in v, Ω, and d and from the different end effects in the different apparatus.

The first break in the curve torque versus rotation rate signals the onset of instability and can be used for a quantitative determination of the critical Taylor number. Such an experiment was made by Donnelly (1958) with an apparatus with a narrow gap ($\eta = 0.95$, $\Gamma = 100$) and a wide gap ($\eta = 0.5$, $\Gamma = 10$). In the narrow gap experiment Donnelly found a spectacular break of the torque curve at a period of the inner cylinder which differed from the theoretically predicted critical period by only 1.1%. This seems to represent the result of only one experiment; however the uncertainty of an experiment can be determined only from a number of repetitions of the experiment. It is unlikely in general that the critical Taylor number can be determined experimentally with an accuracy better than $\pm 2\%$, because the viscosities of the fluids are hardly known better than $\pm 0.5\%$, which contributes 1% of the uncertainty of $\mathcal{T}$, and the uncertainties of Ω^2 and d^4 will add at least another percent to the uncertainty of $\mathcal{T}$. From Donnelly's experiment it follows that the uncertainty of Ω^2 was 2%, and with the uncertainties in v^2 and d^4 the uncertainty of $\mathcal{T}_c$ in his experiments amounts to at least 4% of $\mathcal{T}_c$.

The magnitude of the torque under supercritical conditions was determined theoretically by Stuart (1958), Davey (1962), Kirchgässner and Sorger (1969), DiPrima and Eagles (1977), and Yahata (1977a,b). In general there is, in the moderately supercritical range, qualitative agreement between the theoretical predictions and the experimental measurements, but there is not enough precision in the data to make a critical comparison. In the paper of DiPrima and Eagles (1977) the axial wave number producing maximum torque at a given $\mathcal{T}$ was determined theoretically for $\eta = 0.95$ and $\eta = 0.5$. It was found that the wave number of maximum torque increases with increased Taylor number if $\eta = 0.5$. We know, however, that the wave number of the vortices in large gaps remains constant and equal to a_c according to the experiments. In other words, the fluid does not realize the motions with maximum torque. This is similar to the heat transfer problem by supercritical convection in which the fluid motions do not realize the solution with maximum heat transfer either.

The torque of axisymmetric vortex flow as a function of the Taylor number was computed by Meyer-Spasche and Keller (1980) for radius ratios $\eta = 0.95$ and 0.5. Good agreement of their calculations with the measurements of Donnelly (1958) was found. Fasel and Booz (1984) also computed the torque of axisymmetric vortex flow up to $100\mathcal{T}_c$ for $\eta = 0.5$. There is good qualitative agreement of their results at low supercritical Taylor numbers with Davey's (1962) analytical results and with Donnelly's (1958) data. There is also good qualitative agreement of the numerical results at high Taylor numbers with Donnelly's measurements, although the measured values fall systematically low. This is likely due to end effects. Donnelly's apparatus had a radius ratio of 0.5, a lower resting end plate, and an upper free, and therefore curved, surface; the aspect ratio was only 10. Fasel and Booz suggest that the discrepancy between their numerical results and the experimental data may be due to a change of the axial wavelength of the vortices. This is unlikely because the size of the end cells, which as Fasel and Booz believe causes the change of the wavelength, is practically constant in the case of a resting as well as a free end according to Burkhalter and Koschmieder (1973). The torque calculated by Fasel and Booz seems to approach asymptotically the torque predicted for very large Taylor numbers by Batchelor (1960). He assumed that the Taylor vortices have, for $R \to \infty$, an inviscid core surrounded by boundary layers. It is interesting that this model comes close to reality.

If we summarize the results of the torque measurements of Taylor vortex flow, we find that they have not been particularly informative as far as the characteristics of the flow are concerned. This approach to the investigation of Taylor vortex flow has consequently not been pursued for quite a while. Experience shows that much more specific and accurate information can be obtained comparatively easily from experiments employing visualization techniques, as we shall see in the next chapter. In order to substantiate our statement about the usefulness of the torque measurements we note that the discovery of the wavy vortices (Coles, 1965) came many years after the first torque measurements by Wendt (1933) and Taylor (1936a). Their experiments dealt with nonlinear flow but took no notice of the wavy vortices which we now consider to be the first fundamental feature of nonlinear flow. We also note that there are a couple of theories which seem to calculate correctly the torque measured experimentally, although theory and experiment cannot agree because the theories assumed axisymmetric flow but the experiments dealt with wavy vortex flow. In other words, the accuracy of the torque measurements was not sufficient to provide decisive criteria for the evaluation of the theories.

13

SUPERCRITICAL TAYLOR VORTEX EXPERIMENTS

13.1 Wavy Taylor Vortices

After the Taylor vortex problem had teetered on the brink of non-linearity for many years, the decisive step into nonlinearity was taken by Coles (1965). This by now classical paper describes the outcome of several years of work with an apparatus with two rotating cylinders, with visualization by suspended aluminum particles, following the example of Schultz-Grunow and Hein (1956), and with fluids (silicone oils) much more viscous than water. These experiments established the existence of the heretofore unknown wavy Taylor vortices and recognized them as the first qualitative change of Taylor vortex flow caused by nonlinear conditions. Wavy Taylor vortices, which were initially often called doubly periodic vortices, are characterized by periodic azimuthal waves superposed on the toroidal Taylor vortices which are the consequence of the linear Taylor vortex instability. The entire pattern of wavy vortices moves with uniform velocity in the azimuthal direction. Since we usually associate with the term "wavy" a motion with a periodic vertical oscillation, we should emphasize that wavy Taylor vortices move in the azimuthal direction as rings which have m fixed sinusoidal upward and downward deformations, m being the integer number of the azimuthal waves. There is no vertical upward and downward motion of the deformations of the ring. However an observer watching the translation of the wavy ring past a fixed point in the gap sees a picture which changes periodically in time in the vertical direction as the crests and troughs of the ring pass by. An example of wavy Taylor vortex flow as observed by Coles is shown in Fig. 13.1; another example of wavy Taylor vortices can be seen in Fig. I.2. Prior to Coles, wavy Taylor vortices were probably seen by Taylor (1923) and Lewis (1928), and certainly by Schultz-Grunow and Hein (1956),

Fig. 13.1. Wavy Taylor vortices with four azimuthal waves and 24 axial vortices. The outer cylinder is at rest. The Reynolds number of the inner cylinder is $R_i = 1185$. After Coles (1965).

but were not recognized as a characteristic new feature of the flow. After Coles' preliminary results became known, wavy vortices were also observed by Nissan et al. (1963). There were also the experiments of Schwarz et al. (1964) in which apparently a nonaxisymmetric mode with $m = 1$ was observed.

In order to create wavy Taylor vortices the critical Taylor number of linear axisymmetric theory has to be exceeded by a small part of the critical rotation rate if the experiment is done in a narrow gap, whereas in experiments with a wide gap the rotation rate required to obtain wavy vortices must be much higher than the critical rotation rate. As we shall see later the critical Taylor number $\mathcal{T}_c'$ for the onset of wavy vortex flow depends also on the aspect ratio of the fluid column. The regime diagram for the onset of wavy Taylor vortex

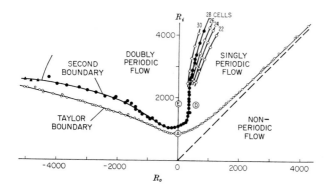

Fig. 13.2. Different regimes in the flow between two rotating cylinders. $\eta = 0.88$, $\Gamma = 14.4$. The fluid is silicone oil of 0.11 cm^2/sec viscosity. Below the Taylor boundary there is circular Couette flow. Above the Taylor boundary there is axisymmetric Taylor vortex flow. Above the second boundary marked by the full circles there is wavy (doubly periodic) Taylor vortex flow. After Coles (1965).

flow as determined by Coles is shown in Fig. 13.2. This figure also provides a splendid confirmation of Taylor's (1923) stability diagram for the onset of axisymmetric Taylor vortices as a function of the rotation rate of both cylinders. Note the large difference in the critical condition for the axisymmetric and the wavy vortices when the outer cylinder co-rotates with the inner cylinder. When both cylinders rotate in the same direction and the Reynolds number $R_o = \Omega r_2^2/\nu$ of the outer cylinder is >400, the wavy vortices appear only when the Reynolds number of the inner cylinder $R_i = \Omega r_1^2/\nu$ is more than two times larger than the critical Reynolds number for axisymmetric vortices at that R_o. On the other hand, in the simple case of a resting outer cylinder the wavy vortices appear when the critical Reynolds number of the inner cylinder has been exceeded by only about 25% (in this apparatus with this η and Γ). For the case of counter-rotating cylinders the difference between the critical linear Reynolds number and the critical Reynolds numbers for wavy flow is between 25 and 50% of R_i according to Fig. 13.2.

As soon as the existence of the wavy Taylor vortices was established, another fundamental aspect of supercritical Taylor vortex flow became apparent, the *nonuniqueness* of the supercritical vortices. Coles writes, "As the speed of the inner cylinder was slowly increased or decreased in the doubly-periodic regime, with the outer cylinder at rest, *the flow pattern was observed to change abruptly, discontinuously, and irreversibly from one state to another at certain well defined and repeatable speeds*" (1965, p. 403). This means that the "state" of wavy Taylor vortex flow, characterized by the

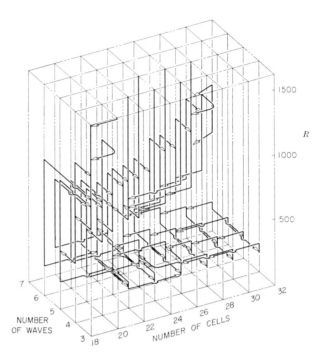

Fig. 13.3. Nonunique state transitions in wavy Taylor vortex flow with the outer cylinder at rest. $\eta = 0.874$, $\Gamma = 27.9$, $\nu = 0.213$ cm^2/sec. Geometry and fluid properties are fixed. Reynolds number is given by $R = \omega_i r_i (r_o - r_i)/\nu$. After Coles (1965).

azimuthal and by the axial wave number, changes as the rotation rate of the inner cylinder is changed. These state changes are *nonunique*. Specifically, the Taylor number at which a particular number of the azimuthal waves decreases by 1, as the Taylor number is increased, is in many cases not the same Taylor number at which the number of the azimuthal waves returns to its previous value when the Taylor number is decreased in turn. There are hysteresis loops or, in other words, there is nonuniqueness, as documented by the fact that at one specific Taylor number one can find two different solutions, depending on whether the Taylor number has been increased or decreased. For an example of such hysteresis loops see Fig. 7 of Coles (1965). There are very many nonunique state changes; the critical Taylor number for these changes were repeatable in most cases within 2 or 3%. A summary of the multitude of possibilities is shown in Fig. 13.3. At any particular Taylor number there exist sometimes 20 or more states, which can also differ in the axial wave number. The state actually observed is determined by the initial conditions of the experiment.

The question then follows whether there is, at a given $\mathcal{T}$, a "preferred state" with a particular azimuthal and axial wave number. That does not mean that a state once established through an experimental procedure will tend to change to the preferred state; once established, a stable state remains in its configuration indefinitely. The notion of a preferred state refers to the mean of the states resulting from a large number of experiments with initial conditions chosen at random so that the preferred state at some $\mathcal{T}$ is given by the statistical mean of the observations. The preferred state may also be the one chosen by the fluid when the fluid is brought quasi-steadily from subcritical conditions to a supercritical Taylor number. If this procedure does not lead to a unique solution, the preferred state would be the mean of the solutions obtained by quasi-steady increases of the Taylor number. The question of the existence of a preferred state was not answered by Coles, nor was it answered later, presumably because of the amount of work required to answer this question. In general, though, it appears from Coles' data that there is a tendency of the wavy Taylor vortices to prefer smaller numbers of azimuthal waves as the Taylor number is increased from moderate supercritical values to higher values of $\mathcal{T}$. The axial number of the vortices also tends to decrease as $\mathcal{T}$ is increased to higher values. Usually the change of the axial wavelength is accomplished by a fairly sudden disappearance of a pair of vortices near the ends of the column. It is a pair of vortices because the horizontal ends of the fluid column determine the direction of flow at the ends and this direction cannot be changed (except in the case of a free top surface). Eliminating only one vortex in the fluid column would produce two adjacent vortices with the same sense of circulation, which is not possible. Although the axial wave number can change this way, the axial wave number in wavy Taylor vortex flow is nevertheless nonunique. The possibility that supercritical Taylor vortex flow is nonunique, as indicated by the stability diagram of linear theory (Fig. 11.1), had found a startling confirmation. The extent to which the axial wavelength of supercritical flow is nonunique is much more evident in the axisymmetric case, as will be shown later.

Finally, something must be said about the tangential velocity of the waves. As mentioned, the entire wave pattern moves with uniform speed in the azimuthal direction. The wave speed was determined by Coles over a very large interval of Reynolds numbers. As the Reynolds number is increased, the tangential velocity decreases gradually; it varies from about 0.5 times the angular velocity Ω_i of the inner cylinder to about 0.35 Ω_i at large values of the Reynolds number. Then the wave speed becomes practically constant. A similar curve can be found in Cognet (1971). More elaborate measurements of the tangential wave speed were described by Fenstermacher et al. (1979) and by King et al. (1984) and will be discussed later. Coles' paper said about all that could be said about wavy Taylor vortex flow at that time.

An innovation in technology, the laser, led to the next advance in knowledge about wavy Taylor vortices. As wavy Taylor vortices move tangentially in the gap between the cylinders, a periodic variation of the velocity components of the flow occurs at a fixed point in the gap. The periodicity observed is a characteristic of the flow. The periodicity depends, however, on the coordinate system of the observer; moving with the waves the flow is actually time-independent. The period of the motion can be determined by laser–Doppler measurements or, more simply, by measurements of the frequency of the variation of the intensity of light scattered at a fixed point by reflecting particles moving with the fluid. The latter technique was introduced by Donnelly et al. (1980). A preliminary report of the results of laser–Doppler measurements of the periodicity of wavy Taylor vortices was published by Swinney (1978); the final results of these studies were published by Fenstermacher et al. (1979). The periodicities of wavy Taylor vortex flow were also measured by Krugljak et al. (1980) and L'vov et al. (1981), and with a very different technique by Bouabdallah and Cognet (1980).

Fenstermacher et al. (1979) measured the radial velocity component of the flow in a pattern with 14 axial vortices and 4 azimuthal waves. Their apparatus had a resting outer cylinder, a radius ratio $\eta = 0.877$, and an aspect ratio $\Gamma = 20$. The pattern was created through suitable initial conditions and was stable under quasi-steady changes of the Reynolds number from $R/R_c = 5.4$ to $R/R_c = 45$. In the range from $R/R_c = 1.2$ to $R/R_c \cong 10$ a primary frequency f_1 was observed which is the consequence of the azimuthal drift of the wavy vortices past a fixed point. The power spectrum of the signal from the radial velocity component contains the frequency f_1 and many higher harmonics of f_1 whose intensity decreases with increased order number n. The power spectrum is shown in Fig. 13.4a, without the higher harmonics because the first harmonic $2f_1$ is already outside of the frequency range in Fig. 13.4a. Note that the linewidth of f_1 is extremely sharp, and that the peak of the signal is about five orders of magnitude above the noise level. That means that the flow is for all practical purposes strictly periodic, in agreement with the smooth appearance of the wavy vortices in the visualization experiments.

So far the results were as expected. However when the Reynolds number of the inner cylinder was increased above $R/R_c = 10.06$, an unexpected second discrete frequency appeared in the power spectrum (Fig. 13.4b) with a power two or three orders of magnitude smaller than the power of f_1. The frequency f_2 is incommensurate with the first frequency f_1; actually the ratio f_2/f_1 increases with increasing Reynolds number. The frequency f_2 marks the appear-

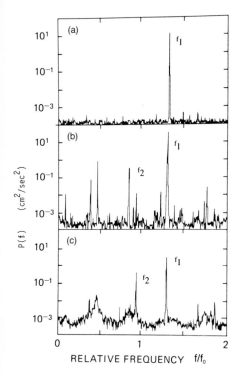

Fig. 13.4. Frequency spectra of the radial velocity component of (a) singly periodic wavy
Taylor vortex flow, (b) modulated wavy vortex flow, and (c) the beginning of the
increase of noise in the spectrum. After Fenstermacher et al. (1979).

ance of modulated wavy vortices which were identified later by Gorman and
Swinney (1982). The signal of the second frequency persisted up to $R/R_c =$
19.3, together with the first frequency, the higher harmonics of f_1 and f_2, and
combinations of f_1 and f_2 (Fig. 13.4c). As the power of the discrete frequencies decreased with increased Reynolds number, the level of noise increased
(compare Figs. 13.4a and 13.4c). The sources of the noise are small transient
disturbances which develop on the wavy vortices; they can be seen with the
naked eye. At $R/R_c = 22$ the first frequency disappeared too, which means
that the azimuthal waves ceased to exist, and they remained absent for all
higher values of the Reynolds number tried. The frequencies f_1 and f_2 were
found to be independent of r and z, but the intensities varied with r and z, as
should be expected. Since the wave speeds are independent of z, the waves
travel in phase.

Fenstermacher et al. wrote that $R = 20R_c$ "should not be identified as the
onset of chaotic flow" (1979, p. 117), rather they believe that the "first

chaotic element'' in the flow appears at $R \cong 12R_c$, with f_1 and f_2 being present. What is missing here is a definition of the term ''chaotic.'' Even if $R = 20R_c$ were defined as the onset of ''chaotic'' flow, the word chosen is a misnomer, because at $R > 20R_c$ the flow definitely retains Taylor vortices, but the word ''chaos'' refers to a state of *complete disorder*. The vortices still present above $20R_c$ can be easily seen with the naked eye, e.g. in Fig. I.4. That means that the flow remains *ordered* at $R > 20R_c$. A definition of ''chaos'' used frequently is that the flow should have a *sensitive dependence on initial conditions*. But this definition does not seem to be particularly fitting for the ''onset of chaos'' at $R \cong 20R_c$ either, because Fenstermacher et al. found that the disappearance of f_2 at $R = 19.3R_c$ was nonhysteretic. What is apparently meant by ''chaotic'' in this paper is disorder in the time dependence, similar to the ''chaotic dynamics'' in the Lorenz (1963) model of convection, to which the authors refer. However the Lorenz equations do not apply to wavy Taylor vortex flow. We shall return to chaotic flow later.

After the discovery of the second sharp frequency, the nature of the flow corresponding to f_2 remained a puzzle; the single point velocity measurements of Fenstermacher et al. were not suited to determining the spatial characteristics of this flow. The motion of the fluid which produces the second frequency f_2 was discovered by Gorman and Swinney (1982). Their experiments were made with an apparatus with $\eta = 0.88$, $\Gamma = 20$, and a resting outer cylinder; the fluid motions were made visible with suspended particles. Mirrors were used to observe the fluid column from all sides. Simultaneous frequency measurements were made with the scattered light technique. It was then found that the frequency f_2 corresponds to a modulation of the azimuthal waves. In the most simple case this amounts to a periodic flattening of the wavy boundaries of the outflow (sources). The best way to observe this phenomenon is with a video camera which rotates around the common axis of the cylinders with the rate of rotation of the (singly) periodic waves. An observer then sees a picture of waves that do not move but flatten periodically. Modulated wavy Taylor vortex flow can occur in several forms; each pattern is characterized by two integers m and k, which define the state m/k. If all waves around the circumference are modulated in phase, or flatten simultaneously, then $k = 0$; if the modulation varies with phase, then $k \neq 0$. For example, if $k = 1$ successive azimuthal waves around the cylinder flatten in sequence. A schematic representation of the modulation patterns for a five-wave state is shown in Fig. 13.5. For the 5/2 state the modulation phase angle is $4\pi/5$. The phase angle satisfies a very simple equation. To each of the m/k states belongs a specific frequency f_2. The f_2 frequency observed by Fenstermacher et al. dealt with one modulated state only, the state 4/1. The depen-

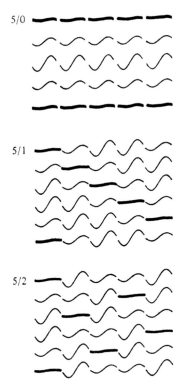

5/0

5/1

5/2

Fig. 13.5. Schematic representation of different types of modulation of a five-wave pattern in a reference system rotating with the waves. Shown are the outflow boundaries. Time increases downward in the graphs. After Gorman and Swinney (1982).

dence of the frequencies f_1 and f_2 of modulated wavy flow on the Reynolds number is so weak that we shall not discuss it here. It appears that the explanation of the second frequency with the modulation of the waves has been accepted; there is no report of a similar experiment in the literature.

The most extensive study of the wave speed – or its equivalent, the frequency – of (singly) periodic waves in Taylor vortex flow was made by King et al. (1984). The wave speed can be a function of all parameters affecting wavy vortex flow: the Reynolds number, the aspect ratio Γ of the column, the radius ratio η of the cylinders, the axial wavelength λ, and the number of waves along the circumference of the fluid column. King et al. found that the wave speed (at a given particular number m of the azimuthal waves, a specific Γ, η, and λ) first decreases with increased Reynolds number, and then reaches a plateau, just as Coles (1965) had found. Additionally they found that the wave speed increases again if $R > 18R_c$ until the waves disappear as identifiable entities. The wave speed was found to be a strong function of the radius ratio, increasing with increasing η, which means that the waves move faster in narrow gaps. The wave speed was also a strong function of the aspect ratio (for a given value of R, η, λ, m), decreasing substantially with increased aspect ratio. This is somewhat paradoxical, because it means that the waves move faster in short fluid columns, although the retarding influence of the horizontal ends of the column should be stronger in short columns. Note, however, that in these experiments the upper end of the fluid column was free. The effects of the average axial wavelength λ and of the number m of the azimuthal waves on the wave speed were small in magnitude and difficult to interpret. Summarizing the data, King et al. state that "to a good approximation the wave speed at large Reynolds number depends only on radius ratio" (1984, p. 375); the word "only" would be better replaced by "mainly." References to other measurements of the wave speed or frequencies can be found in King et al., which we cannot pursue here in detail. To summarize we can say that the results of Coles (1965) concerning the wave speed and the measurements of Fenstermacher et al. (1979) concerning the frequency of (singly) periodic waves in Taylor vortex flow have been found to hold over a wide range of R, Γ, and η and for different values of λ and m. Whereas there is a wealth of data about the frequencies or the wave speed of wavy Taylor vortex flow, there is only one paper (Bust et al., 1985) that is concerned with the amplitudes of the wavy vortices. One should, however, expect that the amplitudes of the wavy vortices are of importance for the dynamics of wavy vortex flow.

We shall now discuss the dependence of the onset of wavy vortex flow on the aspect ratio and the radius ratio. The onset of wavy flow in columns of different Γ ranging from 1 to 107 was studied by Cole (1976) with torque measurements and by visual observations as well. It was found, by both methods, that the critical Taylor number for the onset of wavy flow is strongly affected by the column length, rising sharply when Γ is reduced below 40, in fluid columns with radius ratios between $\eta = 0.954$ and $\eta = 0.894$. This is in contrast to the onset of axisymmetric vortex flow, which was found to occur at virtually the same Taylor number for all $\Gamma > 8$. Cole also noted a surprising dependence of the number of the azimuthal waves at the onset of wavy vortex flow on the aspect ratio, large columns having fewer waves than short columns. In the range of η studied by Cole the radius ratio had little effect on the onset of waves. It is, however, known from various experiments that the onset of wavy flow in columns with radius ratios of 0.7 or smaller occurs at Taylor numbers very much higher than the critical Taylor number. The dependence of the onset of wavy vortex flow on the radius ratio has not yet been studied systematically in one set of experiments. The onset of wavy vortex flow in an apparatus with radius ratio $\eta = 0.507$ and aspect ratios ranging from 10 to 17 was studied experimentally by Lorenzen et al. (1983). The onset of wavy Taylor vortex flow in finite fluid columns has been studied numerically by Edwards, Beane, and Varma (1991). Their comparison of detailed numerical and experimental studies for aspect ratios between 8 and 34 and radius ratio 0.87 showed convincing agreement of the critical Reynolds numbers, the azimuthal wave numbers, and the wave speeds.

The variation of the axial wavelength in a finite column of wavy Taylor vortices with a fixed azimuthal wave number has also been studied. Some data on this aspect of wavy vortex flow can be found in the papers of Ahlers et al. (1983), King and Swinney (1983), and Park et al. (1983). From these glimpses into a complex problem one cannot expect to find the outline of a general solution to a nonlinear problem that depends on several parameters. Walgraef (1986) treated this problem with a very elaborate amplitude equation and could establish qualitative agreement with some of the experimental results. But since the solution of his model depends on the choice of five parameters, one wonders how general these results are.

Finally we come to chaotic flow, in which there is, as mentioned, still a good deal of order. But the term "chaotic flow" will not disappear, so we want to reserve it for the range of Taylor numbers over which the flow exhibits irregular time dependence; such a flow, however, can as well be called weakly turbulent. There are, actually, very few data about the characteristics of chaotic flow, such as the azimuthal and axial wavelengths or their mean values and their standard deviations, as well as the mean wave speed and its standard deviation. Brandstater and Swinney (1987) have determined the dimension of the strange attractor of weakly turbulent Taylor vortices. The range of chaotic flow is bounded at the lower end by laminar wavy vortex flow and laminar modulated wavy vortex flow, which have regular time dependence. Chaotic flow is bounded at the upper end by turbulent Taylor vortex flow, which, as we shall see, is highly organized and for which therefore, the designation "chaotic" is simply not fitting. Chaotic flow ranges from about $100\mathcal{T}_c$ to about $1000\mathcal{T}_c$; there are, of course, no precise limits because even at the lower end the appearance of irregular time dependence depends very much on the aspect ratio, the radius ratio, and the experimental procedure. A picture of chaotic flow is shown in Fig. I.3. It is obvious in this photograph that there is still organization in the flow. The azimuthal waves are still present; their wave number is 2. The axial wavelength can also be seen easily, as well as the irregularities of the flow. The velocity signal from a fixed point in the gap can now no longer have sharp lines, but must have broader lines, and contains rapid fluctuations caused by transient disturbances moving by the point of observation, which means noise in the power spectrum. At higher values of the Taylor number ($<1000\mathcal{T}_c$) the picture remains similar: the irregularities increase and become more rapid, but the azimuthal waves and axial vortices persist. A picture of wavy, nearly turbulent flow can be found in Koschmieder (1979). If the Taylor number is increased further, the two remaining azimuthal waves gradually straighten out, and at around $1000\mathcal{T}_c$ turbulent vortices remain, which are in the mean axisymmetric. The velocity at a particular point then experiences rapid random fluctuations. Rapid random fluctuations of the velocity are usually considered a characteristic feature of turbulence. We shall discuss turbulent Taylor vortices later.

13.2   Axisymmetric Supercritical Vortices

The discovery of wavy supercritical vortex flow by Coles (1965) made the analytical investigation of nonlinear Taylor vortex flow extremely difficult, because the problem had become three-dimensional. It was, therefore, desirable to learn more about supercritical axisymmetric flow, which is certainly much more simple theoretically. The most interesting feature of supercritical flow is the axial wavelength or the size of the vortices. The first to address the question of the supercritical size of the vortices was Stuart (1958), who introduced the so-called shape assumption. From Taylor's (1923) comments concerning the flow under supercritical conditions Stuart concluded that the shape, i.e. the size, of the vortices remains unchanged above the critical Taylor number. The assumption that the wavelength of supercritical axisymmetric vortices remains constant for $\mathcal{T} > \mathcal{T}_c$ has proved to be true for a range of Taylor numbers far exceeding the slightly supercritical conditions Stuart had in mind. Supercritical axisymmetric Taylor vortex flow was later studied theoretically by Kogelman and DiPrima (1970). This study, which we shall discuss in Chapter 14, was related to the predictions of Eckhaus (1965) about nonlinear stability and dealt with the range of nonunique wavelengths at a given supercritical Taylor number.

The size or the wavelength of supercritical axisymmetric Taylor vortices was determined experimentally by Burkhalter and Koschmieder (1973) for Taylor numbers ranging from $\mathcal{T}_c$ to $80\mathcal{T}_c$. The opportunity to investigate axisymmetric supercritical flow at Taylor numbers much larger than $\mathcal{T}_c$ was indicated by Coles' (1965) observation that the onset of wavy flow occurs at much higher Taylor numbers if the radius ratio of the apparatus is about 0.7 or smaller. The experiments of Burkhalter and Koschmieder were made with an apparatus with a resting outer cylinder, with fluid columns of radius ratios $\eta = 0.727$ and $\eta = 0.505$, and with aspect ratios of either $\Gamma = 51$ or $\Gamma = 28$. A large aspect ratio is essential in order to reduce the unavoidable consequences of the ends of the fluid column on the size of the vortices. The result of the experiments was very simple. It turned out that the wavelength of axisymmetric supercritical Taylor vortices is independent of the Taylor number up to Taylor numbers of about $80\mathcal{T}_c$ in fluid columns of radius ratios 0.727 and 0.505, if the Taylor number is increased quasi steadily from subcritical Taylor numbers on.

In order to determine the wavelength in fluid columns of necessarily finite length the end effects of the column have to be taken into account. There are only two simple arrangements of the ends of a fluid column between two concentric cylinders: the end plates (which usually extend over nearly the entire gap) are attached either to the resting outer cylinder or to the rotating inner cylinder. The ends, resting or rotating, determine the direction of the flow at the boundaries. With a resting boundary the radial component of the flow is inward over the boundary; with a rotating boundary the radial component of the flow is directed outward. The size of the Taylor vortex adjacent to a resting boundary is slightly larger than the size of the critical Taylor vortices and does not vary with Taylor number, within the error of measurement. The size of the Taylor vortex over a rotating boundary increases markedly with the Taylor number, and is proportional to $(\mathcal{T}/\mathcal{T}_c)^{1/2}$. Rotating ends, therefore, increasingly compress the other Taylor vortices away from the boundaries, and make it appear that their size or wavelength decreases; for an example of this effect see Fig. 2 of Burkhalter and Koschmieder. However this apparent decrease of the wavelength of the vortices is caused only by the boundary condition. The rotating end boundary is, therefore, not well suited for Taylor vortex experiments and has fallen into disuse.

If in a fluid column with, say, resting end boundaries the Taylor number is slowly increased and the number of vortices remains constant, the wavelength has been found to be constant within the experimental error, because the average wavelength in a finite fluid column, including the end vortices, is determined by the formula $\lambda_{av} = 2L/Nd$, where L is the length of the column and N the number of all vortices. Such measurements have a systematic uncertainty because of the quantization condition, i.e. the condition that only integer numbers of vortices can fill the column. The systematic uncertainty $\Delta\lambda$ of the wavelength in percent is given by the formula $\Delta\lambda/\lambda_{av} = \pm 2 \times 100/N$, where N is again the number of the vortices. The systematic uncertainty of the wavelength in the experiments of Burkhalter and Kosch-mieder with the aspect ratio $\Gamma = 51$ was $\pm 4\%$. There is, additionally, an uncertainty of λ resulting from the standard deviation of the measurements of the sizes of all vortices, which was of the order of 1% in these experiments. The main contribution to this uncertainty comes from the deviation of the size of the end vortices from the size of all other vortices, which are very nearly all of the same size.

Burkhalter and Koschmieder concluded that the wavelength of axisymmetric vortices with large radius ratios is independent of the Taylor number in fluid columns of *infinite* length. One arrives at that conclusion by extrapolation of measurements of the vortex sizes as a function of $\mathcal{T}$ in fluid columns of different length. Burkhalter and Koschmieder observed that, the longer the column, the less the vortex size varied, simply because the contribution of the end effects decreases. Extrapolated, this means that in columns of infinite length the size of the vortices does not vary with $\mathcal{T}$. The simple fact that the wavelength of supercritical axisymmetric Taylor vortices in infinite fluid columns is independent of the Taylor number if the Taylor number is increased quasi steadily has not yet been explained theoretically.

Two earlier measurements of the wavelength of supercritical Taylor vortices have been reported. Donnelly and Schwarz (1965) found that the wavelength increases with $\mathcal{T}$ but did not realize that they were dealing with wavy vortex flow. Their apparatus had small gaps with radius ratios of 0.95 to 0.85, and under these circumstances wavy vortex flow develops at slightly supercritical Taylor numbers. As Coles (1965) observed, the (nonunique) axial wavelength of wavy vortex flow tends to increase with increased Taylor number and that is what Donnelly and Schwarz observed also. Snyder (1969) reported a very small decrease in the wavelength of the order of $d\lambda/d(\mathcal{T}/\mathcal{T}_c) \cong 10^{-4}$. This result was due to an erroneous interpretation of the data. Snyder's results apply to constant numbers of vortices, so the wavelength in his experiments did not vary, within the principal accuracy of his measurements.

The independence of the wavelength from the Taylor number in supercritical Taylor vortex flow is in marked contrast to the increase of the wavelength with increased Rayleigh number in supercritical Rayleigh–Bénard convection. We see here the first example of the difference in the nonlinear behavior of Taylor vortex flow and Rayleigh–Bénard convection. Previously we found that in the linear domain both stability problems are completely analogous. But in the nonlinear regime both problems go separate ways, as we pointed out before in Fig. 6.4. We note, however, that the wavelength of Taylor vortices is constant only as long as the flow is axisymmetric. As Coles (1965) observed, the axial wavelength of wavy vortices can increase, but the increase is nonunique. On the average the wavelength of wavy vortex flow does indeed increase with $\mathcal{T}$, and increases up to $\lambda \cong 4$ at Taylor numbers of around $100\mathcal{T}_c$ (see Fig. 13.9). Turbulent Taylor vortices also have wavelengths which are larger than the critical wavelength. For the wavelength of Taylor vortices to increase it is apparently necessary that the flow be three-dimensional. This is not so with Rayleigh–Bénard convection; the wavelength of axisymmetric convection rolls increases, as is shown in Fig. 6.2.

The question of the uniqueness or nonuniqueness of supercritical flow is, after the wavelength, the second most important aspect of nonlinear axisymmetric Taylor vortex flow. As we have seen wavy Taylor vortex flow is nonunique. Nonuniqueness of axisymmetric supercritical flow was first proved to exist by Snyder (1969). He worked with an apparatus with two independently rotating cylinders and showed that the number of vortices, and hence the wavelength, established when the outer cylinder rotated remained unchanged when the rotation rate of the outer cylinder was reduced to zero. Since the critical wavelength or the number of vortices in a given column is, according to linear theory, a function of the ratio of the rotation rate of both cylinders (see Fig. 11.5) one can, at the onset of instability, create different numbers of vortices in the same column at different values of $\mu = \Omega_2/\Omega_1$. Then, by slowly reducing Ω_2, one can bring the pattern to $\mu = 0$, and Snyder found that the number of vortices did not change when this was done. According to linear theory another number of vortices is formed in the same column at the onset of instability with $\mu = 0$, when the Taylor number is slowly increased from subcritical values on. Twenty-two vortices created when the outer cylinder was co-rotating with the inner cylinder remained thus 22 vortices when $\mu \rightarrow 0$, although 23 vortices were the solution with $\mu = 0$ when the rotation rate of the inner cylinder was increased quasi-steadily to the critical condition and then to supercritical Taylor numbers. Snyder showed also that 26 or 24 vortices created when the cylinders counter-rotated remained either 26 or 24 vortices when the rotation rate of the outer cylinder was reduced to zero. So he showed that, at one particular supercritical Taylor number of the inner cylinder with $\mu = 0$, he could produce 26, 24, 23, or 22 vortices in one and the same fluid column, depending on the initial conditions he used. That is proof of nonuniqueness.

Another set of experiments on the nonuniqueness of supercritical axisymmetric Taylor vortex flow was made by Burkhalter and Koschmieder (1974). They determined the range of nonuniqueness of the wavelength over a large interval of supercritical Taylor numbers and compared their results with the range of nonuniqueness predicted by Kogelman and DiPrima (1970).

Burkhalter and Koschmieder (1974) used an apparatus with a resting outer cylinder and created wavelengths shorter than the critical wavelength through sudden starts of the inner cylinder. In this procedure the rotation rate of the inner cylinder is brought from rest to a final steady supercritical value within a second. The fluid then has a choice in the selection of the wavelength, which no longer has to be the critical wavelength, but can be any wavelength unstable according to the stability diagram of nonlinear theory. The steady wavelengths after sudden starts are usually much shorter than the critical wavelength. The wavelength once selected by the fluid after a sudden start remains indefinitely. The time-dependent formation of the vortices at the inner cylinder during a sudden start was studied experimentally by Kirchner and Chen (1970).

The wavelength established after time-dependent starts is a function of the Taylor number at which the start is made and of the acceleration rate of the inner cylinder. The shortest wavelength at a given supercritical $\mathcal{T}$ is produced by a stepwise increase of Ω (sudden start). With different acceleration rates of the cylinder, all the wavelengths between the wavelength resulting from the stepwise increase of Ω at one particular $\mathcal{T}$ and the critical wavelength resulting from a quasi-steady increase of Ω can be reached. For a graph of the possible wavelengths as a function of $\mathcal{T}$ see Fig. 14.1. The maximal width of the range of nonunique wavelengths produced by sudden starts extends from $\lambda \cong 1.5$ to $\lambda_c = 2.014$ at a Taylor number of about $10\mathcal{T}_c$. With infinite cylinders this range of nonuniqueness would be filled with a continuum of possible steady states, corresponding to the prediction of a continuum of supercritical states by linear theory (Fig. 11.1) or by nonlinear theory (Fig. 14.1). With cylinders of finite length only integer numbers of vortices are accessible because of the quantization condition; the longer the cylinders the finer will be the sequence of possible states.

Just as the critical wavelength does not change when the Taylor number is increased slowly after the onset of instability, so does the wavelength once selected by the fluid after a sudden start not change if the rotation rate of the inner cylinder is changed slowly afterward, as shown in Fig. 9 of Burkhalter and Koschmieder (1974). The wavelength once picked changes only when certain stability limits are reached, which are, on the lower side of $\mathcal{T}$, the Eckhaus boundary, which we shall discuss in Chapter 14, and, on the upper side of $\mathcal{T}$ the wavy vortex instability.

The stability diagrams of linear and weakly nonlinear theory predict the existence not only of a continuum of supercritical vortices with wavelengths $<\lambda_c$, but also a continuum of wavelengths $>\lambda_c$ (see Fig. 11.1 for the stability diagram of linear theory). Wavelengths longer than λ_c were created by

Fig. 13.6. Illustration of the nonuniqueness of supercritical axisymmetric Taylor vortex flow at $\mathcal{T} = 9.1\mathcal{T}_c$ in an apparatus with $\eta = 0.727$ and $\Gamma = 51$ and resting end plates. (*Left*) $\lambda = 1.53$ after a sudden start to $\mathcal{T} = 9.1\mathcal{T}_c$; (*middle*) $\lambda = 2.02$ after a quasi-steady increase of the rotation rate from $\mathcal{T} < \mathcal{T}_c$ to $9.1\mathcal{T}_c$; (*right*) $\lambda = 2.405$ after a filling experiment at $\mathcal{T} = 9.1\mathcal{T}_c$. Visualization with aluminum powder. The intense dark lines mark the sinks, the location of inward radial motion; the weak dark lines mark the sources, the location of radial outward motion. After Burkhalter and Koschmieder (1974).

Burkhalter and Koschmieder with filling experiments, in which the fluid slowly filled the gap while the inner cylinder rotated with a steady supercritical rotation rate. The Taylor vortices are stretched when they are formed, and this elongation of the vortices is preserved when the column is filled and the filling flow is stopped. The long vortices remained indefinitely. Since nonuniqueness is such an important aspect of nonlinear Taylor vortex flow, we illustrate it in Fig. 13.6.

The interval of nonunique wavelengths observed at Taylor numbers ranging from $\mathcal{T}_c$ to about $80\mathcal{T}_c$ is much smaller than the interval of nonunique wavelengths predicted by linear theory and also the smaller interval predicted by the weakly nonlinear theory of Kogelman and DiPrima (1970). We shall compare the experimental results with the theoretical findings in Chapter 14, Fig. 14.1. For Taylor numbers $>80\mathcal{T}_c$ sudden-start experiments produce wavelengths which are equal to or greater than the critical wavelength, because the flow is then three-dimensional. The wavelengths resulting from sudden starts were also observed with a very different technique by Takeda et al. (1990) and are in agreement with the earlier observations. Numerical studies of time-dependent Taylor vortex formation by Neitzel (1984) provided results similar to the experimental results.

We have mentioned before that the wavelength of supercritical Taylor vortices and that of supercritical Rayleigh–Bénard convection vary in completely different ways when the control parameter is increased quasi steadily. The unambiguous nonuniqueness of supercritical Taylor vortices, whether they are axisymmetric or wavy, is also in clear contrast to the apparent uniqueness of supercritical Rayleigh–Bénard convection. As discussed in Section 6.1 the wavelength of moderately supercritical time-independent convection seems to be independent of the initial conditions, although definitive proof of this fact has not yet been provided. We recall also that supercritical surface-tension-driven Bénard convection seems to have a unique wavelength; on the other hand, the variation of its wavelength is opposite to the variation of the wavelength of Rayleigh–Bénard convection. These observations taken together show that the nonlinear behavior of the three instabilities differs from case to case, and that it is unlikely that they can be treated theoretically by one general nonlinear stability theory.

13.3   End Effects

The horizontal boundaries of the fluid column, whose effects on nonlinear flow we referred to earlier, also make themselves felt at subcritical values of the Taylor number. Vortices adjacent to the end plates form far below $\mathcal{T}_c$. When the subcritical Taylor numbers are gradually increased, additional vortices form on top of each other until the vortices coming from symmetric top and bottom plates meet in the middle of the column. The Taylor number is then at its critical value. However in long columns the formation of the vortices in the center section of the column occurs practically simultaneously at or very near the critical Taylor number, in essential agreement with linear theory of the instability. In other words, the results of linear theory of the Taylor vortex instability with infinite columns can be reproduced with good accuracy if only the column is made long enough. The spontaneous onset of the Taylor vortices in the entire column does, however, not occur in the finite columns experiments are made with. The end plates are finite disturbances in circular Couette flow and destabilize the fluid first. This corresponds completely to the observations of the onset of Rayleigh–Bénard convection in bounded fluid layers. Although the premature formation of Taylor vortices at the end plates had been noted in several early experiments (e.g. Coles, 1965), the formation of the vortices in finite columns was really elucidated in a numerical study by Alziary de Roquefort and Grillaud (1978). The sequential appearance of the vortices is convincingly demonstrated in one of their figures, which we reproduce here as Fig. 13.7. Note that the finite amplitude vortices in the last of the graphs come in pairs. Larger values of the radial velocity are where the flow is directed outward, and smaller radial velocities are where the flow is directed inward. The appearance of pairs of vortices in finite amplitude flow is very characteristic and can be seen easily, e.g. in Fig. 13.6. Alziary de Roquefort and Grillaud were also able to show that "by using different initial conditions it was possible to obtain, for the same problem, several steady state solutions differing by the number of cells" (1978, p. 267). The essence of these numerical results was verified experimentally with $\Gamma \approx 18$ and $\eta = 0.506$ by Pfister and Rehberg (1981) and with $\Gamma \approx 20$ and $\eta = 0.747$ by Heinrichs et al. (1986).

If the length of the fluid column is not close to an even integer multiple of the gap width, the vortices may find it difficult to match when the column of vortices coming from the bottom boundary meets the column of vortices coming from the top boundary, as was observed experimentally by Park and Donnelly (1981). Walgraef et al. (1982) have described this with a time-dependent Ginzburg–Landau model. In the end, however, the column is always filled with a perfectly stable pattern of vortices of the same size, except for the end vortices. In short columns similar effects were observed first by Benjamin (1978b) and later by Mullin (1982). When the column length deviates slightly from an integer multiple of the critical vortex length and the Taylor number is increased slowly then, as Mullin wrote:

> the cells develop from the ends, but fail to mesh together in the central region which appears to consist of stagnant fluid. . . . As the speed is further increased, one observes either the sudden appearance of a pair of cells in the stagnant region or the sudden disappearance of the region altogether. Then

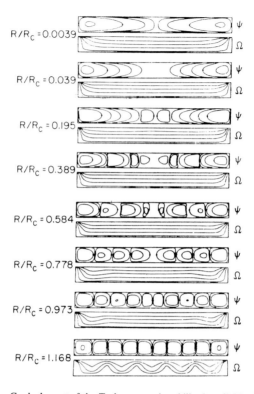

Fig. 13.7. Gradual onset of the Taylor vortex instability in a fluid column with aspect ratio $\Gamma = 10$ and radius ratio $\eta = 0.933$, resting outer cylinder and rotating end plates. Shown are the stream function ψ and the angular velocity Ω of the fluid. The pattern finally consists of 10 vortices. The fluid columns are actually vertical. After Alziary de Roquefort and Grillaud (1978). By permission of Pergamon Press, Inc.

there exists a definite set of cells along the entire length of the cylinders. Upon reduction of speed the structure reverts to its original form abruptly at a lower speed. Thus a definite hysteresis is measured. (p. 211)

This experimental result was illustrated in various bifurcation diagrams which have a cusp. One wonders whether it is correct to draw these diagrams with the cusps, because three very different initial conditions are represented by one curve. If one sticks with one particular initial condition, e.g. the bifurcation to six vortices in Fig. 2 of Mullin (1982), this curve must extend to $\Gamma = 6$, because with this aspect ratio six vortices form under quasi-steady conditions as well as after sudden starts at $\mathcal{T}_c$. Similarly in the same diagram the bifurcation to four vortices must extend to $\Gamma = 4$, because four vortices form at this aspect ratio at the critical condition.

The shorter the fluid columns are, the stronger should be the influence of the end plates. Blennerhassett and Hall (1979) showed theoretically that, in the linear case, the critical Taylor number increases substantially as the length of the column is reduced to aspect ratios of order 1, just as with the critical Rayleigh number in very small containers in Rayleigh–Bénard convection (Davis, 1967). Hall (1980a) proved the same in the weakly nonlinear case. In very short columns unexpected effects were found by Benjamin (1978a,b), in particular the so-called anomalous modes. These are axisymmetric Taylor vortices of *odd* number which fill short fluid columns although the top and bottom end plates are of the same (nonrotating) type; an example of such a flow is shown in Fig. 13.8b. The fluid flow in this case is anomalous because it is asymmetric with respect to the horizontal plane through the middle of the fluid column, although the boundary conditions are symmetric about the midplane. There is also another type of anomalous flow, namely flow with *even* numbers of vortices with a circulation which is *opposite* to the normal circulation of even-numbered vortices in a column with nonrotating top and bottom end plates (see Fig. 13.8c). This flow is anomalous because the direction of motion in the midplane is opposite to the direction the flow should have in a column with resting end plates. Normally the flow over nonrotating end plates is inward (Burkhalter and Koschmieder, 1973), but with anomalous even-numbered vortices the flow near the nonrotating end plates is outward.

Before we discuss anomalous Taylor vortex flow further, we should be specific about the conditions under which anomalous vortices can be established. Anomalous vortices in short fluid columns with nonrotating end plates can be created by sudden starts at supercritical Taylor numbers, but not by quasi-steady increases of the Taylor number from subcritical values. Anomalous vortices have not been created in fluid columns with rotating end plates; the rotating end boundary condition is so strong that vortices with a circulation opposite to the normal circulation over rotating ends evidently are not possible. For a recent study of Taylor vortex flow in short fluid columns with rotating end plates and a wide gap see Tavener et al. (1991). On the other hand, the resting end boundary condition is apparently so weak that a reversal of the direction of flow over the end is possible under certain circumstances. It has been said (Mullin, 1982) that fluid columns with odd numbers of vortices created by changes of the length (or the aspect ratio Γ) of the fluid column are anomalous, but in these cases the odd number of vortices is not really anomalous because the column is not symmetric about the midplane when the length of the column is varied by a one-sided motion of one end plate. It is perfectly normal for the column to be filled with odd numbers of vortices if the boundary conditions at the ends are not the same. If the length of a column of Taylor vortices is varied, the fluid has really no choice in the selection of the wavelength of the motion, but after a sudden start the fluid has a choice in the selection of the wavelength out of the continuous spectrum of possible wavelengths.

Benjamin (1978b) worked with a fluid column of aspect ratio $\Gamma = 3.25$ and radius ratio $\eta = 0.615$, a resting outer cylinder, and resting top and bottom end plates. In this apparatus two vortices appeared when the Reynolds number was increased quasi steadily from subcritical on. Two vortices also appeared after sudden starts to low supercritical values of the Reynolds number. At around $3.4R_c$ the axisymmetric anomalous three-vortex pattern appeared when the inner cylinder was started suddenly. The vortices are then of unequal size; there is a pair of vortices of the same size and a larger vortex with outflow near one end plate, as shown in Fig. 13.8b. There are two types of three-vortex patterns which should appear with the same probability; the anomalous vortex with outflow near a resting end plate is either over the bottom plate or under the top plate. The anomalous three-vortex patterns are perfectly stable. The narrow range of Reynolds numbers in which three-vortex flow could be created was followed by a wider range in Reynolds numbers in which anomalous four-vortex flow was created, which in turn was followed by another range of anomalous three-vortex flow. At even higher Reynolds numbers Benjamin's data are sketchy. But it is clear from these experiments that the anomalous vortices could be created only in certain limited ranges of highly supercritical flow and through sudden starts. These experiments were limited to one aspect ratio and one radius ratio. One should, however, expect a strong dependence of the conditions for the formation of anomalous vortex flow on Γ and η. Experiments on the dependence of anomalous flow on the aspect ratio were made by Bielek and Koschmieder (1990). Working with an apparatus with $\eta = 0.605$ they found anomalous three-vortex flow with aspect ratios of 3.0, 3.25, and 3.75, but no anomalous vortices at all with an aspect ratio of 3.5. It was also learned that the formation of the anomalous three-vortex flow was closely linked to the existence of the anomalous four-vortex flow. The three-vortex patterns never formed directly after a sudden start, but were the decay product of an anomalous four-vortex pattern which formed initially after the start. This is supported by the observation that the three-vortex flows form at Reynolds numbers just below or just above the range of Reynolds numbers for the anomalous four-vortex flow. The real puzzle is the anomalous four-vortex flow without which there may be no anomalous three-vortex flows. Without going into detail we note that it was also found that in fluid columns with a significantly smaller gap ($\eta = 0.885$) no anomalous vortices were found for aspect ratios between 3 and 4.5.

There has been strong interest in the anomalous vortices, which was fueled in part by the interpretation of these experiments as showing that the conventional theory of the Taylor vortex instability was inadequate. To quote Benjamin (1978b), "A necessary part of the task has been to play down the standard theory of the Taylor experiment, . . . whose shortcomings, particularly its conceptional limitations, need to be recognized" (p. 42). Or to quote Benjamin and Mullin (1982), "In our view the presence of the ends bears crucially on what happens in a Taylor apparatus, however long" (p. 220), or Lorenzen and Mullin (1985), "the finite length of the cylinders . . . has been recognized as a dominating effect" (p. 3463). I do not share this view. For the classical linear Taylor vortex problem in a narrow, long gap the ends of the column do not bear crucially on what happens in the column. We recall that Taylor (1923) worked with fluid columns of aspect ratios of 100 or more and verified the onset of instability with an accuracy of the order of a percent. We should be able to do a little better now, but even a percent deviation from the theoretical prediction cannot be called a crucial difference. Similarly it has been found that the critical wavelength of the motions in finite columns is in the percent range in agreement with theory for infinite columns. Anomalous Taylor vortices, which are not described by the standard theory of Taylor vortex flow, actually occur only under very particular supercritical conditions. At present we do not have a theoretical prediction for the value of the wavelength at a given supercritical Taylor number. However the observed wavelengths of anomalous supercritical Taylor vortices fall well into the range of wavelengths possible according to the weakly nonlinear theory of Kogelman and DiPrima (1970), which deals with columns of infinite length. As all experiments with axisymmetric supercritical Taylor vortices show, the wavelength of the flow in finite but long columns is very nearly the same over the entire length of the column except for the vortices adjacent to the ends. That can be easily seen with the naked eye (see, e.g., Fig. 13.6). Measured individual vortex sizes in columns with 34 or 50 vortices can be found in Table 2 of Burkhalter and Koschmieder (1973). The uniformity of the size of the vortices means that the end boundaries dominate only the end vortices, but are insignificant for the rest of the column. To put this in other words, away from the ends the standard theory of Taylor vortex flow describes what is happening very well.

The significance of the ends will certainly increase in short fluid columns, and anomalous vortices can then be observed. However Lorenzen and Mullin

(1985) maintain that anomalous modes can also be created in long columns with aspect ratios up to 47. These anomalous vortices were created in fluid columns with rotating ends which produced, as is usual for rotating ends, end vortices with outflow over the end plates. The rotation of the end plates was then gradually brought to a halt while the rotation rate of the inner cylinder was held constant. Nevertheless the vortices adjacent to the ends maintained their previous direction of circulation, which means that their circulation had become "anomalous" with respect to the usual direction of circulation over a resting end. It is testimony to the astonishing stability of supercritical axisymmetric Taylor vortices that the flow in the column had not changed in spite of the drastic change of the boundary condition at the ends, but this does not prove something anomalous; the fluid was symmetric about the midplane of the column before and after the change of the end conditions. Under these conditions the flow is anomalous only if one makes the direction of the flow over the resting ends the criterion for anomalous flow, but this criterion is misleading because, as we shall see later, there are always minor vortices at the ends which make the direction of motion over the ends ambiguous. The criterion using the direction of the flow at the boundary deals exclusively with a complication due to end effects and is of little significance for the Taylor vortex problem as a whole. Columns with rotating ends are never anomalous but satisfy just as well the theoretical results for infinite fluid columns. So the anomalous flows depend critically on the presence of *resting* ends (and wide gaps), which means they are a complication due to specific end effects. To summarize, it appears that in long fluid columns of aspect ratio of the order of 10 or more the Taylor vortex instability is the dominant feature and that the onset of instability and supercritical flow can be described successfully by studies of the flow between rotating cylinders of infinite length.

A fairly large number of analytical as well as numerical studies of Taylor vortex flow in short fluid columns have been made. This work began with Benjamin's (1978a) qualitative discussion of the possible bifurcations and the dependence of the number of vortices on the length of the column. This line of reasoning was continued by Schaeffer (1980). He introduced a parameter τ by which, when $\tau = 0$, free horizontal boundaries are represented for which the problem has its well-known solution, whereas $\tau = 1$ describes the case of realistic fixed boundaries for which we have no solution. Schaeffer assumed that small values of τ ($\tau \ll 1$) would suffice to show the qualitative properties of the problem with fixed boundaries. Hall (1980b) studied Schaeffer's model in the narrow gap case with asymptotic methods in order to find whether two or four cells (i.e. vortices) develop in short columns when the Reynolds number is increased quasi steadily. Hall (1982) continued this work for a wide range of radius ratios using a perturbation expansion that could handle time-dependent perturbations. In addition to flows with even numbers of cells he also investigated asymmetric flows, as exemplified by the three-vortex pattern. His results suggest that such flows are probably not accessible by slowly increasing rotation rates of the inner cylinder, as is borne out by the experiments.

These studies were followed by numerical investigations. Cliffe (1983) studied single-cell and two-cell flows, of which the single cell is necessarily anomalous, because the direction of the circulation at both boundaries cannot be the same, although both boundaries are of the same type. Lücke et al. (1984) studied Taylor vortex flow with aspect ratio $\Gamma = 1.05$, numerically and experimentally. An experimental investigation in a similar aspect ratio range was made by Pfister et al. (1988). Cliffe and Mullin (1985) studied two-, three-, and four-cell flows numerically and experimentally. There seem to be noticeable differences between the photographs of the flows and the streamlines computed in this paper. It appears from the photographs that there is strong coupling of the flow in pairs of vortices, which is, however, not apparent in the streamlines. The bifurcations of one to eight cell patterns were studied experimentally by Nakamura et al. (1990). Four- and six-cell flows were studied by Cliffe (1988), and flows with cells ranging from two to six were computed by Bolstad and Keller (1987). The latter paper makes it very obvious that anomalous vortices, whether odd-numbered or anomalous even-numbered, are accompanied by minor vortices at the horizontal boundaries of the column, where the flow seems to be outward, but should be inward. The minor vortices correct this by a circulation which is directed inward at the (resting) boundary. Considering the existence of the minor vortices, it follows that "all suggestions of 'wrong' direction of rotation or of odd numbers of vortices disappear" (p. 231). Two examples of anomalous Taylor vortices with minor vortices computed by Bolstad and Keller, together with the streamlines of a normal two-cell flow, are shown in Fig. 13.8. The minor vortices appear regularly in experiments and can be seen easily. The first mention of minor vortices occurred in a one-line sentence of Benjamin (1978b). They could also be seen in Cliffe (1983) with a one-cell flow, in Lücke et al. (1984), and in Cliffe and Mullin (1985) with three-, four-, and five-cell solutions. It is of importance to realize the existence of these vortices; they ensure that in all cases the direction of motion of the fluid over a resting end plate is inward if the inner cylinder rotates. On the other hand, the minor vortices are usually so small that one is justified in disregarding them when the number of vortices in the column is counted. The existence of the minor vortices does not change the anomalous character of a pattern; e.g. a

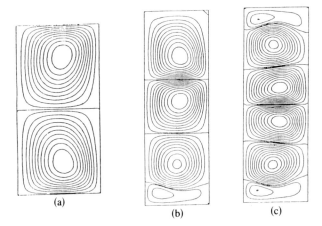

Fig. 13.8. Computed streamlines of Taylor vortex flow in short fluid columns. (a) Normal two-cell mode, $R = 240$, $\Gamma = 2$, $\eta = 0.615$. (b) Anomalous three-cell mode with a minor vortex at the bottom, $R = 240$, $\Gamma = 3.0$, $\eta = 0.615$. (c) Anomalous four-cell mode with minor vortices at top and bottom. $R = 283.9$, $\Gamma = 2.886$. After Bolstad and Keller (1987).

three–vortex pattern is asymmetric about the midplane of the column regardless of whether the minor (fourth) vortex is considered or not. As I see it, the results of the numerical investigations support the view that anomalous Taylor vortices are not an intrinsic feature of Taylor vortex flow but the consequence of a complication of the problem brought about solely by the presence of the horizontal boundaries of the fluid column.

13.4   Turbulent Taylor Vortices

Before we discuss turbulent Taylor vortices we must first define what we mean by turbulent. Turbulence, when discussed in the following, refers to a flow which is characterized by rapid, random variations of the velocity at any point in the fluid. Rapid means on a time scale that is short compared with the viscous relaxation time d^2/v. To put this in another way, we consider motions at very high values of the control parameter, either the Taylor number or the Reynolds number, with Taylor numbers of the order of, say, $1000\mathscr{T}_c$ or more. Under these conditions it might be possible to simplify the Navier–Stokes equations significantly with approximations using very high values of the control parameter. The first indication that ordered Taylor vortices can still exist at very high Taylor numbers came from Pai's (1943) experiment. He worked with air in an apparatus with $\Gamma = 20$ or 10 with a gap with $\eta = 0.937$ or 0.881 in which very high Taylor numbers were reached by high rotation rates of the inner cylinder. He measured the velocity distribution and the pressure field, much as Wendt (1933) did. He also determined qualitatively the form of the motions and found "that ring-shape vortices still exist at Reynolds numbers as high as several hundred times the critical Reynolds number" (p. 10). From the number of vortices that he mentions it follows that the size of the Taylor vortices that he observed must have been substantially larger than the size of critical Taylor vortices. Pai also observed some nonuniqueness of the flow. Another indication of the existence of turbulent Taylor vortices appeared in the photographs of Schultz-Grunow and Hein (1956), but no notice was taken of the implication of these pictures. The existence of turbulent Taylor vortices was then established by Koschmieder (1979) and by Barcilon et al. (1979).

It is possible to reach very high Taylor numbers with modest rotation rates via the dependence of the Taylor numbers on d^4. Turbulent Taylor vortices obtained at modest rotation rates in large gaps are afflicted by the same problems which we have encountered with convection experiments made at very high Rayleigh numbers in deep fluid layers. What one actually observes in either case is weak turbulence. In order to obtain well-organized turbulent Taylor vortices one has to work with comparatively narrow gaps ($\eta > 0.5$) and relatively large rotation rates. The results concerning turbulent Taylor vortices to be discussed in the following apply strictly only to experiments with narrow gaps.

The experiments of Koschmieder (1979) were made with a long apparatus with a narrow gap and a resting outer cylinder, with $\Gamma = 123$ and $\eta = 0.896$, and extended from the just critical to Taylor numbers of the order of $10^3 \mathcal{T}_c$, as shown in Figs. I.1–I.4. Two types of experiments were made: either the rotation rate of the inner cylinder was increased slowly with a steady acceleration rate of 7×10^{-4} rad/sec^2 (or less), or the rotation rate was brought through a sudden start to a final steady value. In the slow-acceleration experiments, steady axisymmetric Taylor vortices were observed at the critical Taylor number (see Fig. I.1). When the Taylor number was increased, wavy Taylor vortices formed (see Fig. I.2). In these experiments the axial wavelength of the motions increased on the average. At $7\mathcal{T}_c$ (Fig. I.2) the axial wavelength was already 20% above the critical wavelength $\lambda_c \cong 2$; however the flow at such values of $\mathcal{T}$ was clearly nonunique, which means that the axial as well as the azimuthal wavelength differed at the same $\mathcal{T}$ depending on the initial conditions, as we know from Coles' (1965) work. When the Taylor number was increased further, the axial wavelength increased further, on the average, and so did the azimuthal wavelength; the latter means that the number of the azimuthal waves decreased. At around $100\mathcal{T}_c$ the axial wavelength reached a maximum at $\lambda \approx 3.4$, much larger than the critical wavelength $\lambda_c \cong 2$. The number of azimuthal waves had then decreased to two, and was apparently unique. Sudden-start experiments at $100\mathcal{T}_c$ produced vortices which had, within the experimental error, the same axial wavelength $\lambda \approx 3.4$ as the vortices after slow acceleration. And after sudden starts at $100\mathcal{T}_c$, the vortices also always had two azimuthal waves.

There is a remarkable similarity between turbulent Taylor vortices, which are, in the mean, axisymmetric, and laminar supercritical axisymmetric vortices. The independence of the wavelength of turbulent vortices from the Taylor number at $\mathcal{T} > 1000\mathcal{T}_c$ is reminiscent of the constancy of the wavelength of laminar axisymmetric supercritical vortices after slow increases of the Taylor number in large gaps, as discussed in Section 13.2. After sudden starts above $1000\mathcal{T}_c$, the axial wavelength of turbulent vortices was noticeably smaller than $\lambda \approx 3.4$. This means that the vortices were nonunique, as shown in Fig. 13.9. This behavior is qualitatively the same as the behavior of laminar axisymmetric supercritical vortices in experiments with large gaps, in which sudden-start experiments produce wavelengths shorter than λ_c, as was shown by Burkhalter and Koschmieder (1974). In an infinite column, a continuum of possible wavelengths of the turbulent vortices would exist, with the longest wavelength being produced by quasi-steady acceleration experiments and the shortest by sudden starts. The wavelengths in between those resulting from sudden starts ($\lambda \approx 2.4$) and those resulting from quasi-steady acceleration experiments ($\lambda \approx 3.4$) can be realized through experiments with different acceleration rates. This is perfectly analogous to the observations with moderately supercritical axisymmetric flow. We also note that the wavelength of turbulent, in the mean axisymmetric, vortices produced by sudden starts remains constant when the Taylor number is increased afterward to $\mathcal{T} > 1000\mathcal{T}_c$, just as is the case when the Taylor number of laminar supercritical axisymmetric Taylor vortices produced by sudden starts is increased.

In the range from $100\mathcal{T}_c$ to $1000\mathcal{T}_c$ two long azimuthal waves remained which gradually straightened out when $\mathcal{T}$ was increased, and at $1000\mathcal{T}_c$ the flow appeared to be in the mean axisymmetric, though turbulent. The sequence of events is best demonstrated by the graph in Fig. 13.9. As we discussed in Section 13.1 the range from around $100\mathcal{T}_c$ to around $1000\mathcal{T}_c$ is most suitably called the chaotic regime.

Above around $1000\mathcal{T}_c$ order emerged from chaos. Although the flow was clearly turbulent in the sense defined at the beginning of this section, on the average, or on a time scale that was long compared with the viscous relaxation time, the mean value of the velocity at a given point in the fluid was constant ($\pm$ the standard deviation of the velocity fluctuations), as docu-

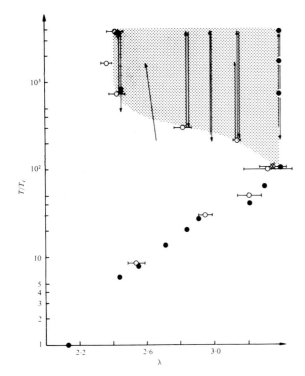

Fig. 13.9. Wavelengths of turbulent Taylor vortices. Solid circles show the results of steady acceleration experiments; open circles show the results of sudden-start experiments. The average value of 10 steady acceleration experiments to $100\mathcal{T}_c$ is marked by the crossed-out circle. The shaded area above $100\mathcal{T}_c$ marks the region of nonunique flow. Arrows in the nonunique area indicate shifts of the Taylor number after a sudden start had been made. After Koschmieder (1979).

mented by the formation of a permanent unchanging pattern. The flow was characterized by an axial wavelength $\lambda \approx 3.4$ after slow-acceleration experiments (see Fig. 5 of Koschmieder, 1979) and an axial wavelength $\lambda \approx 2.4$ after sudden starts (see Fig. I.4). In either case the wavelength once established was independent of the Taylor number when the Taylor number was increased further, and the wavelength was uniform over the greatest part of the column away from the (resting) end plates, with an accuracy of $\pm 4\%$ for a column with 82 vortices. We note specifically that even after sudden starts the wavelength of the turbulent vortices was larger than the critical wavelength of linear theory, and after slow-acceleration experiments the wavelength of the turbulent vortices was 1.7 times as large as the critical wavelength. This says that the size of Taylor vortices increases if the Taylor number is increased to very high values. The increase of the wavelength of

Taylor vortex flow at these extremely high Taylor numbers is reminiscent of the increase of the wavelength of Rayleigh–Bénard convection which already occurs, however, at slightly supercritical values of the Rayleigh number. The difference in the condition for the growth of the wavelength may be understood if one considers that in Rayleigh–Bénard convection the disturbances grow in an originally resting fluid, whereas in Taylor vortex flow the disturbances grow in Couette flow, with which the disturbances can apparently compete only when the Taylor number is very high.

The experiments of Mobbs described in Barcilon et al. (1979) were made with an apparatus with aspect ratio $\Gamma = 65$ and radius ratio $\eta = 0.908$, which had a resting outer cylinder and end plates attached to the rotating inner cylinder. As shown by the photographs in Barcilon et al. the flow at very high Taylor numbers exhibited a periodic structure of Taylor vortices which were in the mean axisymmetric. They wrote, "The striking feature of the flow patterns for increasing T/T_c from about 400 upwards to at least 80 000 . . . is the stability of the form of the visible Taylor cells" (p. 460). This is qualitatively the same result as in Koschmieder (1979); however the wavelength of the vortices was not measured. Emphasis in Barcilon et al. is on the "herringbone"-like patterns of streaks at the outer wall converging at the sinks of the Taylor vortices. Such herring-bone patterns were already seen in the photographs of Schultz-Grunow and Hein (1956). These patterns do not appear in the photographs of Koschmieder (1979) (Fig. I.4), because the time exposure of his photographs was shorter than the exposure of Mobbs' pictures. There is a high-frequency periodicity in the Taylor vortices which will in all likelihood not be a sharp periodicity. The streak width was measured by Mobbs and is shown as a function of the Taylor number in Fig. 7 of Barcilon et al. The frequency of these motions was also measured and is shown in Fig. 2 of Barcilon and Brindley (1984). This frequency appears to be too sharp and is far too low to be caused by the herring-bone pattern. Barcilon et al. (1979) conjecture that the herring-bone pattern is caused by Görtler vortices on the inside of the concave outer wall. The authors "postulate that the Taylor vortex flow redistributes the mean zonal velocity profile so as to create, near the cylindrical walls, boundary layers of thickness δ in which centrifugal instabilities cause Görtler vortices to form on scale related to δ" (p. 455). The authors also argue that the persistence and stability of both Taylor and Görtler vortex structures over a large range of Taylor numbers suggest "that each is in a certain sense marginally stable" (p. 455), without explaining what is meant by marginal stability in this case. One wonders whether the fine-scale structure is not merely caused by small-scale vortices within the turbulent Taylor vortices. Small vortices with a size distribution seem to be a natural part of a turbulent flow. The suggestions of Barcilon et al. were backed up by a linear analysis of Barcilon and Brindley (1984) which is based on the assumption that the scales of motion of the turbulent Taylor vortices and the Görtler vortices are near marginal stability. Time will tell whether these hypotheses are fruitful.

At even higher Taylor numbers in the range from $4 \times 10^4 \mathcal{T}_c$ to $2 \times 10^6 \mathcal{T}_c$ toroidal Taylor vortices highly uniform in size and intensity still exist, as was shown by Smith and Townsend (1982) and Townsend (1984). Their experiments were made with a horizontal apparatus with a large gap ($\eta = 0.666$) and an aspect ratio $\Gamma = 23.5$. The outer cylinder was at rest. The ends of the gap between both cylinders were partially closed with end plates of different outer diameter attached to the inner cylinder. This arrangement induced a slow axial flow in the system which moved the Taylor vortices with uniform velocity over the fixed hot wire anemometers, and in this way made it possible to determine the velocity field of the toroidal vortices. This method worked fine in the lower range of the Taylor numbers but caused problems at the higher end of the Taylor numbers when the vortices changed from the toroidal to a helical form or possibly to segmented helical vortices. The theoretical understanding of turbulent Taylor vortex flow is not sufficiently advanced to evaluate the implications of the many velocity profiles measured in these experiments. But the experiments demonstrate clearly that, even at around a million times the critical Taylor number, turbulent Taylor vortices still exist.

We shall end this section on turbulent Taylor vortices with a discussion of an astonishing phenomenon discovered by Coles (1965), the so-called spiral turbulence (Fig. 13.10). Spiral turbulence consists of a helical band of turbulence wrapped around the inner cylinder, alternating with laminar flow in between. Spiral turbulence occurs primarily when both cylinders counterrotate, but also with the inner cylinder at rest or even when both cylinders co-rotate. The spiral band of turbulence can be right-handed or left-handed, and each direction occurs with the same probability when the flow is established from rest. The helical pattern rotates around the axis at about the mean angular velocity of the two cylinders over a wide range of cylinder speeds. An anemometer mounted on either one of the cylinders measures therefore an intermittent signal consisting of alternating periods of laminar and turbulent flow, as shown in Fig. 2 in Van Atta's (1966) paper. Van Atta showed also that spiral turbulence is created by a catastrophic transition when the rotation rate of the inner cylinder is increased slowly in a direction opposite to the motion of the outer cylinder, over a large range of Reynolds numbers up to $-100,000$ for the outer cylinder and $20,000$ for the inner cylinder. When, on

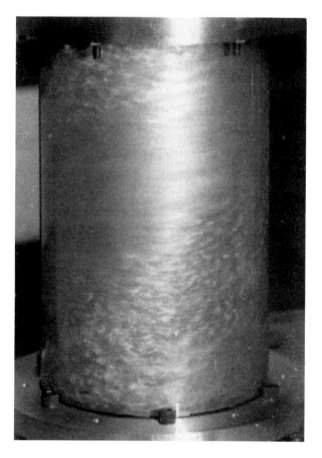

Fig. 13.10. Spiral turbulence between two rotating cylinders. $R_i = 5250$, $R_o = -15,880$, aspect ratio $\Gamma = 14$, radius ratio $\eta = 0.88$. After Coles (1965).

the other hand, the rotation rate of the inner cylinder was decreased after spiral turbulence had been established, it disappeared at a significantly lower Reynolds number of the inner cylinder; there was hysteresis. Laminar and turbulent flow in spiral turbulence are not separated on radial lines, but as Van Atta showed, the interface between laminar and turbulent flow has a leading edge inside from the outer cylinder, and the turbulent flow has a tail near the inner cylinder. So far no theoretical explanation for the formation of spiral turbulence has been advanced.

Let us summarize the section on turbulent Taylor vortices. There is the surprising fact that at extremely high values of the Taylor number Taylor vortices exist at all, are in the mean axisymmetric, and have a size or wavelength only slightly larger than the critical vortices, and that the wavelength of the turbulent vortices is independent of the Taylor number. Just as astonishing is the fact that the size of the turbulent Taylor vortices depends on the initial conditions, or that they are nonunique. It is remarkable that the qualitative behavior of turbulent Taylor vortices (nonuniqueness and independence of the size of the vortices from the Taylor number) resembles so closely the qualitative behavior of laminar axisymmetric Taylor vortex flow. There is a message in this which we have not grasped yet. Turbulent Taylor vortex flow seems to offer an excellent opportunity for further experimental investigation of turbulence, because the experimental conditions are very simple and can be very well controlled, so that theoretical assumptions can be approximated with great accuracy. Besides, turbulent Taylor vortices and spiral turbulence are perfect examples of the much-discussed coherent structures in turbulence.

13.5 Experiments with Two Rotating Cylinders

The stability diagram of linear theory of the Taylor vortex instability (Fig. 11.4) deals also with the case of two independently rotating cylinders. Taylor's (1923) experiments confirmed that this diagram predicted accurately the onset of instability when the cylinders either co-rotate or counter-rotate. The region above the linear stability curve remained unexplored until Coles (1965) ventured into this area and discovered the wavy vortices and spiral turbulence; the latter was studied in more detail by Van Atta (1966). Supercritical Taylor vortex flow between two rotating cylinders was studied again by Snyder (1970). In this paper experiments with a long fluid column and radius ratios ranging from 0.2 to 0.959 are described and the stability diagrams for the different radius ratios are given. Since Snyder did not employ an efficient flow visualization technique, the waveforms he observed appear to be uncertain. Later Andereck et al. (1986) worked with only one radius ratio, $\eta = 0.883$, and primarily one aspect ratio, $\Gamma = 30$. They found a multitude of flow forms, which reflect the influence of the many parameters that affect the two-rotating-cylinder problem. These parameters are the Reynolds numbers of the inner and outer cylinders, the radius ratio, the aspect ratio, the boundary conditions at the end of the fluid column, and the initial conditions of the experiments.

The experimental stability diagram of Andereck et al. (Fig. 13.11) is based on one particular experimental procedure which was followed at all Reynolds numbers in order to obtain consistent results. That means that Fig. 13.11 does not show the total complexity of Taylor vortex flow between two rotating

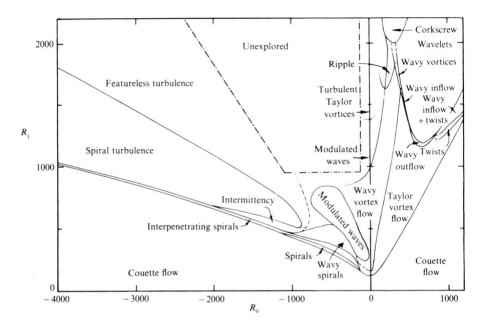

Fig. 13.11. Experimental stability diagram of supercritical Taylor vortex flow between two rotating cylinders for one particular experimental procedure. The average axial wavelength $\lambda_{av} = 2.00$, $\eta = 0.833$, $\Gamma = 30$. After Andereck et al. (1986).

cylinders. The procedure in these experiments was first to accelerate slowly the outer cylinder from rest to a final rotation rate with the inner cylinder at rest. The fluid is then always stable, except at very high rotation rates. Then the inner cylinder was slowly accelerated, instability occurred, but the increase of the rotation rate was continued until a final value was reached. This procedure does not permit the observation of hysteresis.

Before we discuss some of the flow forms, we note that there is not always agreement of the stability diagram in Fig. 13.11 with the results of earlier observations. For counter-rotating cylinders ($R_o < 0$) Taylor's stability diagram, which we have reproduced in Fig. 11.4, and Coles' stability diagram, which we have reproduced in Fig. 13.2, indicate that the onset of instability takes place in the form of regular axisymmetric Taylor vortices to Reynolds numbers of the outer cylinder $R_o = -2000$ in Taylor's experiments and to $R_o = -4000$ in Coles' experiments. We note, however, that in a footnote in Krueger et al. (1966) it is stated that Coles did observe a change from closed vortex rings to a weak helical structure on the Taylor boundary for rotation rates $-0.80 < \mu < -0.75$ in an apparatus with radius ratio $\eta = 0.87$. On the other hand, Andereck et al. found the onset of instability in the form of spiral vortices in the range $-1200 < R_o < -150$. Spiral vortices in this case are Taylor vortices of *equal* size which wind around the inner cylinder in the axial direction, *not* the spiral vortices of different size observed by Taylor (1923) which we showed in Fig. 11.6. Spiral vortex flow will be discussed soon. Spiral vortices at the onset of instability for R_o smaller than about -100 were also observed by Snyder (1970). The onset of instability in the form of nonaxisymmetric flow for $\mu < -0.78$ was predicted theoretically by Krueger et al. (1966). Wavy Taylor vortices are clearly absent in Fig. 13.11 for $R_o < -150$. Coles and Andereck et al. followed the same experimental procedure, but the aspect ratios and end conditions differed. Spiral turbulence, which is very prominent in Fig. 13.11, was found by Coles even with co-rotating cylinders, although that occurred apparently with a different experimental procedure. We find it odd that the closed region of modulated wavy vortex flow in Fig. 13.11 is severed from modulated wavy vortex flow with a resting outer cylinder. Concerning the topic of counter-rotating cylinders, we want to point out the extended region in the stability diagram of Fig. 13.11 in which featureless turbulent flow occurs when the control parameters $|R_o|$ and R_i are large. The absence of vortices under these conditions is in contrast to the well-organized turbulent Taylor vortices which form at equally large values of R_i when the outer cylinder is at rest. The different reaction of the fluid to these conditions is the consequence of the larger shear in the fluid when the cylinders counter-rotate.

Spiral Taylor vortex flow between counter-rotating cylinders which winds with small slope around the inner cylinder as adjacent vortices of *equal* size was first discussed theoretically by Krueger et al. (1966). They mention that, "when k $\neq$ 0 it is impossible, within the framework of linear stability theory, to distinguish between a solution of the disturbance equations corresponding to a wave traveling in the θ-direction but standing in the z-direction $(\exp[i(\omega t + m\theta)]\cos \lambda z,)$ and a wave traveling in both the θ- and z-directions $(\exp[i(\omega t + m\theta + \lambda z)])$'' (p. 533). The latter form characterizes the spiral vortices, whereas the first form characterizes wavy vortex flow. Spiral flow can be left-handed or right-handed and comes with an arbitrary phase angle. Note that the term "standing [wave] in the z-direction'' in the comment of Krueger et al. can be misinterpreted, because in general a standing wave oscillates in the vertical direction at the location of the antinodes, whereas the crests and troughs of wavy Taylor vortices do not move in the vertical direction. A thorough study of the linear aspects of spiral flow determining the condition for the onset of spiral vortex flow of different azimuthal wave number m can be found in Langford et al. (1988). A substantial effort has been

Fig. 13.12. Twisted Taylor vortices between two co-rotating cylinders. $R_i = 1040$, $R_o = 720$. Direction of rotation of the cylinders is from left to right. After Andereck et al. (1968).

devoted to the nonlinear theory of spiral vortices, which was started by DiPrima and Grannick, (1971) and continued by Babenko et al. (1982), Demay and Iooss (1984), Chossat and Iooss (1985), Golubitsky and Stewart (1986), and Golubitsky and Langford (1988). Experimental data on spiral flow are few; the critical conditions for the onset of spiral flow were observed by Snyder (1970), and a careful comparison of experimentally measured wave speeds of nonlinear spiral vortices with the predictions of a cubic order amplitude equation has been made by Edwards, Tagg, Dornblaser, and Swinney (1991). The characteristics of a modified form of spiral flow, the so-called ribbons, were investigated by Tagg et al. (1989).

When the cylinders co-rotate, the effects of the end plates become very conspicuous. As mentioned in Section 13.2 the size of the vortices at rotating column ends increases as $(\mathcal{T}/\mathcal{T}_c)^{1/2}$. If, however, as Andereck et al. (1986) found, at a particular value of $R_o > 0$ the Reynolds number of the inner cylinder is much larger than the critical value at this R_o, vortices are shed from the enlarged end vortex when the Reynolds number of the inner cylinder is increased. That means that under highly supercritical conditions the effects of rotating end plates begin to decrease.

Several new types of vortex flow were observed between co-rotating cylinders, of which we reproduce here, for its aesthetic appeal, a picture of the twisted Taylor vortices (Fig. 13.12). Note how well the twists are in phase over the length of the column. The twisted vortices fill only a small portion of the unstable range in the stability diagram of Fig. 13.11. Twisted vortices were predicted theoretically by Demay and Iooss (1984), Chossat and Iooss (1985), and Golubitsky and Stewart (1986), and were also discussed by Iooss (1986) and Nagata (1986). Most of the unstable range of supercritical flow between co-rotating cylinders in Fig. 13.11 is filled with regular axisymmetric Taylor vortices, and at higher Reynolds numbers of the inner cylinder with the so-called wavelets, a type of flow in which the azimuthal waves on the sinks of the vortices and the azimuthal waves on the sources of the vortices move with different angular velocities and have different azimuthal wavelengths. This does not make for a very regular pattern. The transition from regular Taylor vortices to wavelets takes place via a number of intermediate flow forms, among them the twists. We shall not discuss all these variants of Taylor vortex flow because we would soon be lost in too much detail; the many parameters involved in these flows render the understanding of these complicated nonlinear flows very difficult. In order to understand them it will be necessary to understand first the less complicated problem of nonlinear flow with the outer cylinder at rest.

14

NONLINEAR THEORY OF TAYLOR VORTICES

14.1 Weakly Nonlinear Axisymmetric Taylor Vortices

In order to explain the multitude of experimental findings discussed in the preceding chapter it is necessary to solve the nonlinear equations for Taylor vortex flow. From the solution of the nonlinear equations we want to learn first of all how the amplitude of the Taylor vortices increases when the critical Taylor number has been exceeded. Clearly the exponential growth of the disturbances considered in linear theory is unrealistic; there must be a finite equilibrium amplitude for supercritical Taylor vortices. Other points of obvious importance are the size or the wavelength of Taylor vortices under supercritical conditions and the question of uniqueness of supercritical flow. Also of particular interest are the critical condition for the onset of wavy vortex flow, the wave speed and the amplitudes of the waves, and their axial and azimuthal wavelengths. Ultimately one would like to be able to describe turbulent Taylor vortices, but without having understood the laminar supercritical problems the chances are slim that we will make much headway with the turbulent problem.

The first foray into the theory of nonlinear Taylor vortex flow was made by Stuart (1958), at a time when an experiment dealing specifically with supercritical flow had not yet been made. Using Taylor's (1923) comments about the fluid flow after the onset of instability Stuart concluded that supercritical flow should be axisymmetric. The governing equations are then

$$\frac{\partial u}{\partial t} + u \frac{\partial u}{\partial r} + w \frac{\partial u}{\partial z} - \frac{v^2}{r} = -\frac{1}{\rho} \frac{\partial p}{\partial r} + \nu \left(\nabla^2 - \frac{1}{r^2} \right) u, \qquad (14.1)$$

$$\frac{\partial v}{\partial t} + u \frac{\partial v}{\partial r} + w \frac{\partial v}{\partial z} + \frac{uv}{r} = \nu \left(\nabla^2 - \frac{1}{r^2} \right) v, \qquad (14.2)$$

$$\frac{\partial w}{\partial t} + u \frac{\partial w}{\partial r} + w \frac{\partial w}{\partial z} = -\frac{1}{\rho} \frac{\partial p}{\partial z} + \nu \nabla^2 w, \tag{14.3}$$

with ∇^2 given by (11.6). The continuity equation in the axisymmetric case is given by (11.7).

Stuart developed the disturbances u', v', w' in Fourier series. For the disturbances he obtained an energy equation which is an exact consequence of the axisymmetric equations (14.1)–(14.3). In equilibrium it is

$$\iint (-\overline{\rho u' v'}) \left(\frac{\partial \overline{v}}{\partial r} - \frac{\overline{v}}{r} \right) r \, dr \, dz = \mu \iint (\xi'^2 + \eta'^2 + \zeta'^2) r \, dr \, dz, \tag{14.4}$$

where the bars denote mean values and the vorticity components ξ', η', ζ' are given by

$$\xi' = -\frac{\partial v'}{\partial z}, \qquad \eta' = \frac{\partial u'}{\partial z} - \frac{\partial w'}{\partial r}, \qquad \zeta' = \frac{1}{r} \frac{\partial}{\partial r} (r v'). \tag{14.5}$$

Equation (14.4) says that in equilibrium the rate of transfer of kinetic energy from the mean flow to the disturbance balances the rate of viscous dissipation of kinetic energy by the disturbance. This equation corresponds to Chandrasekhar's (1961) balance theorem for Rayleigh–Bénard convection [equation (2.45)]. The consequences of Stuart's balance theorem for axisymmetric Taylor vortex flow have received just as little attention as the consequences of the balance theorem for Rayleigh–Bénard convection.

Stuart presented an approximate solution of the nonlinear stability problem for the narrow gap case; he assumed that under slightly supercritical conditions the Taylor vortices would have the same shape as the critical vortices, that the generation of their harmonics could be ignored, and that the vortex flow was close to equilibrium. Collectively, these three assumptions came to be known as the "shape assumption." He also required that the outer cylinder be at rest. Stuart derived an equation for the equilibrium amplitude A_e of the supercritical vortices; he found that A_e is proportional to $R^{-1} \sqrt{(1 - \mathcal{T}_c/\mathcal{T})}$ For values of $\mathcal{T}$ close to $\mathcal{T}_c$, i.e. in the weakly nonlinear regime, it follows that the equilibrium amplitude is proportional to $\sqrt{(\mathcal{T} - \mathcal{T}_c)/\mathcal{T}_c} = \sqrt{\varepsilon}$, as Landau (1944) had anticipated. The $1/R$ factor in the formula for the equilibrium amplitude is, however, not compatible with the continued increase of the amplitude at larger values of the Taylor number. Using this formula for the amplitudes of the velocity components Stuart computed the torque required to maintain the rotation of the inner cylinder and found good agreement with Taylor's (1936a) torque measurements, not only in the weakly nonlinear regime, but also under moderately supercritical conditions up to about $6\mathcal{T}_c$. This agreement is fortuitous because at less than $1.5\mathcal{T}_c$ wavy Taylor vortices form in a narrow gap. The wavy vortices invalidate the assumptions made by Stuart and change the magnitude of the torque, as was found in the experiments of Debler et al. (1969).

The investigation of weakly nonlinear axisymmetric Taylor vortex flow between infinite cylinders was continued by Davey (1962). He developed the velocity components of the disturbances in Fourier series in harmonics of the wave number a, writing

$$u' = \sum_{n=1}^{\infty} u_n(r,t)\cos naz,$$

$$v = \bar{v} + v' = \bar{v}(r,t) + \sum_{n=1}^{\infty} v_n(r,t)\cos naz, \qquad (14.6)$$

$$w' = \sum_{n=1}^{\infty} w_n(r,t)\sin naz.$$

Davey eliminated p and w from equations (14.1)–(14.3) and the continuity equation and tried a solution of the remaining two differential equations with

$$u_n(r,t) = A^n\left\{u_n(r) + \sum_{m=1}^{\infty} A^{2m}u_{nm}(r)\right\}, \qquad n \geq 1$$

$$v_n(r,t) = A^n\left\{v_n(r) + \sum_{m=1}^{\infty} A^{2m}v_{nm}(r)\right\}, \qquad n \geq 1 \qquad (14.7)$$

$$\bar{v} = \bar{v}_l + \sum_{m=1}^{\infty} A^{2m}f_m(r),$$

following the example of Stuart (1960) and Watson (1960). The time-dependent amplitudes are determined from the equation

$$\frac{1}{A}\frac{dA}{dt} = \sum_{m=0}^{\infty} a_m A^{2m}, \qquad a_0 = s. \qquad (14.8)$$

From this series only the first two terms on the right hand side were retained, because the amplitude was supposed to be small. That means that the amplitude equation became

$$dA/dt = sA + a_1 A^3. \tag{14.9}$$

This is the equation originally proposed by Landau (1944), which we encountered before in the theory of Rayleigh–Bénard convection [equation (7.14)]. A solution of (14.9) which satisfies the condition that the amplitude increases as e^{st} as $A \to 0$ was given by Davey as

$$A^2 = Kse^{2st}/(1 - a_1 Ke^{2st}), \tag{14.10}$$

from which it follows that in the first approximation the equilibrium amplitude is given by

$$A_e^2 = -s/a_1, \tag{14.11}$$

the sign of a_1 determining whether the disturbances are subcritical or supercritical. Amplitude equations with a fifth order term in A were studied by Eagles (1971) and DiPrima and Eagles (1977). No significant differences in the results with the third order or fifth order amplitude equations were found.

Using numerical methods Davey determined the critical wave number and the amplitude of the vortices in the case of a wide gap ($\eta = 0.5$) with the outer cylinder at rest, in the case of a narrow gap with the outer cylinder at rest, and in the case of a narrow gap when both cylinders rotate with nearly the same speed, assuming that the wave number, which is equal to a_c of linear theory for the particular η, is fixed at a_c. For the wide gap Davey found that for $\mu = 0$ and $\eta = 0.5$ the equilibrium amplitude varies as

$$A_e^2 = 0.09017(1 - \mathcal{T}_c/\mathcal{T}). \tag{14.12}$$

This means, as it is commonly expressed, that the amplitude increases in a first approximation as $\varepsilon^{1/2}$, where $\varepsilon = (\mathcal{T} - \mathcal{T}_c)/\mathcal{T}_c$. The relationship that the amplitude is proportional to $\varepsilon^{1/2}$ has also been found in the wide gap case ($\eta = 0.5$ and 0.6122) by Yahata (1977a) using a Galerkin method and by Yahata (1977b) with a mode-coupling approach. We note finally that Davey also confirmed Stuart's balance theorem [equation (14.4)].

When the amplitude of the motion is known, the torque can be calculated. In the range from R_c to $1.35R_c$ Davey found good agreement of the calculated torque with the torque measured by Donnelly (1958) with an apparatus with $\eta = 0.5$ and $\Gamma = 10$. The agreement between theory and experiment actually extends over a wider range than could be expected from a weakly nonlinear theory. A similar increase of the equilibrium amplitude was found for the narrow gap problem. The torque calculated for the narrow gap problem in the case of a resting outer cylinder was compared with the measurements of Taylor (1936a), made with an apparatus with $\eta = 0.97$ and $\Gamma = 767$. Agreement was fairly good for Reynolds numbers up to about $3.7R_c$. On the other hand, when compared with the measurements of Donnelly (1958) with an apparatus with $\eta = 0.95$ and $\Gamma = 100$, the slopes of the theoretical and experimental curves differed from the beginning. We note, however, that in the narrow gap case the torque measurements can hardly be expected to agree with a theory which assumes axisymmetric flow, because at around $1.1R_c$ wavy vortices form, which invalidates the assumption of axisymmetric flow. Yahata (1977a,b) similarly computed the torque in the slightly supercritical range.

The validity of the amplitude equation (14.9), the principal result of Davey's (1962) study, was to some extent supported by the torque measurements. It was, nevertheless, of importance to verify directly the variation of the amplitude of the Taylor vortices as a function of the Taylor number. The first such measurements were made by Donnelly and Schwarz (1965). They observed that the square of the amplitude of the vortices was a linear function of the Taylor number between $\mathcal{T}_c$ and $\approx 1.2\mathcal{T}_c$ for $\eta = 0.95$ and $\eta = 0.90$, and between $\mathcal{T}_c$ and $\approx 1.4\mathcal{T}_c$ for $\eta = 0.85$. The amplitude reached a maximum at $1.2\mathcal{T}_c$ for $\eta = 0.95$ and at $1.4\mathcal{T}_c$ for $\eta = 0.90$ and decreased abruptly when the Taylor number was increased further, the decrease being almost as rapid as the initial increase. Donnelly and Schwarz wrote that "it appears possible that the decrease . . . is due to a 'wavy' disturbance" (p. 544). Although the technique with which the amplitude was measured permitted only relative measurements, and although the three curves showing the variation of the amplitude for the three values of η are not consistent, these experiments confirmed in a qualitative way the validity of the amplitude equation for Taylor vortex flow.

The amplitude of the radial velocity component of steady axisymmetric Taylor vortices was measured directly by Gollub and Freilich (1976) in the center of the gap of an apparatus with radius ratio $\eta = 0.61$ and aspect ratio $\Gamma \cong 30$ using the laser–Doppler technique. The basic harmonic of the flow was found to increase as $A_1 = (0.158 \pm 0.008)\varepsilon^{0.50\pm0.03} + (0.029 \pm 0.005)\varepsilon^{1.42\pm0.05}$ cm/sec for $(\mathcal{T} - \mathcal{T}_c)/\mathcal{T}_c$ up to 0.5, in good agreement with the predictions of Davey (1962), according to which the amplitude is proportional to $\sqrt{\varepsilon}$. At larger values of the Taylor number a second and third harmonic of the flow increased in importance. The amplitude of the second harmonic increased as $A_2 = (0.066 \pm 0.004)\varepsilon^{0.76\pm0.05} - (0.021 \pm 0.001)\varepsilon^{1.7\pm0.3}$ cm/sec. It is interesting that the second harmonic did not increase as ε, but it will require much more substantiation to raise this observation to a rule. Additional measurements of the amplitude of the velocity components in fluid columns of different radius ratios and at different locations in the gap are necessary in order to establish the validity of Landau's equation for Taylor vortex flow in general. The measurement of the amplitude at different locations in the gap will be of particular importance for the determination of the role of the higher harmonics. The numerical study of Fasel and Booz (1984) shows that there are very strong local variations of the amplitude increase in a wide gap. It would also be of great interest to determine the variation of the amplitude of the velocity components of wavy vortex flow as a function of the Taylor number.

The experimental confirmation of the validity of the Landau equation in Rayleigh–Bénard convection (Bergé, 1975; Dubois and Bergé, 1978), as well as in the Taylor vortex problem (Gollub and Freilich, 1976), and theoretical successes with the Ginzburg–Landau equation in convection (Newell and Whitehead, 1969; Segel, 1969) and in Taylor vortex flow with finite fluid columns (Blennerhassett and Hall, 1979; Graham and Domaradzki, 1982) have created the impression that the Landau equation or the Ginzburg–Landau equation is a panacea for the nonlinear problems in Rayleigh–Bénard convection and the Taylor vortex instability. Whether that is really so is doubtful. The derivation of the Landau equation in Rayleigh–Bénard convection and in Taylor vortex flow is valid only for two-dimensional, weakly nonlinear flow. A successful description of weakly nonlinear flow should lead to a theory of moderately supercritical flow of which weakly nonlinear flow is the most simple special case, but the Landau equation does not do that. It deals only with the amplitude of the motions, but the amplitude is likely not to be the only parameter which determines the characteristics of nonlinear flow. Likewise, the Ginzburg–Landau equation applies only in the weakly nonlinear regime, and the amplitude is supposed to vary only slowly in the direction perpendicular to the axis of the rolls. The Ginzburg–Landau equation does apparently match reality in bounded fluid layers in the weakly nonlinear regime, but the bounded layers lack the generality of the problems in infinite layers. It does not appear that the crucial nonlinear problems in Rayleigh–Bénard convection and Taylor vortex flow depend on the presence of lateral boundaries.

Concerning the validity of the experimental verification of the Landau equation one has to keep in mind that the results of Dubois and Bergé (1978) apply to the case of supercritical convective motions with a constant wavelength, whereas in general the wavelength of the motions in Rayleigh–Bénard convection increases when $\mathfrak{R} > \mathfrak{R}_c$. In the Taylor vortex problem the amplitude equation was verified by Gollub and Freilich (1976) only in an apparatus with a wide gap, where the wavelength of the motions is constant when $\mathcal{T} > \mathcal{T}_c$, as was assumed in the derivation. There are simply not enough data to prove that the Landau equation holds under a variety of conditions. The scarcity of data supporting the Landau equation does not seem to justify the optimistic belief that with the Landau equation most of the outstanding nonlinear problems in Rayleigh–Bénard convection and Taylor vortex flow can be solved.

In the 1960s a number of analytical studies of supercritical Taylor vortex flow were published. The existence of steady secondary flow between concentric cylinders was proved rigorously by Velte (1966). The exact number of branching solutions beyond the critical eigenvalue was determined by Kirchgässner and Sorger (1969), and the nonuniqueness of the stationary problem, the stability of the stationary solutions, and the selection of a distinct wave number were discussed in Kirchgässner and Sorger (1968). They found that, in the weakly nonlinear case, "the only possible stable branching solution" is that with the critical wave number. This agrees with the experimental observation that the wave number of supercritical vortices is the critical wave number if the Taylor number in increased quasi steadily from subcritical values on. On the other hand, *stable* supercritical vortices with wave numbers that are different from the critical wave number can be produced easily in the experiments by different initial conditions. Kirchgässner and Sorger stated also that their results apply analogously to the Bénard convection problem, which is, however, in contradiction with the experimental facts. Other analytical investigations of slightly supercritical Taylor vortex flow are in Iudovich (1966), Ivanilov and Iakovlev (1966), and Ovchinnikova and Iudovich (1974), in which again a connection between the nonlinear Bénard problem and the Taylor vortex problem was found.

14.2   Theory of Wavy Taylor Vortices

When the preliminary results of Coles' experiments with wavy Taylor vortices became known, DiPrima (1961) began the theoretical investigation of wavy Taylor vortex flow. In his paper DiPrima outlined a program for the study of wavy vortices, suggesting that "one should examine the stability of the steady motion consisting of the distorted original velocity distribution plus the Taylor vortices," and he noted that "such an analysis clearly presents a formidable problem" since the amplitude of the axisymmetric supercritical vortices alone was an extremely difficult problem (p. 751). Therefore he pursued first a more simple task, studying only the stability of the velocity of Couette flow to infinitesimal disturbances which are nonaxisymmetric. DiPrima considered wavy azimuthal disturbances of the form

$$u(r,\theta,z,t) = u(r)\cos kz \, e^{i(st+m\theta)}, \tag{14.13}$$

$$v(r,\theta,z,t) = V(r) + v'(r)\cos kz \, e^{i(st+m\theta)}, \tag{14.14}$$

$$w(r,\theta,z,t) = w(r)\sin kz \, e^{i(st+m\theta)}, \tag{14.15}$$

with the azimuth angle θ and the integer number m of the azimuthal waves. The growth rate s can, in general, be complex. When the analysis is completed, only the real part of the disturbances is considered. The *linearized* Navier–Stokes equations for a fluid column of infinite length in the neutral state lead to a sixth order system for the velocity components u and v, which together with the no-slip boundary condition at the cylinder walls leads to a rather complicated eigenvalue problem. The problem was therefore simplified by using the narrow gap approximation, with the requirement that both cylinders rotate in the same direction ($\mu \geq 0$) and by replacing the radial distribution of the angular velocity $\Omega(r)$ by a linear profile. Neglecting further two small imaginary terms in the differential equations, DiPrima finally arrived at the equations

$$(DLD - a^2L)u = -0.5a^2\mathcal{T}(1 + \mu)v, \tag{14.16}$$

$$Lv = u, \tag{14.17}$$

with the wave number a, Taylor number $\mathcal{T}$, and the operator

$$L = (D^2 - a^2) - iR(\beta + mf(x)), \tag{14.18}$$

with $D = d/dx$, x replacing the radial direction, and with $\beta = \omega_r/\Omega_1$, where ω_r is the angular velocity of the azimuthal waves. These equations have to be solved with the boundary conditions

$$u = v = Du = 0 \quad \text{at} \quad x = \pm\tfrac{1}{2}. \tag{14.19}$$

The eigenvalue problem posed by (14.16), (14.17), and (14.19) was solved with a variational method from which it followed (in the first approximation) that the critical Reynolds numbers for the onset of wavy vortex flow increase slightly and monotonically (up to $m = 3$) and are larger than the critical Reynolds number for the axisymmetric mode ($m = 0$), in the cases $\mu = 0$ and $\mu = 0.5$ and for radius ratio $\eta = 0.89$. This is the essential result of this study, meaning that azimuthal disturbances can indeed cause azimuthal waves, even in linear theory; nevertheless the onset of instability takes place in the form of axisymmetric vortices because they have the lowest critical Reynolds number. Actually, the critical Reynolds numbers for $m = 0$ and $m = 1$ were found to be very close together, a result not supported by the experiments of Coles, who observed the onset of wavy vortices at a few percent above the onset of axisymmetric instability. This difference between theory and experiment is likely to be due to the fact that the wavy vortices in Coles' experiments formed in nonlinear flow, whereas DiPrima's theory is linear. It also followed from DiPrima's calculations that the dimensionless angular velocity of the waves is 0.5 for all wave numbers m studied, in qualitative agreement with Coles' observations. DiPrima's result that the critical Reynolds number is an increasing function of the number of azimuthal waves was confirmed by Roberts (1965) in the case $\mu = 0$, for four different radius ratios.

The consequences of nonaxisymmetric disturbances on Couette flow between counter-rotating cylinders ($\mu < 0$) were studied by Krueger et al. (1966) in the narrow gap case with linear theory. Using a sixth order system very similar to (14.16)–(14.17) they arrived at a complicated eigenvalue problem depending on μ, a, m, $\mathcal{T}$ that was solved numerically in the neutral case. From these calculations emerged a principal new feature, namely the finding that the *onset* of instability between counter-rotating cylinders can occur in the form of nonaxisymmetric motions if $\mu < -0.78$. That means that for $\mu < -0.78$ the critical Taylor number for nonaxisymmetric modes is smaller than for the axisymmetric mode. The critical, i.e. minimal, Taylor number of instability should be reached with azimuthal waves of larger wave number m when μ is decreased below -0.78; e.g. at $\mu = -1.25$ the instability should occur with vortices with five azimuthal waves. It should also be possible to increase the azimuthal wave number at the critical condition by decreasing the gap width. Finally it was shown that the axial wavelength should be larger than the axial wavelength of axisymmetric vortices if the onset of instability occurs in the form of a nonaxisymmetric flow.

The onset of instability in a small gap between counter-rotating cylinders was observed by Coles (1965) to take place in the form of axisymmetric vortices (see Fig. 13.2), but there is a footnote in Krueger et al. (1966) saying that there may have been spiral vortices at $\mu < -0.78$. On the other hand, Snyder (1970) and Andereck et al. (1986) found the onset of instability in nonaxisymmetric (spiral) form at $\mu < -0.78$ (see Fig. 13.11), confirming the prediction of Krueger et al. The most likely cause for the differences in the results of the experiments are different boundary conditions at the column ends. Neither Snyder nor Andereck et al. observed azimuthal wavy vortices at the onset; they observed instead spiral vortices, i.e. Taylor vortices which wind around the inner cylinder in the vertical direction and rotate around the inner cylinder with about the mean rotation rate of the cylinders. This is not the spiral flow observed by Taylor (1923) shown in Fig. 11.6, which is characterized by spirals of unequal size. The spiral flow observed by Snyder and Andereck et al. was discussed in Section 13.5. The prediction of Krueger et al. that instability should occur with vortices with larger azimuthal wave numbers if μ is decreased to values lower than -0.78 has not been confirmed so far; although the experiments have been extended to values of μ substantially below -0.78, the onset of instability has always been in the form of spiral flow.

The conditions for the formation of nonaxisymmetric wavy *supercritical* Taylor vortices were investigated by Davey et al. (1968) using the results of the preceding studies of Davey (1962) and Krueger et al. (1966). The paper of Davey et al. deals with finite amplitude motions $(\mathcal{T} > \mathcal{T}_c)$ between infinitely long cylinders with the outer cylinder at rest, with the fluid being contained in a narrow gap $(\eta \to 1)$, and with a fixed wave number $a = a_c$. We note that the narrow gap approximation, which plays an important role in the calculations made in this study, is almost a necessary condition for the study of wavy Taylor vortices because wavy vortices are prominent mainly in the narrow gap case; in wide gaps the vortices remain axisymmetric up to fairly high Taylor numbers. The nonlinear Navier–Stokes equations can now no longer be reduced to a single sixth order differential equation as in (11.10). In the nonlinear case one has to deal with three partial differential equations for the velocity components which incorporate nonlinear terms. In order to solve this problem the velocity components of the disturbances were developed in Fourier series. If the azimuthal velocity component is periodic with period $2\pi/a$ in the nondimensional vertical direction ζ and periodic with the period $2\pi/k$ in the azimuth ϕ, then it is in general

$$v(x,\zeta,\phi,\tau) = \sum_{q=-\infty}^{\infty} \left\{ v_{0q}(x,\tau) + \sum_{n=1}^{\infty} v_{cnq}(x,\tau)\cos na\zeta \right.$$
$$\left. + v_{snq}(x,\tau)\sin na\zeta \right\} e^{iqk\phi}, \tag{14.20}$$

where x is the nondimensional radial coordinate, τ the nondimensional time, and $k = m\Omega_0 d^2/\nu$, a nondimensional version of the azimuthal wave number m. The subscripts c and s refer to the cosine and sine functions. Formulas similar to (14.20) apply to the u and w velocity components. Note that the terms $\cos na\zeta$ and $\sin na\zeta$ represent a phase change by $\pi/2$ in the vertical direction. The choice of the phase between infinite cylinders is arbitrary, but the phase poses a problem in finite, though possibly long, columns where the phase is fixed by the ends of the column. Since wavy vortices are actually observed in finite fluid columns, it seems that the instability leading to wavy vortices cannot depend critically on the phase. From equation (14.20) only the four fundamental terms

$$v_{c10}(x,\tau)\cos a\zeta, \qquad v_{s10}(x,\tau)\sin a\zeta,$$

from the axisymmetric motions and

$$v_{c11}(x,\tau)\cos a\zeta \, e^{ik\phi}, \qquad v_{s11}(x,\tau)\sin a\zeta \, e^{ik\phi},$$

from the nonaxisymmetric motions were retained. This means that only the interactions of two axisymmetric disturbances shifted in phase by $\pi/2$ with two nonaxisymmetric disturbances, also shifted in phase by $\pi/2$, were studied. All four fundamental modes have the same axial wave number.

The velocities were then expanded in powers and products of the amplitudes $A_c(\tau)$, $A_s(\tau)$, $B_c(\tau)$, $B_s(\tau)$, the amplitudes B belonging to the nonaxisymmetric motions. In order to be consistent with the differential equations for u, v, w the amplitudes must satisfy a system of ordinary nonlinear differential equations,

$$\frac{dA_c}{d\tau} = a_0 A_c + a_1 A_c^3 + a_1 A_c A_s^2 + a_3 A_c |B_c|^2 + a_4 A_c |B_s|^2$$
$$+ a_5 A_s B_c B_s^* + a_5^* A_s B_c^* B_s, \qquad (14.21)$$

$$\frac{dB_c}{d\tau} = b_0 B_c + b_1 B_c |B_c|^2 + b_2 B_c |B_s|^2 + b_3 B_c A_c^2 + b_4 B_c A_s^2$$
$$+ (b_3 - b_4) B_s A_c A_s + (b_1 - b_2) B_c^* B_s^2,$$

and two similar equations for A_s and B_s, with the asterisk indicating complex conjugate. The coefficients a_n and b_n are functions of μ, a, k, and $\mathcal{T}$. The amplification rates a_0 and b_0 are considered so small that the amplitude equations can be truncated at third order terms. Relations between the different coefficients a_n and b_n determine the stability of the different types of flow; e.g. for wavy vortex flow it is

$$A_s = B_c = 0, \qquad A_c^2 = A_e^2 = \frac{a_0 b_{1r} - a_4 b_{0r}}{a_4 b_{4r} - a_1 b_{1r}}, \qquad B_s = \beta_e e^{i\omega(\tau - \tau_s)}$$

$$\beta_e^2 = \frac{a_1 b_{0r} - a_0 b_{4r}}{a_4 b_{4r} - a_1 b_{1r}}, \qquad \omega = b_{0i} + b_{1i}\beta_e^2 + b_{4i}A_e^2, \qquad (14.22)$$

with the subscripts r and i standing for real and imaginary, and the subscript e for equilibrium.

The values of the coefficients a_n and b_n were computed and it was found that the system of equations has a solution for stable laminar Couette flow, a range of parameters for which laminar Couette flow is unstable but Taylor vortices exist which are stable, and a range of parameters for which Taylor vortices are unstable against perturbations which are periodic in the azimuth and have the same axial wave number as the Taylor vortices but are shifted in axial phase by $\pi/2$. The critical Taylor number $\mathcal{T}_c'$ for the onset of wavy vortex flow was found to be about 8% above the critical Taylor number for the onset of axisymmetric vortex flow. $\mathcal{T}_c'$ was found to depend weakly on the azimuthal wave number m, $m = 1$ having the lowest $\mathcal{T}_c'(m)$. The neighboring wavy cell boundaries were found to be in phase. There was also a solution for a spiral mode. We summarize and note that wavy Taylor vortex flow was found to be the consequence of the interaction of one axisymmetric Taylor vortex mode of finite amplitude with one finite amplitude (not infinitesimal) nonaxisymmetric mode which is shifted in phase by $\pi/2$.

The results of Davey et al. are in qualitative agreement with the experimental findings. A quantitative agreement is elusive because the experimental data come from experiments with finite fluid columns. According to Cole (1976) the critical Taylor number $\mathcal{T}_c'(m)$, as well as the azimuthal wave number m at the onset of wavy flow (p. 10), depends on the aspect ratio of the column. Coles (1965) observed the onset of wavy vortex flow with $m = 4$, but his fluid column was relatively short. At the onset of wavy flow the boundaries of neighboring vortices are in phase, as can easily be seen in Fig. 19a of Coles; only at higher values of the Reynolds number are the vortex boundaries no longer in phase.

Davey et al. also discussed the physical mechanism that causes the formation of wavy vortex flow. They discussed qualitatively the stability of the shear flow between neighboring vortices following a suggestion of Meyer (1967). This shear instability, or jet instability as it is also called, has been discussed a number of times in the literature. Since a definitive answer of the question posed has not yet emerged, we shall not pursue this matter.

Following Davey et al. (1968) wavy vortex flow was studied numerically by Eagles (1971). He dropped the narrow gap approximation and took fifth order terms in the amplitude equation into account. No significant differences in the results of his study as compared with those of Davey et al. appeared.

Similar calculations for selected values of $\eta \leq 0.95$ were made by DiPrima et al. (1984). Nakaya (1975) also studied wavy vortex flow, dropping the requirement $\mu = 0$ and the condition that the wave number of the axisymmetric and nonaxisymmetric modes be the same. He found, however, that wavy vortex flow develops most rapidly when both types of disturbances have the same axial wave number. Yahata (1981, 1983) studied the transition to wavy vortex flow with a 56-mode model. His (1983) result, that the motions change from periodic to quasi-periodic motions with three frequencies and then to a motion with two frequencies followed by nonperiodic motions when the Reynolds number is increased, lacks experimental confirmation as far as the three-frequency motion is concerned. Jones (1981) determined numerically the stability boundary for small nonaxisymmetric disturbances with wave numbers $m = 1, 2, 3$, finding again that the wave number $m = 1$ has the lowest eigenvalue. Moser et al. (1983) treated a few cases of nonaxisymmetric flow numerically. Marcus (1984a,b) made extensive numerical studies of wavy vortex flow, reproducing with accuracy the azimuthal wave speed measured in several experiments. He also investigated the mechanism of instability and concluded that an instability of the radial outflow boundary is responsible for the (singly) periodic wavy vortex flow, and that the second frequency of the modulated wavy vortices is associated with an instability of the inflow boundary. Since the inflow jet is weaker than the outflow jet, the transition to modulated vortex flow occurs at a higher Reynolds number. Walgraef et al. (1984) studied the onset of wavy vortex flow in finite fluid columns with an amplitude equation. Jones (1985) studied the transition to wavy vortices in wide gaps $(0.5 < \eta < 0.8)$ and for different axial wavelengths. He also discussed the physical mechanism that makes the vortices wavy. He found that the increased dissipation in wide gaps delays the onset of wavy flow to higher Taylor numbers. Marx and Haken (1989) studied wavy vortex flow by using a generalized Ginzburg–Landau equation.

The multitude of these theoretical and numerical investigations have contributed to the understanding of some aspects of wavy vortex flow. But it seems that we have still a long way to go in order to explain convincingly the wavy vortex problem so clearly expressed in the figure in Coles (1965), which we have reproduced in Fig. 13.3. Wavy vortex flow is, of course, a formidable problem because it not only is nonlinear but also depends on a multitude of parameters, the Reynolds number, the azimuthal wave number, the axial wave number, the radius ratio, and the aspect ratio, and very strongly on the initial conditions, and one can also consider the dependence on the ratio of the rotation rates of the inner and outer cylinders. Since the nonlinear theory of Taylor vortices between two rotating cylinders is presented in detail by Chossat and Iooss (1992) we shall not pursue this topic further.

14.3 Moderately Supercritical Axisymmetric Taylor Vortices

We shall now look at the problem of moderately supercritical axisymmetric Taylor vortex flow. For the investigation of supercritical axisymmetric flow the narrow gap approximation has to be dropped, because supercritical flow in narrow gaps remains axisymmetric only for a very small range of Taylor numbers. After the amplitude of axisymmetric vortices had been determined by Davey (1962), the stability of axisymmetric supercritical vortices with respect to axisymmetric disturbances in an infinite column was studied by Kogelman and DiPrima (1970). They applied Eckhaus' (1965) stability theory to a broad class of nonlinear partial differential equations which contained as special cases axisymmetric Taylor vortex flow, two-dimensional Rayleigh–Bénard convection rolls, and Poiseuille flow between parallel walls. Kogelman and DiPrima arrived at a formula for the neutral stability curve of axisymmetric nonlinear Taylor vortices which is the same as equation (7.10). In terms of the wavelength the interval of stable supercritical wavelengths is given by

$$\lambda_{nl} = \frac{\lambda_c \lambda_l \sqrt{3}}{\lambda_c + \lambda_l (\sqrt{3} - 1)}, \tag{14.23}$$

where λ_{nl} is either the maximal or the minimal wavelength of stable vortex flow at a given supercritical Taylor number according to weakly nonlinear theory, λ_l is either the maximal or minimal wavelength which follows from linear theory at a given $\mathcal{T} > \mathcal{T}_c$ according to equation (11.17), and finally λ_c is the critical wavelength at the critical Taylor number. Either (7.10) or (14.23) shows that there is a continuum of nonunique stable solutions for supercritical axisymmetric Taylor vortex flow. The nonlinear range of stable flow is not as wide as the linear range; both ranges contain wavelengths either smaller or larger than the critical wavelength. The linear as well as an extrapolated nonlinear stability range is shown in Fig. 14.1. The reality of stable, nonunique supercritical axisymmetric Taylor vortices with wavelengths different from the critical wavelength was demonstrated in Fig. 13.6.

The range of the wavelengths of stable axisymmetric supercritical Taylor vortex flow was investigated experimentally by Burkhalter and Koschmieder (1974) over a range of Taylor numbers far exceeding the weakly nonlinear regime (see Fig. 14.1). In these experiments Taylor vortices with wavelengths shorter than the critical wavelength were established through the sudden-start procedure; wavelengths longer than the critical wavelength were produced by the filling experiments. All wavelengths between the minimal wavelength reached at a particular supercritical $\mathcal{T}$ by sudden starts and between the

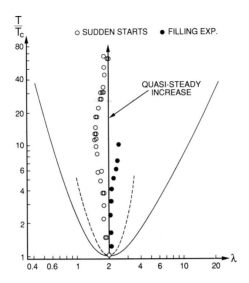

Fig. 14.1. Range of the wavelength of stable axisymmetric supercritical Taylor vortex flow according to theory and experiments. The continuous curve represents the neutral curve of linear theory for an infinite fluid column in a narrow gap. The dashed curved represents an extrapolated Eckhaus stability diagram of weakly nonlinear theory. The circles mark the experimental results with a fluid column of radius ratio $\eta = 0.727$ and aspect ratio $\Gamma = 53$. After Burkhalter and Koschmieder (1974).

critical wavelength reached by quasi-steady increases of $\mathcal{T}$ can be realized either by starts with different acceleration rates of the inner cylinder or by increases or decreases of the Taylor number of a particular pattern established at another Taylor number. The wavelengths that can be established in experiments are, however, subject to the quantization condition; there is not a continuum of possible wavelengths at a given $\mathcal{T}$ but rather a sequence of discrete values of the size of the vortices which fill the finite fluid column. As Fig. 14.1 shows, the range of stable supercritical axisymmetric Taylor vortices observed in experiments is much smaller than the range of stable supercritical flow predicted by the Eckhaus instability or by Kogelman and DiPrima, already under slightly supercritical conditions, $\mathcal{T} = 1.5\mathcal{T}_c$.

Nakaya (1974) determined another nonlinear stability diagram for supercritical axisymmetric vortex flow in an infinite fluid column of radius ratio $\eta = 0.5$. He used the amplitude equation and took fifth order terms into account. His calculations extend up to $3R_c$. The stability diagram from the fifth order approximation is not as wide as that of Kogelman and DiPrima but still wider than the range of nonunique flows observed in the experiments. Also of interest is Nakaya's computation of the equilibrium amplitude of the vortices. He found that the wave numbers of the maximum amplitude increase when the Reynolds number is increased, but we know from the experiments that the wave numbers of the vortices remain constant with increased R, which means that the flow does not follow the maximum amplitude.

Another calculation of the range of stability of nonlinear axisymmetric vortex flow with respect to axisymmetric disturbances was made for three aspect ratios (η = 0.892, 0.75, 0.5) by Riecke and Paap (1986) using a Galerkin procedure to solve the Navier–Stokes equations in the axisymmetric case, assuming that the outer cylinder is at rest and that the fluid layer is of infinite length. Their stability diagram is shown in Fig. 14.2 for the case η = 0.75, which we shall use because this diagram was verified experimentally by Dominguez-Lerma et al. (1986). As this figure shows, the Reynolds numbers of the calculated stability diagram go up to about $2R_c$, i.e. above the weakly nonlinear regime. Figure 14.2 shows also that in the weakly nonlinear range, $R \leq 1.1R_c$, the stability diagram of Riecke and Paap coincides with the Eckhaus instability, but that for larger R the range of stable supercritical flow becomes noticeably smaller. Also it was found that the width of the stable range at a given R decreases when the radius ratio was decreased.

The stability of axisymmetric Taylor vortices in an apparatus of radius ratio η = 0.75 and aspect ratio $\Gamma \cong 40$ was investigated experimentally by Dominguez-Lerma et al. (1986). The procedure of these experiments was as follows: First an integer number of vortex pairs with the critical wavelength was created by increasing R slowly from subcritical values to about $2R_c$. Then the column was either expanded at constant R by moving the upper end plate of the column upward or compressed by moving the end plate inward. If the number of the vortices does not change during this procedure, the wavelength of the vortices decreases as the column is compressed and increases as the column is expanded, because the wavelength is proportional to the column length ($\lambda = L/Nd$), assuming that the vortices of the column are of uniform size, as they are except for the end vortices. In this way the wavelength was changed at $R \cong 2R_c$ by a maximum of about 25% by expansion and a maximum of about 33% by compression. After a pattern with a wavelength different from λ_c had been established this way at $\cong 2R_c$, the rotation rate or the Reynolds number was decreased slowly. At a particular value of the Reynolds number the pattern became unstable; the wavelength changed by the formation of an additional pair of vortices when λ was too long or the loss of a pair of vortices when λ was too short. The points of instability of the flow are marked as the solid circles in Fig. 14.2. Within the error of the experiment

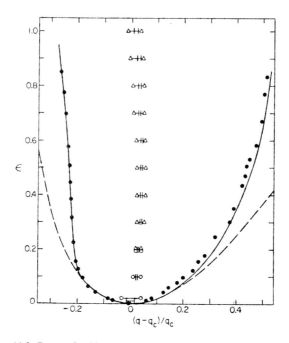

Fig. 14.2. Range of stable wave numbers q of axisymmetric Taylor vortices with radius ratio $\eta = 0.75$ as a function of $\varepsilon = (R - R_c)/R_c$. The dashed curve is the Eckhaus boundary. The solid curve is the nonlinear boundary according to Riecke and Paap. The solid circles show the experimental stability limits. The stability limits in the presence of spatial ramps of two different slopes are marked by triangles and open circles. After Dominguez-Lerma et al. (1986).

the agreement of theory and experiment is remarkable. The experiments also showed that the boundary conditions at the stationary ends were virtually inconsequential for the stability of the fluid in columns of this apect ratio.

Since we are dealing here with the best quantitative agreement of an experiment with a theoretical study of moderately supercritical Taylor vortex flow, we must be specific about the experimental uncertainty. The principal uncertainty of the wavelength (or the wave number of q) is given by the quantization condition. From $\Delta\lambda/\lambda = \pm 2 \times 100/N$ it follows that the uncertainty of the wavelength was of the order of 5%. There is also an uncertainty in the average wavelength following from the size distribution of the vortices. Dominguez-Lerma et al. give this as 0.1%, an underestimate because they omit from consideration the three vortex pairs at the column ends, which contribute most to the uncertainty of the average wavelength. The uncertainty of the Reynolds number is given as 0.5%. This value follows from an experimental determination of Ω_c, which the authors make since they apparently do not know the viscosity of their water–glycerol fluid mixture. Since the viscosity of water–glycerol mixtures is hardly known better than 1%, we must assume that the uncertainty of the Reynolds number was of the order of at least 1%.

Figure 14.2 also presents the results of the experiments of Dominguez-Lerma et al. (1986) concerning wave number selection by a spatial ramp, which is said to culminate in the selection of a unique wave number for supercritical Taylor vortex flow. The existence of a unique supercritical wave number in supercritical Taylor vortex flow would be of a fundamental importance and therefore these experiments deserve careful attention. The notion of a unique supercritical wave number contradicts the consensus that supercritical Taylor vortex flow is nonunique, which consensus was built on a long history of nonuniqueness going back to Coles' (1965) paper. Nonuniqueness has since then been observed innumerable times, as discussed in Chapter 13.

In the experiments on Taylor vortices with a spatial ramp Dominguez-Lerma et al. used an apparatus consisting of two different sections. The first section of length L had a gap of uniform width and radius ratio 0.75 and aspect ratios between 30 and 40, and was connected to a second section in which the radius of the outer cylinder decreased linearly in the axial direction, so the gap between both cylinders decreased steadily. The ramping angles were between 0.002 and 0.03. When the rotation rate of the inner cylinder was gradually increased, axisymmetric vortices formed in the straight section, not in the ramped cylinder section in which the local Reynolds number is smaller because the gap width is smaller. We disregard in the following some small end effects.

When the conditions in the straight section were supercritical, the wavelength of a pair of vortices in the straight section was measured. Then the length L of the column and with it the wavelength of the vortices was varied by moving the end plate of the straight section either inward or outward in order to study the range $\Delta\lambda$ of the variation of the wavelength. As L is varied the wavelength should vary because the wavelength in a column depends on L according to the formula $\lambda = L/Nd$. Using the ramped section as one end of the fluid column in the straight section introduces a flexible end condition, because the vortices can penetrate into the ramped section. Nevertheless the range $\Delta\lambda$ measured when L was varied at a given supercritical R did not differ much from the range given by the formula $\Delta\lambda/\lambda = \pm 2 \times 100/N$, which follows from the quantization condition for flow between two rigid ends. But now, with a ramp at the end of the straight section, the column could no longer be compressed (or expanded) as was possible in the case of a uniform gap with two rigid ends, the procedure which led to the experimental points marked by the solid circles in Fig. 14.2. The maximal or minimal wavelengths or their difference $\Delta\lambda$ observed with a ramped end at various R are marked at the center of Fig. 14.2. What the authors observed with the ramped end was that the wavelength of supercritical Taylor vortices in the straight section was practically constant and, as a function of ε, was equal to the critical wavelength. That is well known (see Fig. 14.1). When $R \geq 1.5R_c$ the observed mean values of the wavelength begin to differ from the critical wavelength λ_c. This may originate from the interaction of the vortices in the straight section with the vortices which begin to form in the ramped section when the condition there becomes critical. Taylor vortex flow in the ramped section is not discussed by Dominguez-Lerma et al., but was studied experimentally by Wimmer (1983) and will be discussed in Section 15.2. Summarizing, Dominguez-Lerma et al. concluded that the results shown in Fig. 14.2 "suggest strongly that a unique wave number $q_s(\varepsilon)$ near the centers of the bands for finite α would be selected in the limit of vanishingly small ramp angle α" (1986, p. 4967).

Much more important for the question of uniqueness than the points in the center of Fig. 14.2, originating from a quasi-steady increase of R, is the trend of the wavelength of supercritical vortices which have a wavelength significantly different from the critical wavelength. Such vortices can, as we have seen in Chapter 13, be produced either through sudden starts or other initial conditions. Dominguez-Lerma et al. address this question in Fig. 15 of their paper, where the time evolution of a pattern with a wavelength 6% larger than λ_c and a pattern with a wavelength 4% smaller than λ_c is shown. They say that "in all cases studied the decay was to a final state having the mean wavelength expected." That seems to prove indeed a tendency to a preferred wavelength. But the evidence is not convincing because these experiments were made at very small supercritical Reynolds number, $R = 1.02R_c$, and in an apparatus with a narrow gap, $\eta = 0.892$, in which wavy vortex flow is imminent at small supercritical Taylor numbers. In order to show unambiguously that there is, after a sudden start, a tendency to a preferred wavelength in an apparatus with a ramped section it must be shown that this tendency occurs in an apparatus with a larger gap, say $\eta = 0.75$, and at definitely supercritical Reynolds numbers, say $1.5R_c$ or more.

The selection of a truly unique wavelength must be independent of the sign of the ramp angle. Recent experiments by Ning et al. (1990) showed that the "unique" wavelength in the case of a diverging ramp on the inner cylinder, or in the case of converging ramps on the outer and inner cylinders, differs dramatically from the critical wavelength and varies in opposite directions. Similar results were obtained theoretically by Paap and Riecke (1991), who calculated a number of curves for the selected wave numbers of Taylor vortices in the homogeneous section of a Taylor vortex apparatus as a function of the Reynolds number, when the homogeneous section is connected to different subcritical ramps. There can be only one unique wavelength, not any number of them. Therefore one has to abandon the idea that the existence of a unique wavelength in supercritical axisymmetric Taylor vortex flow can be proved using such ramps. Anyway, the idea of the existence of a unique wavelength for supercritical Taylor vortices contradicts the experimental evidence, which has shown so convincingly that supercritical axisymmetric (as well as wavy vortex flow) is nonunique. Nonuniqueness of supercritical Taylor vortex flow can be considered a certainty. By using a particular boundary condition at one column end which differs from the boundary condition at the other column end the wavelength of the vortices in a column of uniform width can be forced, as the experiments of Dominguez-Lerma et al. and Ning et al. show. But that is irrelevant to the Taylor vortex problem in general. In an infinite column all vortex pairs experience the same forces from above or below and therefore do not expand or contract at a given $\mathcal{T}$, and the same applies in a fluid column with resting end plates on the top and bottom. All that a ramp at one end does is to introduce a one-sided force on the vortices. This is not part of the Taylor vortex problem that we discuss when we face the problem of nonuniqueness of Taylor vortex flow.

The idea that a spatial variation of a control parameter may cause the selection of a preferred wave number was originally proposed by Kramer et al. (1982), who used an amplitude equation with an ad hoc forcing term, was elaborated on by Hohenberg et al. (1985), and also applied to the Rayleigh–Bénard convection problem by Kramer and Riecke (1985). The results of the experiments with ramped Taylor vortices or with nonuniformly heated (ramped) Rayleigh–Bénard convection do not seem to support the idea that a unique wave number for the unbounded problem can be found this way. In Rayleigh–Bénard convection the wavelength seems to be unique to begin with.

We shall now look at the results of the numerical studies of supercritical axisymmetric Taylor vortex flow. Modern numerical studies of nonlinear axisymmetric Taylor vortices were made by Meyer-Spasche and Keller (1980), who were mainly interested in the torque and the determination of the critical Reynolds number, by Jones (1981, 1985), who was mainly interested in the transition to wavy vortex flow, and by Fasel and Booz (1984), who focused on axisymmetric vortices at very high Taylor numbers. They studied the flow in an infinite fluid column of radius ratio $\eta = 0.5$ with a resting outer

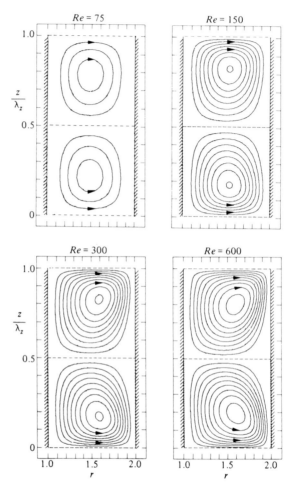

Fig. 14.3. Computed streamlines of axisymmetric supercritical Taylor vortices in a gap with
η = 0.5 for R = 75, 150, 300, 600 with R_c = 69. After Fasel and Booz (1984).

cylinder and a fixed wavelength λ = 2.0, starting with Reynolds numbers
below critical and then increasing the Reynolds number up to R = 690 or
$100\mathcal{T}_c$. They give a detailed description of the flow field at various Reynolds
numbers, of which we reproduce here the figure showing the streamlines of
the flow at various Reynolds numbers (Fig. 14.3). This figure makes the increasing concentration of the outgoing flow very obvious. Since according to
Fig. 14.3 the sources and sinks are not of the same intensity, the vortices appear to be arranged in pairs, as one can see easily in experiments with supercritical Taylor vortices (see Fig. 13.6). Another feature apparent in the

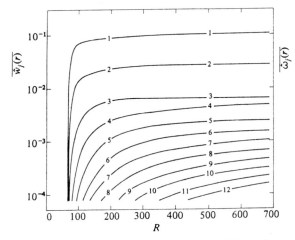

Fig. 14.4. Amplitudes of the harmonic components of the vertical velocity component of axisymmetric Taylor vortices averaged over the gap, as a function of the Reynolds number. After Fasel and Booz (1984).

flow fields computed by Fasel and Booz is the increasing concentration of the azimuthal and vertical motions in boundary layers at the cylinder walls; this means that "at large R the fluid mass in the center portion of each cell moves with almost constant velocity in the azimuthal direction" (1984, p. 40).

Fasel and Booz also determined the amplitudes of the higher harmonics of the flow variables by Fourier decomposition of the flow fields at different Reynolds numbers. There are strong local variations of the amplitudes of the variables in the gap, but when averaged over the gap it was found that "the lower the harmonic the stronger is its initial growth. But more important, beyond the [initial] strong amplification there is a rapid levelling off with an obvious asymptotic behaviour. . . . For large Re the lower harmonics remain practically constant while higher harmonics still rapidly grow with Re" (p. 50). So it turns out that a large number of harmonics contribute to the total solution at large R. In the strongest example, the $n = 12$ component of vorticity at $R = 690$ was found to have still more than 20% of the amplitude of the $n = 3$ component. The variation of the amplitudes of the averaged harmonics of the vertical velocity component is shown in Fig. 14.4. This figure has a startling qualitative similarity to the amplitude curves of the first four harmonics of the shear measured at the inner wall by Snyder and Lambert (1966) in an apparatus with $\eta = 0.5$.

If we now look back and summarize what we have learned about the theory of supercritical Taylor vortices, we find that, in spite of the progress made, we are not yet able to explain why the wavelength of supercritical axisymmetric vortices is independent of the Taylor number and equal to the critical wavelength when the Taylor number is increased slowly from subcritical values. Nor do we understand why supercritical axisymmetric vortices can be nonunique. As far as the more difficult wavy vortex problem is concerned, we do not have an explanation for the nonuniqueness of the azimuthal wave number, nor can we explain the complexities of Coles' stability diagram. Basic questions still await answers.

15

MISCELLANEOUS TOPICS

15.1   Taylor Vortices between Eccentric Cylinders

The basic arrangement of the Taylor vortex experiments with two circular concentric cylinders can be varied in several ways, one of which is to make the cylinders eccentric. A very substantial amount of theoretical as well as experimental work has been devoted to the eccentric Taylor vortex problem. As soon as the cylinder axes do not coincide the Taylor vortex problem is no longer axisymmetric, which brings about a significant increase in the mathematical difficulty of the problem. Taylor vortices between eccentric cylinders have practical implications; they cause the so-called superlaminar regime in journal bearings in which the results of classical lubrication theory do not hold any longer although the flow is still laminar. Journal bearings are characterized by very small clearance ratios of the order of 10^{-3}. The clearance ratio is defined by $\delta = d/r_i$, where d is the gap width in the concentric case, or the mean gap width in the eccentric case, and r_i the radius of the inner cylinder. The clearance ratio is related to the radius ratio in the concentric case by $\delta = (1 - \eta)/\eta$; δ can be larger than 1. A new parameter of the eccentric problem is the eccentricity ε, defined as the ratio of the distance of the axes of the cylinders divided by the mean gap width, with $0 \le \varepsilon \le 1$.

The eccentric arrangement of the cylinders with a wide gap on the one side and a narrow gap on the opposite side raises the question of whether the onset of the Taylor vortex instability will occur at a Taylor number smaller than the critical Taylor number in the concentric case, i.e. whether the side with the wider gap determines the onset of instability where the Taylor number is larger because of the dependence of the Taylor number on d^4 or whether the rotation rate of the inner cylinder has to be increased beyond the critical rotation rate of the concentric case, i.e. whether the narrow gap determines the onset of instability. One would like to assume that the Taylor vortices still

Fig. 15.1. Eccentric axisymmetric Taylor vortices in a fluid column with $\Gamma = 38$, $\varepsilon = 0.371$, and $\delta = 0.375$ at $\mathcal{T} = 1.32\mathcal{T}_c$. The wide gap is to the left; rotation of the inner cylinder is from right to left. The entire column is filled with vortices of the same size and appearance, except for the end vortices. The heavy dark lines are the sinks, the location of inward flow at the outer cylinder; the fine dark lines are the sources. The intensification of the sinks on the left side of the column indicates separation. Without separation axisymmetric eccentric Taylor vortices cannot be distinguished visually from axisymmetric Taylor vortices between concentric cylinders. After Koschmieder (1976).

appear in their usual form if the eccentricity is small, but one wonders whether large eccentricities will not modify the form of the Taylor vortices or eliminate the Taylor vortices altogether.

The first experiments on Taylor vortex flow between eccentric cylinders were made by Cole, whose results, including his much earlier preliminary studies, were published in Cole (1967). Since very small clearance ratios are difficult to work with, he tried clearance ratios between 0.138 and 0.478. Supplementary measurements with two additional clearance ratios are in Cole (1969). His cylinders were of moderate length; he used oil and air as test fluids. He found first of all that in the eccentric case "the Taylor vortex pat-

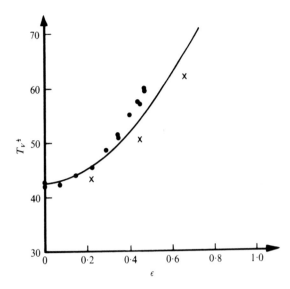

Fig. 15.2. Increase of the critical condition for the formation of eccentric Taylor vortices as a function of the eccentricity ε, according to the measurements of Vohr with δ = 0.099 (circles) and Kamal with δ = 0.09 (crosses). The curve represents the theoretical prediction of Eagles et al. The ordinate is in Reynolds numbers with a correction for δ and ε². After Eagles et al. (1978).

tern still appears'' (p. 1200), even at eccentricities as large as about 0.5. Figure 15.1 illustrates Taylor vortices between eccentric cylinders. Cole observed also that the critical Taylor number in the eccentric case increases rapidly with eccentricity; in other words, eccentricity was found to be stabilizing. He determined the critical condition from the break of the torque versus rotation rate curve. Results of two experiments similar to his but with smaller clearance are shown in Fig. 15.2 together with a theoretical prediction.

Cole's results were confirmed by Kamal (1966), who studied primarily the separation of the circumferential flow from the outer cylinder which occurs at eccentricities larger than 0.3. We shall return to separation soon. In order to support his theoretical results Kamal made a number of experiments measuring the onset of instability. He too found that eccentricity is stabilizing in the case δ = 0.09 and Γ = 35.4. As Fig. 15.2 shows, his measurements are in close agreement with Vohr's results. Vohr (1968) made another set of experiments with eccentric Taylor vortices, combining torque measurements with flow visualization experiments in a long apparatus with δ = 0.099 and Γ ≅ 62. He found practically the same rapid increase of the critical Taylor number with eccentricity that Cole and Kamal had observed. In the visualization experiments Vohr also noted wavy Taylor vortices between eccentric cylinders to which Kamal already had referred in a one-line sentence. Eccentric wavy Taylor vortices form under supercritical conditions.

Although the stabilizing effect of eccentricity on the onset of the Taylor vortex instability should have been established by the three experiments discussed above, a period of uncertainty about the consequences of eccentricity for the Taylor vortices followed. This seems to have resulted from the so-called local theory of instability between eccentric cylinders, which predicted that eccentricity may be destabilizing. We shall not pursue the "local" theory of the instability because it was superseded by a better "nonlocal" description of the flow in which the whole flow field is taken into account. The experiments of Castle and Mobbs (1968) indicated that the onset of instability may occur at Taylor numbers smaller than the critical Taylor number in the concentric case $\mathcal{T}_c$(conc). Castle and Mobbs observed a "first instability" at values $\mathcal{T} < \mathcal{T}_c$(conc) which did not extend over the entire gap. The "first instability" was made visible with dye released at the inner cylinder. On the other hand, torque measurements made with a similar apparatus indicated a stabilizing effect of eccentricity, similar to the earlier results of Cole, Kamal, and Vohr. Versteegen and Jankowski (1969) studied the effect of eccentricity on the flow between two independently rotating cylinders. They found that eccentricity could be stabilizing or destabilizing depending on the speed of the outer cylinder, and that the destabilizing effect became more pronounced as Ω_2 increased. Frêne and Godet (1971) worked with an apparatus with very small clearance ($\delta = 0.005$) and an aspect ratio $\Gamma = 1000$ and observed the onset of instability at $\mathcal{T} < \mathcal{T}_c$(conc) when $\varepsilon \cong 0.33$.

The stabilizing effect of eccentricity was confirmed again by Castle et al. (1971) in a paper that was mainly concerned with wavy Taylor vortices between eccentric cylinders, a topic that we shall not pursue here because the wavy vortices complicate substantially a problem that is already made difficult by the eccentricity. We note, however, that wavy eccentric vortex flow was investigated theoretically by Weinstein (1977a,b). Castle et al. (1971) found the onset of plane eccentric Taylor vortices at values of $\mathcal{T} > \mathcal{T}_c$(conc) using torque and visual measurements; this means that they found a stabilizing influence of eccentricity. Koschmieder (1976) used an apparatus with $\delta = 0.375$ and $\Gamma \cong 50$ and observed the onset of instability at Taylor numbers which were practically at the same increasing values that Cole had found. Apparently no other experimental determination of the critical Taylor number as a function of eccentricity has been made. The consensus seems to be that eccentricity has a stabilizing effect on the onset of Taylor vortex flow.

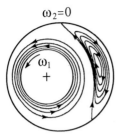

$\omega_2 = 0$

ω_1
+

Fig. 15.3. Streamlines of separated two-dimensional flow between two eccentric cylinders. The inner cylinder is rotating. $\varepsilon = 0.5$, $\delta = 1.0$, $R_i = 40$. After San Andres and Szeri (1984).

In Koschmieder's (1976) paper is a curve showing the variation of the axial wavelength of the vortices at the onset as a function of eccentricity. Whereas up to $\varepsilon \cong 0.3$ the critical wavelength is practically independent of ε, and within the experimental uncertainty equal to λ_c of the concentric case, for $\varepsilon > 0.4$ the wavelength decreases noticeably. This observation is in qualitative agreement with a sentence in Cole (1967) and a curve in Castle and Mobbs (1968) for their so-called second instability, which extended throughout the gap. The decrease of the critical wavelength with increased eccentricity seems to indicate that the characteristics of eccentric Taylor vortices are determined by the narrow section of the gap, in agreement with the stabilizing influence of eccentricity. There is an interesting sentence in Kamal (1966) according to which "the wavelength seemed to decrease with increase in speed" of the inner cylinder (p. 717). This, when confirmed in detail, will be of importance for nonlinear flow between eccentric cylinders.

We shall now turn to the question of whether the eccentricity brings about a substantial modification of the basic flow around the inner cylinder and with it of the Taylor vortices when the fluid is unstable. The first theoretical study of the flow between eccentric cylinders with the inner cylinder rotating was made by Wannier (1950) using classical lubrication theory (Reynolds, 1886). Wannier calculated the streamlines of the flow and showed an example of the separation of the flow around the inner cylinder. In this case the fluid forms, besides the circumferential circulation around the inner cylinder, an eddy in the wide gap with a circulation opposite to the circulation around the inner cylinder. For an example of separation in the flow between eccentric cylinders see Fig. 15.3.

Separation between eccentric cylinders was treated again by Kamal (1966), also with the approximations of lubrication theory. He gave a formula for the azimuthal velocity component of the basic flow as a function of the azimuth angle θ. The azimuth angle θ is measured at the center of the inner cylinder from the line connecting the axes of both cylinders, with $\theta = 0$ pointing in the direction of the large gap. In the case in which the clearance between both cylinders is very small and the Reynolds number is small it is

$$v_\theta = \frac{1}{2\mu R} \frac{\partial p}{\partial \theta}(r - h)r + \frac{r}{h}v_r, \tag{15.1}$$

where h is the variable gap width given by

$$h = d(1 + \varepsilon \cos \theta), \tag{15.2}$$

with the gap width $d = r_o - r_i$. In (15.1) μ is the dynamic viscosity, R is the Reynolds number $R = v_\theta d/v$, r is measured from the inner cylinder, and ε is the eccentricity. The pressure p between eccentric rotating cylinders was determined by Sommerfeld (1904). Using his formula for the pressure and the criterion for separation $dv_\theta/dr = 0$, Kamal arrived at a formula for the angle at which separation begins at the outer cylinder. It is

$$\cos \theta_s = \frac{1 - 4\varepsilon^2}{\varepsilon(2 + \varepsilon^2)}. \tag{15.3}$$

According to this formula separation does not occur for eccentricities $\varepsilon < 0.3$. Also it follows that the angles of separation and of reattachment of the circumferential flow to the outer cylinder are symmetric with respect to the line through the centers of the two cylinders. And (15.3) says that the angle of separation is, in this approximation, independent of the clearance δ and the Reynolds number.

Introducing a bipolar coordinate system Kamal (1966) solved the Stokes equation, i.e. the Navier–Stokes equations without inertial terms, and calculated the streamlines of an example of flow with separation. His solution is qualitatively similar to Wannier's solution, which Kamal was not aware of. In Kamal's solution the streamline of separation $\Psi = 0$ intersects the outer cylinder at right angles, whereas in Wannier's solution this angle is acute, as it is in reality. Since we discuss here Taylor vortices between eccentric cylinders, we shall not pursue the theory of separated flow between eccentric cylinders in detail, other than to refer to Ballal and Rivlin (1976) and San Andres and Szeri (1984), who have treated this problem in great detail. There does not seem to exist an experimental confirmation of the results of these studies.

What is of importance here is that separation, in other words the presence of a large eddy with a circulation opposite to the circumferential circulation

Fig. 15.4. Top view of separated flow in the presence of Taylor vortices. Rotation of the inner cylinder is clockwise. $\varepsilon = 0.556$, $\delta = 0.375$, $\mathcal{T} = 1.05\mathcal{T}_c$. Visualization by aluminum powder floating on the fluid surface. After Koschmieder (1976).

around the inner cylinder, has only a marginal effect on the formation and stability of the Taylor vortices between both cylinders. As Fig. 15.1 shows separation leaves only a weak mark on the Taylor vortices, which extend virtually unaltered from the narrow gap to the wide gap, where the flow is separated. One particular consequence of separation is that, near the point of separation at the outer cylinder, the fluid moves in a direction *opposite* to the direction of the inner cylinder! This is another example of the astounding stability of Taylor vortices. In this case the vortices form in spite of the presence of the separated eddy and actually force their form upon the separated flow. The horizontal streamlines of separated flow in the presence of Taylor vortices as observed from above are shown in Fig. 15.4. As seen from the side the flow in this case is very similar to the flow in Fig. 15.1. Note that reattachment does not leave a clear mark in Fig. 15.4, in agreement with the observation that reattachment does not leave a noticeable mark when observed from the side. As far as the occurrence of separation is concerned, it has been confirmed experimentally that $\varepsilon > 0.3$ is indeed required to observe separation in Taylor vortex flow between eccentric cylinders under critical conditions. It is, on the other hand, possible to observe the consequences of separation at values $\varepsilon < 0.3$ if the Taylor vortices are supercritical. It has also been observed that the angle of separation θ_s for a specific ε increases if the Taylor number is increased above critical.

The basic flow between two eccentric, rotating, circular cylinders of small eccentricity and clearance and infinite length was investigated theoretically by DiPrima and Stuart (1972a). They derived the Sommerfeld pressure distribution and the associated flow as a solution of the Navier–Stokes equations by an expansion in the clearance ratio and the Reynolds number. The formula for the pressure distribution is rather long, so we shall refer here to equation (91) of DiPrima and Stuart. Separation was found to occur only if $\varepsilon \geq 0.30278 + 0.03818\delta$. The stability of this basic flow with respect to infinitesimal disturbances of the Taylor vortex type with only the inner cylinder rotating was then studied in the linear and nonlinear cases by DiPrima and Stuart (1972b, 1975), and their results were summarized and corrected by Eagles et al. (1978).

Assuming slightly supercritical conditions and that the clearance ratio, the eccentricity, and the Taylor vortex amplitude are small, and working with a constant (critical) wave number of the vortices, Eagles et al. found after long calculations that the critical Taylor number of the flow between eccentric cylinders is given by

$$\mathcal{T}_{ac} = 1694.97(1 + 1.1618\delta)(1 + 2.6185\varepsilon^2) + O(\delta^2, \delta\varepsilon^2, \varepsilon^4), \qquad (15.4)$$

using for the Taylor number the formula

$$\mathcal{T}_a = (q_1 r_1/\nu)^2 \delta^3, \qquad (15.5)$$

where q_1 is the velocity of the inner cylinder and r_1 the radius of the inner cylinder. The formula (15.4) for the critical Taylor number as a function of ε and δ says that eccentricity is stabilizing. That is in agreement with the experimental results (see Fig. 15.2). Also of interest are the flow and the pressure field. In the linear approximation the Taylor vortices have, to the lowest order, the form of Taylor vortices between concentric cylinders modified by a multiplicative factor $B(\phi)$ which varies the strength of the vortices in azimuthal direction. It is, in the linear case, and for small ε and δ,

$$B(\phi) = \text{const} \exp\left[\frac{\Gamma}{kT_0^{1/2}}(\sin\phi - 1)\right], \qquad (15.6)$$

where $\Gamma = 23.09$ is a constant, $T_o = 1695$, and k relates the clearance ratio and eccentricity. ϕ is the bipolar angle which is in the first order in ε the same as the azimuthal angle θ in polar coordinates. In the nonlinear case $B(\phi)$ is much more complicated; $B(\phi)$ is plotted as a function of ϕ in Fig. 5 of Eagles et al. (1978) for one set of parameters. Similar to the velocity field the pressure between eccentric cylinders depends on the azimuthal angle. This causes a plane Poiseuille flow to develop between the cylinders. A pressure distribution calculated for slightly supercritical flow is plotted in Fig. 6 of Eagles et al. With the study of Eagles et al. the investigation of Taylor vortex flow between eccentric cylinders seems to have come to a temporary end. Pending in particular is an experimental investigation of the nonlinear aspects of Taylor vortex flow between eccentric cylinders. Although we have measurements of the torque between eccentric cylinders which extend into the nonlinear regime, we do not know how the wavelength of the vortices varies under supercritical conditions, when wavy vortex flow sets in, and whether or not the supercritical flow is unique or nonunique. The theoretical investigation of Taylor vortex flow between eccentric cylinders is taken up in Chapter 5 of Chossat and Iooss (1992).

15.2 Taylor Vortices between Conical Cylinders

Another simple way to change the traditional arrangement of the Taylor vortex experiments is to make the coaxial cylinders conical. There are a number of possible variations of the form of either the inner or the outer cylinder or both, varying the slope of the cylinder wall or the sign of the slope. As soon as there is a slope in a cylinder wall the Reynolds number becomes a local Reynolds number because on a sloping inner wall the Reynolds number varies as the radius of the inner cylinder. It is not worthwhile to pursue all possibilities, because it appears that in all cases the flow that develops is similar in that it is, up to moderately supercritical conditions, the superposition of a basic three-dimensional Couette flow upon the toroidal Taylor vortices caused by the instability of the basic flow with regard to disturbances of the Taylor vortex type. The Taylor vortex problem between conical cylinders was investigated experimentally by Wimmer (1983) and numerically by Abboud (1988).

The most simple case of flow between conical cylinders is that in which the slopes of the inner and the outer cylinder are the same, so that the gap width between the cylinders is constant. An example of a stationary flow in this geometry with a resting outer cylinder is shown in Fig. 15.5. Clearly visible

Fig. 15.5. Taylor vortices of alternating larger and smaller size between a conical rotating inner and a conical resting outer cylinder. The gap width is constant, $d = 1.0$ cm, the slope of the cylinder walls $8.15°$, axial length 12.5 cm, $r_{\mathrm{imax}} = 4$ cm, $R = r_{\mathrm{imax}}^2 \omega / v = 1037$. Steady state, resting top and bottom plates. After Wimmer (1983).

in this figure are Taylor vortices of alternating large and small size, similar to the convection rolls of alternately large and small size on a nonuniformly heated plate in Fig. 8.3. In the large Taylor vortices the circulation of the vortices is of the same sense as the overall three-dimensional basic circulation which extends throughout the length of the gap; in the small vortices the circulation is opposite to the overall circulation. The overall circulation is one long loop going up as a spiral along the inner cylinder and coming down as a spiral along the outer cylinder. This circulation forms because of the dependence of the centrifugal force on r, which makes the radial pressure gradient on top of the fluid larger than at the bottom of the column. This determines the direction of the overall circulation. Also, the overall three-dimensional Couette circulation present in the flow in Fig. 15.5 decreases in amplitude in the upward direction, because the area of a horizontal section through the gap increases upward. Since the velocity components of the overall circulation are large at the bottom of the column, the bottom vortex in Fig.

15.5, which turns with the overall circulation, is much larger than the second vortex from below, which turns in the opposite direction, meaning opposite to the overall circulation. The bottom vortex is also much larger than the second vortex from the top, which turns with the overall circulation. This is so because of the decrease of the amplitude of the basic circulation in the upward direction. The wavelength of the bottom vortex is about 2.2 times larger than the wavelength of the second vortex from the top. The vertical extension of the second vortex from the top is about equal to the gap width. As can be seen in Fig. 15.5 all five vortices turning against the overall circulation are nearly of the same size, and they are nearly as long as the gap is wide.

If the Reynolds numbers are smaller than in the case just discussed, the basic flow can coexist side-by-side with Taylor vortices. There can then be, for example, a single vortex pair on top of the column, whereas the rest of the column is filled with the basic circulation slightly modified by a weak vortex adjacent to the single vortex pair on top. Streamlines of such a flow were computed by Abboud (1988). Wimmer also observed that different numbers of vortices can fill the fluid column at the same rotation rate of the inner cylinder, depending on the acceleration rate of the cylinder. That means that Taylor vortex flow between conical cylinders is nonunique and corresponds in this respect to the flow between coaxial right circular cylinders. As we shall see the interaction of an overall circulation with a local Taylor vortex instability that occurs between conical cylinders also occurs in the flow between two concentric rotating spheres.

15.3 Taylor Vortices between Rotating Spheres

So many papers have been written about Taylor vortex flow between concentric spheres that a summary of Taylor vortex flow would be incomplete without reference to this topic. Taylor vortex flow between two concentric rotating spheres is an academic problem which is much more complex than the Taylor vortex problem between concentric cylinders but is of interest for applied mathematics. The problem has, in theory, the advantage of a fluid layer between two spheres having no ends, meaning that Taylor vortices can form without interference of the troublesome end conditions. In reality there is, however, always the shaft which turns the inner sphere and introduces an end condition. But one hemisphere of the fluid layer is in a good approximation free from end effects. It is obvious that in some configurations of two concentric spheres, with the inner one rotating, the Taylor vortex instability must occur, because between the two tangential cylindrical surfaces at the equator of the spheres the stability problem is just the same as in the usual Taylor vortex experiment. The early papers on the instability of the flow between two rotating spheres were the analytical studies of Bratukhin (1961), Haberman (1962), Ovseenko (1963), and Yakushin (1969, 1970), all limited to low values of the Reynolds number. The first report of an experimental investigation of the instability of a thin fluid layer between a rotating inner and resting outer sphere was that of Khlebutin (1968). This short paper confirmed the occurrence of instability and noted the absence of instability in larger gaps when $\sigma = (r_2 - r_1)/r_1 > 0.19$. The fluid motions were made visible with aluminum powder, and it was observed that annular vortices formed at the critical condition in the equatorial region. These Taylor vortices were stable and stationary.

The first thorough experimental investigation of Taylor vortices between concentric spheres was that of Sawatzki and Zierep (1970). From their pictures we reproduce here a photograph showing the basic flow between a rotating inner and resting outer sphere (Fig. 15.6). The three-dimensional axisymmetric basic flow is caused by different radial pressure gradients at the equator and the poles, which differ because of the dependence of the centrifugal force on r. The motion is therefore outward at the equator, continues along the outer sphere with an azimuthal velocity component, is inward at the poles, returns along the inner cylinder, and is symmetric about the equatorial plane. The streamlines are spirals, as is obvious in Fig. 15.6; the slope of the spirals is determined by the rotation rate. There is no simple formula for the basic flow as was the case for Couette flow between concentric cylinders [equation (10.3)]; spherical Couette flow can be calculated only in approximations. When in the experiments the rotation rate was increased beyond a critical value, which depends in particular on the gap width, Taylor vortices formed near the equator; the basic flow, however, remained over a large part of the gap. There is only a local critical Reynolds number because the basic flow between two rotating spheres depends on the latitude, whereas between concentric cylinders the basic flow is independent of z. An example of instability near the equator is shown in Fig. 15.7. It cannot be seen in Fig. 15.7, but the vortices are of alternately larger and smaller size, because every second vortex turns in a direction opposite to the overall circulation of the basic flow. The size difference of two adjacent vortices has been measured by Sawatzki and Zierep and is obvious in the computed velocity fields of Bartels (1982). The number of vortices in a given gap depends on the rotation rate and in particular on the gap width. If the gap width is small, many vortices can appear consecutively when the rotation rate is increased after the critical

Fig. 15.6. Basic flow between a rotating inner and resting outer sphere. The gap width is 5 mm, r_o = 80 mm, σ = 0.0667, $R = r_i^2\omega/v$ = 2250, $\mathcal{T} \approx \mathcal{T}_c \cong$ 44. Courtesy J. Zierep.

condition has been reached at the equator. There will nevertheless be a remnant of the basic flow near the poles.

If in a given apparatus the rotation rate is increased sufficiently the vortices become wavy and in the end turbulent, just as was the case between concentric cylinders. Turbulent vortices were clearly identifiable vortices which covered the entire gap from the equator to the poles, but were of spiral form converging at the pole; see, e.g., Fig. 11 in Sawatzki and Zierep, taken at a Reynolds number $R = r_1^2\,\Omega/v$ = 1.58 × 10^5. In these experiments at least five different states and additional substates were observed over the entire range of rotation rates. We shall not discuss all these states because of their confusing multitude. These experiments were continued by Wimmer (1976), who pointed out that different initial conditions lead to different modes of flow at the same Reynolds number, as one would expect from our experience

Fig. 15.7. Taylor vortices in the fluid between a rotating inner and resting outer sphere. The gap width is 3.5 mm, r_o = 80 mm, σ = 0.0458, R = 7600, $\mathcal{T}$ = 74. Courtesy J. Zierep.

with Taylor vortices between concentric cylinders. A long series of experiments investigating the stability of spherical Couette flow was made by Yavorskaya and her colleagues (Yavorskaya et al., 1975, 1977, 1978; Yavorskaya and Belyaev, 1991). Of particular interest is their observation (Belyaev et al., 1978) that for $\sigma > 0.23$, i.e. for large gaps, Taylor vortices did not form at all. Munson and Menguturk (1975) made experiments with three different gaps, σ = 0.135, 1.273, and 2.289. Their photographs show the basic flow in these gaps. With σ = 2.289 they observed no transition until the flow became turbulent. Wimmer (1981) made a series of experiments in which the outer sphere rotated also. It is startling to find in this paper a stability diagram for the flow between two rotating spheres which is practically the same as Taylor's (1923) stability diagram (Fig. 11.4) for the flow between two rotating cylinders, rotating either in the same direction or in opposite

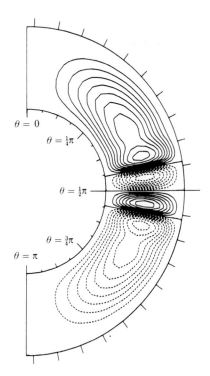

$\theta = 0$

$\theta = \frac{1}{4}\pi$

$\theta = \frac{1}{2}\pi$

$\theta = \frac{3}{4}\pi$

$\theta = \pi$

Fig. 15.8. Meridional streamlines of two-vortex flow between a rotating inner and resting outer sphere at $R = 90$ in a gap with $\sigma = 0.18$. After Marcus and Tuckerman (1987a).

directions. A thorough experimental study extending from the basic flow to turbulence was made by Nakabayashi (1983), who observed 14 different flow regimes. In many of the experiments discussed above torque measurements were made which confirmed the existence of the transitions observed visually. We cannot pursue this topic here.

After the early analytical studies of the stability of the flow between two rotating spheres, the topic was studied again by Munson and Joseph (1971a), who investigated the basic flow and studied its stability (Munson and Joseph, 1971b) with the energy method. The stability of the basic flow in the linear approximation was investigated by Munson and Menguturk (1975), by Walton (1978) for narrow gaps and a resting outer sphere, and by Soward and Jones (1983) for spheres rotating with different angular velocities. The results of these studies are in agreement with the experimental observations of the onset of instability. In narrow gaps the value of the critical Taylor number between concentric spheres approaches the critical Taylor number between concentric cylinders, as it must. The flow between concentric spheres was also studied in a number of numerical investigations, beginning with Bonnet and Alziary de Roquefort (1976) and followed by Bartels (1982), Dennis and Quartapelle (1984), Schrauf (1986), and Marcus and Tuckerman (1987a,b). These elaborate studies reproduced many of the features observed in the experiments, mostly for gap widths around $\sigma = 0.15$ in order to compare the results with Wimmer's (1976) measurements. For this value of σ the numerical technique is superior to the analytical studies because too many terms in the basic flow have to be taken into account when the gap is wide. To demonstrate the numerical results we reproduce here the streamlines of a two-vortex flow (Fig. 15.8). In view of the expense required to make such studies one wonders whether it is practical to pursue this line of investigation much further. Recently the interest has shifted to Taylor vortex flow between concentric spheres which is asymmetric about the equatorial plane. Although the basic flow is symmetric about the equator, an asymmetric pair of vortices has been found numerically and experimentally in a small range of supercritical Reynolds numbers. Both types of asymmetric pairs, either with the larger vortex in the upper hemisphere or with the larger vortex in the lower hemisphere, have been observed in the experiments of Bühler (1990) with equal probability. They were created either by strong acceleration or by slow deceleration.

We have learned from these studies many features of Taylor vortex flow between rotating spheres, and we have found many similarities to the Taylor vortex problem between concentric cylinders. But this problem has remained a problem of interest for applied mathematics only, and has not furthered our understanding of the Taylor vortex problem between concentric cylinders, which was our prime concern.

REVIEW

Looking back on what we have learned about Taylor vortices since Taylor's (1923) original paper, what seems to be most impressive is the tremendous progress that has been made in the experimental field. Foremost among the modern experimental discoveries is the observation of wavy vortices by Coles (1965). This started the experimental investigation of nonlinear Taylor vortex flow. Since then nonlinear Taylor vortices have been explored in great detail, ending up with the observation of well-organized turbulent Taylor vortices. Our knowledge about Taylor vortices extends now over the entire range of Taylor numbers, from the critical Taylor number to Taylor numbers of the order of a million times $\mathcal{T}_c$. The basic experimental observations of Taylor vortex flow seem to have been made.

The theoretical description of the Taylor vortices builds on Taylor's (1923) linear theory of the instability of Couette flow. In the 30 years following Taylor's paper the theoretical effort was mainly concerned with, to quote DiPrima, "more elegant and practical techniques for solving the eigenvalue problem" (Krueger et al., 1966, p. 521) for a larger range of parameters without using the narrow gap approximation. It appears that the linear investigation of the Taylor vortex instability has been completed. At the end of the 1950s the theoretical investigation of weakly nonlinear Taylor vortex flow began. This effort culminated in the derivation of the amplitude equation and the determination of the critical condition for the onset of wavy vortex flow between a rotating inner and resting outer cylinder. The analytical theory of weakly nonlinear flow has been extended into the moderately supercritical range in only a very few studies. In the moderately supercritical range nonunique wavy vortex flow blossoms, as is exemplified by Coles' stability diagram. We certainly cannot claim to understand Coles' diagram of the nonunique states. Further up on the Taylor number scale is modulated wavy vortex flow, which has not been explained yet, and even further up is irregular or chaotic flow, which has not been explained rigorously although chaotic Taylor vortex flow is considered a prime example of "chaos." It is too much to ask to continue this theoretical line to turbulent Taylor vortices. But actually the theoretical description of turbulent Taylor vortices may be less difficult than the description of chaotic Taylor vortex flow because extremely high Taylor numbers may make it possible to simplify the Navier–Stokes equations significantly. One may ask whether it is worth the effort to pursue these obviously very difficult nonlinear aspects of the Taylor vortex problem. The answer to this question seems to be yes, because in the case of Taylor vortex flow we can pursue the formation of turbulence from laminar flow to full turbulence with great precision in all detail through a number of very characteristic stages. In other words, basic theoretical work can be done on this problem, which may, in the end, rank as high as the pioneering studies that we have examined in this book.

REFERENCES

Abboud, M. (1988). Ein Beitrag zur theoretischen Untersuchung von Taylor-Wirbeln im Spalt zwischen Zylinder/Kegel-Konfigurationen. *Z. angew. Math. Mech.* **68**, T 275–277.

Ahlers, G. (1974). Low-temperature studies of the Rayleigh-Bénard instability and turbulence. *Phys. Rev. Lett.* **33**, 1185–1188.

Ahlers, G., & Behringer, R.P. (1978). Evolution of turbulence from the Rayleigh-Bénard instability. *Phys. Rev. Lett.* **40**, 712–716.

Ahlers, G., Cannell, D.S., & Dominguez-Lerma, M.A. (1983). Possible mechanism for transitions in wavy Taylor-vortex flow. *Phys. Rev. A* **27**, 1225–1227.

Ahlers, G., Cannell, D.S., & Steinberg, V. (1985). Time-dependence of flow patterns near the convective threshold in a cylindrical container. *Phys. Rev. Lett.* **54**, 1373–1376.

Ali, M., & Weidman, P.D. (1990). On the stability of circular Couette flow with radial heating. *J. Fluid Mech.* **220**, 53–84.

Alziary de Roquefort, T., & Grillaud, G. (1978). Computation of Taylor vortex flow by a transient implicit method. *Comput. Fluids* **6**, 259–269.

Andereck, C.D., Liu, S.S., & Swinney, H.L. (1986). Flow regimes in a circular Couette system with independently rotating cylinders. *J. Fluid Mech.* **164**, 155–183.

Avsec, D. (1939). Tourbillons thermoconvective dans l'air: Application a la météorology. Thesis, University of Paris.

Babenko, K.I., Afendikov, A.L., & Yur'ev, S.P. (1982). Bifurcation of Couette flow between rotating cylinders in the case of a double eigenvalue. *Dokl. Akad. Nauk SSSR* **266**, 73–78; *Sov. Phys. Dokl.* **27**, 706–709.

Ball, K.S., & Farouk, B. (1988). Bifurcation phenomena in Taylor–Couette flow with buoyancy effects. *J. Fluid Mech.* **197**, 479–501.

Ball, K.S., & Farouk, B. (1989). A flow visualization study of the effects of buoyancy on Taylor vortices. *Phys. Fluids A* **1**, 1502–1507.

Ballal, B.Y., & Rivlin, R.S. (1976). Flow of a Newtonian fluid between eccentric rotating cylinders: Inertial effects. *Arch. Rat. Mech. Anal.* **62**, 237–294.

Barcilon, A., & Brindley, J. (1984). Organized structures in turbulent Taylor–Couette flow. *J. Fluid Mech.* **143**, 429–449.

Barcilon, A., Brindley, J., Lessen, M., & Mobbs, F.R. (1979). Marginal instability in Taylor–Couette flows at very high Taylor number. *J. Fluid Mech.* **94**, 453–463.

Barenblatt, G.I., Iooss, G., & Joseph, D.D. (1983). *Nonlinear Dynamics and Turbulence.* Pitman, Boston.

Bartels, F. (1982). Taylor vortices between two concentric spheres. *J. Fluid Mech.* **119**, 1–25.

Batchelor, G.K. (1960). A theoretical model of the flow at speeds far above the critical. *J. Fluid Mech.* **7**, 416–418.

Behringer, R.P. (1985). Rayleigh–Benard convection and turbulence in liquid helium. *Rev. Mod. Phys.* **57**, 657–687.

Behringer, R.P., & Ahlers, G. (1982). Heat transport and temporal evolution of fluid flow near the Rayleigh-Bénard instability in cylindrical containers. *J. Fluid Mech.* **125**, 219–258.

Belyaev, Yu.N., Monakhov, A.A., & Yavorskaya, I.M. (1978). Stability of spherical Couette flow in thick layers when the inner sphere revolves. *Izv. Akad. Nauk SSSR, Mekh. Zh. i Gaza* **13**(2), 9–15; *Fluid Dyn.* **13**, 162–168.

Bénard, H. (1900). Les tourbillons cellulaires dans une nappe liquide. *Rev. Gén. Sciences Pure Appl.* **11**, 1261–1271, 1309–1328.

Bénard, H. (1901). Les tourbillons cellulaires dans une nappe liquide transportant de la chaleur par convection en régime permanent. *Ann. Chim. Phys.* **23**, 62–144.

Bénard, H. (1908). Formation de centres de giration à l'arriére d'un obstacle en mouvement. *Compt. Rend.* **147**, 839–842.

Bénard, H. (1928). Sur les tourbillons cellulaires, les tourbillons en bandes et la théorie de Rayleigh. *Bull. Soc. Franc. Phys.* No. 266, 112S–115S.

Bénard, H. (1930). *Proc. 3rd. Int. Congr. Appl. Mech.* Vol. 1, p. 120.

Bénard, H., & Avsec, D. (1938). Travaux récents sur les tourbillons cellulaires et les tourbillons en bandes, applications a l'astrophysique et la météorology. *J. Phys. Radium* **9**, 486–500.

Benguria, R.D., & Depassier, M.C. (1987). Oscillatory instabilities in the Rayleigh–Bénard problem with a free surface. *Phys. Fluids* **30**, 1678–1682.

Benguria, R.D., & Depassier, M.C. (1989). On the linear stability theory of Bénard–Marangoni convection. *Phys. Fluids A* **1**, 1123–1127.

Benjamin, T.B. (1978a). Bifurcation phenomena in steady flows of a viscous fluid. I. Theory. *Proc. Roy. Soc. London A* **359**, 1–26.

Benjamin, T.B. (1978b). Bifurcation phenomena in steady flows of a viscous fluid. II. Experiments. *Proc. Roy. Soc. London A* **359**, 27–43.

Benjamin, T.B., & Mullin, T. (1982). Notes on the multiplicity of flows in the Taylor experiment. *J. Fluid Mech.* **121**, 219–230.

Bensimon, D. (1988). Pattern selection in thermal convection: Experimental results in an annulus. *Phys. Rev. A* **37**, 200–206.

Berdnikov, V.S., Getling, A.V., & Markov, V.A. (1990). Wavenumber selection in Rayleigh–Bénard convection: Experimental evidence for the existence of an inherent optimal scale. *Exp. Heat Transfer* **3**, 269–288.

Bergé, P. (1975). Rayleigh–Benard instability: Experimental findings obtained by light scattering and other optical methods. In *Fluctuations, Instabilities and Phase Transitions*, pp. 323–352, ed. T. Riste. Plenum Press, New York.

Berkovsky, B.M., & Fertman, V.E. (1970). Advanced problems of free convection in cavities. *4th. Int. Heat-Transf. Conf. Paris*, Vol. 4, NC 2.1, 1–12. Elsevier, Amsterdam.

Bestehorn, M., & Haken, H. (1991). Associative memory of a dynamical system: The example of the convection instability. *Z. Phys. B* **82**, 305–308.

Bestehorn, M., & Pérez-Garcia, C. (1987). Coexistence of patterns with different symmetries in Bénard–Marangoni convection. *Europhys. Lett.* **4**, 1365–1370.

Bhattacharyya, S.P., & Nadoor, S. (1976). Stability of thermal convection between non-uniformly heated plates. *Appl. Sci. Res.* **32**, 555–570.

Bielek, C.A., & Koschmieder, E.L. (1990). Taylor vortices in short fluid columns with large radius ratio. *Phys. Fluids A* **2**, 1557–1563.

Blennerhassett, P.J., & Hall, P. (1979). Centrifugal instabilities of circumferential flows in finite cylinders: Linear theory. *Proc. Roy. Soc. London A* **365**, 191–207.

Block, M.J. (1956). Surface tension as the cause of Bénard cells and surface deformation in a liquid film. *Nature* **178**, 650–651.

Bolstad, J.H., & Keller, H.B. (1987). Computation of anomalous modes in the Taylor experiment. *J. Comput. Phys.* **69**, 230–251.

Bonnet, J.-P., & Alziary de Roquefort, T. (1976). Ecoulement entre deux sphères concentriques en rotation. *J. Mécanique* **15**, 373–397.

Bouabdallah, A., & Cognet, G. (1980). Laminar–turbulent transition in Taylor–Couette flow. In *Laminar–Turbulent Transition*, pp. 368–377, eds. R. Eppler & H. Fasel. Springer, Berlin.

Boubnov, B.M., & Golitsyn, G.S. (1986). Experimental study of convective structures in rotating fluids. *J. Fluid Mech.* **167**, 503–531.

Boussinesq, J. (1903). *Théorie Analytique de la Chaleur*, Vol. 2, p. 172. Gauthier-Villars, Paris.

Brandstater, A., & Swinney, H.L. (1987). Strange attractors in weakly turbulent Couette–Taylor flow. *Phys. Rev. A* **35**, 2207–2220.

Bratukhin, Iu.K. (1961). On the evaluation of the critical Reynolds number for the flow of fluid between two rotating spherical surfaces. *Prikl. Math. Mekh.* **25**, 858–866; *J. Appl. Math. Mech.* **25**, 1286–1299.

Brown, S.N., & Stewartson, K. (1977). On thermal convection in a large box. *Stud. Appl. Math.* **57**, 187–204.

Brown, S.N., & Stewartson, K. (1978). On finite amplitude Bénard convection in a cylindrical container. *Proc. Roy. Soc. London A* **360**, 455–469.

Brown, W. (1973). Heat–flux transitions at low Rayleigh number. *J. Fluid Mech.* **60**, 539–559.

Buell, J.C., & Catton, I. (1983a). The effect of wall conduction on the stability of a fluid in a right circular cylinder heated from below. *J. Heat Transf.* **105**, 255–260.

Buell, J.C., & Catton, I. (1983b). Effect of rotation on the stability of a bounded cylindrical layer of fluid heated from below. *Phys. Fluids* **26**, 892–896.

Buell, J.C., & Catton, I. (1986a). Wavenumber selection in large-amplitude axisymmetric convection. *Phys. Fluids* **29**, 23–70.

Buell, J.C., & Catton, I. (1986b). Wavenumber selection in ramped Rayleigh–Bénard convection. *J. Fluid Mech.* **171**, 477–494.

Bühler, K. (1990). Symmetric and asymmetric Taylor vortex flow in spherical gaps. *Acta Mech.* **81**, 3–38.

Bühler, K., Kirchartz, K.R., & Oertel, H. (1979). Steady convection in a horizontal fluid layer. *Acta Mech.* **31**, 155–171.

Bühler, K., & Oertel, H. (1982). Thermal cellular convection in rotating rectangular boxes. *J. Fluid Mech.* **114**, 261–282.

Burkhalter, J.E., & Koschmieder, E.L. (1973). Steady supercritical Taylor vortex flow. *J. Fluid Mech.* **58**, 547–560.

Burkhalter, J.E., & Koschmieder, E.L. (1974). Steady supercritical Taylor vortices after sudden starts. *Phys. Fluids* **17**, 1929–1935.

Busse, F.H. (1967a). On the stability of two-dimensional convection in a layer heated from below. *J. Math. Phys.* **46**, 140–149.

Busse, F.H. (1967b). The stability of finite amplitude cellular convection and its relation to an extremum principle. *J. Fluid Mech.* **30**, 625–649.

Busse, F.H. (1978). Nonlinear properties of thermal convection. *Rep. Prog. Phys.* **41**, 1929–1967.

Busse, F.H., & Clever, R.M. (1979). Instabilities of convection rolls in a fluid of moderate Prandtl number. *J. Fluid Mech.* **91**, 319–335.

Busse, F.H., & Whitehead, J.A. (1971). Instabilities of convection rolls in a high Prandtl number fluid. *J. Fluid Mech.* **47**, 305–320.

Bust, G.S., Dornblaser, B.C., & Koschmieder, E.L. (1985). Amplitudes and wavelengths of wavy Taylor vortices. *Phys. Fluids* **28**, 1243–1247.

Castaing, B., Gunaratne, G., Heslot, F., Kadanoff, L., Libchaber, A., Thomae, S., Wu, X-Z., Zaleski, S., & Zanetti, G. (1989). Scaling of hard turbulence in Rayleigh–Bénard convection. *J. Fluid Mech.* **204**, 1–30.

Castillo, J.L., & Velarde, M.G. (1982). Buoyancy–thermocapillary instability: The role of interfacial deformation in one- and two-component fluid layers heated from below or above. *J. Fluid Mech.* **125**, 463–474.

Castle, P., & Mobbs, F.R. (1968). Hydrodynamic stability of the flow between eccentric rotating cylinders: Visual observations and torque measurements. *Proc. Inst. Mech. Engrs.* **182**, 41–52.

Castle, P., Mobbs, F.R., & Markho, P.H. (1971). Visual observations and torque measurements in the Taylor vortex regime between rotating eccentric cylinders. *J. Lub. Tech.* **93**, 121–129.

Catton, I. (1970). Convection in a closed rectangular region: The onset of motion. *J. Heat Transf.* **92**, 186–188.

Cerisier, P., Jamond, C., Pantaloni, J., & Charmet, J.C. (1984). Deformation de la surface libre en convection de Bénard–Marangoni. *J. Physique* **45**, 405–411.

Cerisier, P., Perez-Garcia, C., Jamond, C., & Pantaloni, J. (1987). Wavelength selection in Bénard–Marangoni convection. *Phys. Rev. A* **35**, 1949–1952.

Cerisier, P., & Zouine, M. (1989). Disorder and wave number selection in Benard–Marangoni instability. *Physicochem. Hydrodyn.* **11**, 659–670.

Chana, M.S., & Daniels, P.G. (1989). Onset of Rayleigh–Bénard convection in a rigid channel. *J. Fluid Mech.* **199**, 257–279.

Chandra, K. (1938). Instabilities of fluids heated from below. *Proc. Roy. Soc. London A* **164**, 231–242.

Chandrasekhar, S. (1952). The thermal instability of a fluid sphere heated within. *Phil. Mag. Ser. 7* **43**, 1317–1329.

Chandrasekhar, S. (1953a). The instability of a layer of fluid heated below and subject to Coriolis forces. *Proc. Roy. Soc. London A* **217**, 306–327.

Chandrasekhar, S. (1953b). The onset of convection by thermal instability in spherical shells. *Phil. Mag. Ser. 7* **44**, 233–241.

Chandrasekhar, S. (1954). The stability of viscous flow between rotating cylinders. *Mathematika* **1**, 5–13.

Chandrasekhar, S. (1957). The thermal instability of a rotating fluid sphere heated within. *Phil. Mag. Ser. 8* **2**, 845–858.

Chandrasekhar, S. (1958). The stability of viscous flow between rotating cylinders. *Proc. Roy. Soc. London A* **246**, 301–311.

Chandrasekhar, S. (1961). *Hydrodynamic and Hydromagnetic Stability*. Oxford, Clarendon Press.

Chandrasekhar, S., & Elbert, D. (1955). The instability of a layer of fluid heated below and subject to Coriolis forces. II. *Proc. Roy. Soc. London A* **231**, 198–210.

Charlson, G.S., & Sani, R.L. (1970). Thermoconvective instability in a bounded cylindrical fluid layer. *Int. J. Heat Mass Transfer* **13**, 1479–1496.

Charlson, G.S., & Sani, R.L. (1971). On thermoconvective instability in a bounded cylindrical fluid layer. *Int. J. Heat Mass Transfer* **14**, 2157–2160.

Chen, M.M., & Whitehead, J.A. (1968). Evolution of two-dimensional periodic Rayleigh convection cells of arbitrary wave-number. *J. Fluid Mech.* **31**, 1–15.

Chossat, P., Demay, Y., & Iooss, G. (1987). Interaction de modes azimutaux dans le problème de Couette–Taylor. *Arch. Rat. Mech. Anal.* **99**, 213–248.

Chossat, P., & Iooss, G. (1985). Primary and secondary bifurcations in the Couette–Taylor problem. *Japan J. Appl. Math.* **2**, 37–68.

Chossat, P., & Iooss, G. (1992). *The Couette–Taylor Problem*. Springer, Berlin.

Christopherson, D.G. (1940). Note on the vibrations of membranes. *Quart. J. Math.* **11**, 63–65.

Chu, T.Y., & Goldstein, R.J. (1973). Turbulent convection in a horizontal layer of water. *J. Fluid Mech.* **60**, 141–159.

Ciliberto, S. (1991). Spatiotemporal chaos in Rayleigh–Bénard convection in an annulus. *Eur. J. Mech. B/Fluids* **10**, 193–198.

Clever, R.M., & Busse, F.H. (1974). Transition to time-dependent convection. *J. Fluid Mech.* **65**, 625–645.

Clever, R.M., & Busse, F.H. (1979). Nonlinear properties of convection rolls in a horizontal layer rotating about a vertical axis. *J. Fluid Mech.* **94**, 609–627.

Clever, R.M., & Busse, F.H. (1987). Nonlinear oscillatory convection. *J. Fluid Mech.* **176**, 403–417.

Cliffe, K.A. (1983). Numerical calculations of two-cell and single-cell Taylor flows. *J. Fluid Mech.* **135**, 219–233.

Cliffe, K.A. (1988). Numerical calculations of the primary-flow exchange process in the Taylor problem. *J. Fluid Mech.* **197**, 57–79.

Cliffe, K.A., & Mullin, T. (1985). A numerical and experimental study of anomalous modes in the Taylor experiment. *J. Fluid Mech.* **153**, 243–258.

Cloot, A., & Lebon, G. (1984). A nonlinear stability analysis of the Bénard–Marangoni problem. *J. Fluid Mech.* **145**, 447–469.

Cognet, G. (1971). Utilisation de la polarographie pour l'étude de l'écoulement de Couette. *J. Mécanique* **10**, 65–90.

Cole, J.A. (1967). Taylor vortices with eccentric rotating cylinders. *Nature* **216**, 1200–1202.

Cole, J.A. (1969). Taylor vortices with eccentric rotating cylinders. *Nature* **221**, 253–254.

Cole, J.A. (1976). Taylor-vortex instability and annulus-length effects. *J. Fluid Mech.* **75**, 1–15.

Coles, D. (1965). Transition in circular Couette flow. *J. Fluid Mech.* **21**, 385–425.

Couette, M. (1890). Etudes sur le frottement des liquides. *Ann. Chim. Phys.* **6**, 433–510.

Croquette, V. (1989a). Convective pattern dynamics at low Prandtl number: Part I. *Contemp. Phys.* **30**, 113–133.

Croquette, V. (1989b). Convective pattern dynamics at low Prandtl number: Part II. *Contemp. Phys.* **30**, 153–171.

Croquette, V., Mory, M., & Schosseler, F. (1983). Rayleigh–Bénard convective structures in a cylindrical container. *J. Physique* **44**, 293–301.

Cross, M.C. (1982). Ingredients of a theory of convective textures close to onset. *Phys. Rev. A* **25**, 1065–1076.

Cross, M.C. (1983). Phase dynamics of convective rolls. *Phys. Rev. A* **27**, 490–498.

Cross, M.C., & Newell, A.C. (1984). Convection patterns in large aspect ratio systems. *Physica D* **10**, 299–328.

Cross, M.C., Daniels, P.G., Hohenberg, P.C., & Siggia, E.D. (1983). Phase-winding solutions in a finite container above the convective threshold. *J. Fluid Mech.* **127**, 155–183.

Cross, M.C., Tesauro, G., & Greenside, H.S. (1986). Wavenumber selection and persistent dynamics in models of convection. *Physica D* **23**, 12–18.

Daniels, P.G. (1977). The effect of distant sidewalls on the transition to finite amplitude Bénard convection. *Proc. Roy. Soc. London A* **358**, 199–221.

Daniels, P.G. (1980). The effect of centrifugal acceleration on axisymmetric convection in a shallow rotating cylinder or annulus. *J. Fluid Mech.* **99**, 65–84.

Daniels, P.G. (1984). Roll-pattern evolution in finite-amplitude Rayleigh–Bénard convection in a two-dimensional fluid layer bounded by distant sidewalls. *J. Fluid Mech.* **143**, 125–152.

Daniels, P.G., Golbabai, A., & Soward, A.M. (1984). Axisymmetric overstable convection in a rotating fluid layer uniformly heated from below. *Q. J. Mech. Appl. Math.* **37**, 597–617.

Daniels, P.G., & Ong, C.F. (1990). Nonlinear convection in a rigid channel uniformly heated from below. *J. Fluid Mech.* **215**, 503–523.

Dauzère, C. (1912). Sur les changements qu'éprouvent les tourbillons cellulaires lorsque la température s'élève. *Compt. Rend.* **155**, 394–398.

Davey, A. (1962). The growth of Taylor vortices in flow between rotating cylinders. *J. Fluid Mech.* **14**, 336–368.

Davey, A., DiPrima, R.C., & Stuart, J.T. (1968). On the instability of Taylor vortices. *J. Fluid Mech.* **31**, 17–52.

Daviaud, F., & Dubois, M. (1991). Spatio-temporal intermittency in 1 D Rayleigh–Bénard convection. *Eur. J. Mech. B/Fluids* **10**, 199–204.

Davies-Jones, R.P. (1970). Thermal convection in an infinite channel with no-slip sidewalls. *J. Fluid Mech.* **44**, 695–704.

Davis, S.H. (1967). Convection in a box: Linear theory. *J. Fluid Mech.* **30**, 465–478.

Davis, S.H. (1969a). Buoyancy–surface tension instability by the method of energy. *J. Fluid Mech.* **39**, 347–359.

Davis, S.H. (1969b). On the principle of exchange of stabilities. *Proc. Roy. Soc. London A* **310**, 341–358.

Davis, S.H. (1971). On the possibility of subcritical instabilities. In *Instability of Continuous Systems*, pp. 222–227, ed. H. Leipholz. Springer, Berlin.

Davis, S.H. (1987). Thermocapillary instabilities. *Ann. Rev. Fluid Mech.* **19**, 403–435.

Davis, S.H., & Homsy, G.M. (1980). Energy stability theory for free-surface problems: Buoyancy–thermocapillary layers. *J. Fluid Mech.* **98**, 527–553.

Deane, A.E., & Sirovich, L. (1991). A computational study of Rayleigh–Bénard convection. Part 1. Rayleigh-number scaling. *J. Fluid Mech.* **222**, 231–250.

Deardorff, J.W., & Willis, G.E. (1967). Investigation of turbulent thermal convection between horizontal plates. *J. Fluid Mech.* **28**, 641–654.

Debler, W., Füner, E., & Schaaf, B. (1969). Torque and flow patterns in supercritical circular Couette flow. In *Proc. 12th Int. Congr. Appl. Mech.* pp. 158–178, eds. M. Hetényi & W.G. Vincenti. Springer, Berlin.

Demay, Y., & Iooss, G. (1984). Calcul des solutions bifurquées pour le problème de Couette–Taylor avec les deux cylindres en rotation. *J. Méc. Théor. Appl.*, special issue, 193–216.

Dennis, S.C.R., & Quartapelle, L. (1984). Finite difference solution to the flow between two rotating spheres. *Comp. Fluids* **12**, 77–92.

Dijkstra, H.A., & van de Vooren, A.I. (1989). Multiplicity and stability of steady solutions for Marangoni convection in a two-dimensional rectangular container with rigid sidewalls. *Num. Heat Transfer* **16**, 59–75.

DiPrima, R.C. (1961). Stability of nonrotationally symmetric disturbances for viscous flow between rotating cylinders. *Phys. Fluids* **4**, 751–755.

DiPrima, R.C., & Eagles, P.M. (1977). Amplification rates and torques for Taylor-vortex flows between rotating cylinders. *Phys. Fluids* **20**, 171–175.

DiPrima, R.C., Eagles, P.M., & Ng, B.S. (1984). The effect of radius ratio on the stability of Couette flow and Taylor vortex flow. *Phys. Fluids* **27**, 2403–2411.

DiPrima, R.C., Eckhaus, W., & Segel, L.A. (1971). Non-linear wave-number interaction in near-critical two-dimensional flows. *J. Fluid Mech.* **49**, 705–744.

DiPrima, R.C., & Grannick, R.N. (1971). A non-linear investigation of the stability of flow between counter-rotating cylinders. In *Instability of Continuous Systems*, pp. 55–60, ed. H. Leipholz. Springer, Berlin.

DiPrima, R.C., & Stuart, J.T. (1972a). Flow between eccentric rotating cylinders. *J. Lub. Tech.* **94**, 266–274.

DiPrima, R.C., & Stuart, J.T. (1972b). Non-local effects in the stability of flow between eccentric rotating cylinders. *J. Fluid Mech.* **54**, 393–415.

DiPrima, R.C., & Stuart, J.T. (1975). The nonlinear calculation of Taylor vortex flow between eccentric rotating cylinders. *J. Fluid Mech.* **67**, 85–111.

DiPrima, R.C., & Swinney, H.L. (1981). Instabilities and transition in flow between concentric rotating cylinders. In *Hydrodynamic Instabilities and the Transition to Turbulence*, pp. 139–180, eds. H.L. Swinney & J.P. Gollub. Springer, Berlin.

Doelman, A. (1989). Slow time-periodic solutions of the Ginzburg–Landau equation. *Physica D* **40**, 156–172.

Dominguez-Lerma, M.A., Cannell, D.S., & Ahlers, G. (1986). Eckhaus boundary and wave-number selection in rotating Couette–Taylor flow. *Phys. Rev. A* **34**, 4956–4970.

Donnelly, R.J. (1958). Experiments on the stability of viscous flow between rotating cylinders. I. Torque measurements. *Proc. Roy. Soc. London A* **246**, 312–325.

Donnelly, R.J., Park, K., Shaw, R., & Walden, R.W. (1980). Early transitions in Couette flow. *Phys. Rev. Lett.* **44**, 987–989.

Donnelly, R.J., & Schwarz, K.W. (1965). Experiments on the stability of viscous flow between rotating cylinders. VI. Finite-amplitude experiments. *Proc. Roy. Soc. London A* **283**, 531–546.

Drazin, P.G. (1975). On the effect of side walls on Bénard convection. *Z. angew. Math. Phys.* **26**, 239–245.

Drazin, P.G., & Reid, W.H. (1981). *Hydrodynamic stability*. Cambridge University Press.

Dubois, M., & Bergé, P. (1978). Experimental study of the velocity field in Rayleigh–Bénard convection. *J. Fluid Mech.* **85**, 641–654.

Dubois, M., & Bergé, P. (1981). Instabilités de couche limite dans un fluide en convection: Evolution vers la turbulence. *J. Physique* **42**, 167–174.

Dubois, M., Bergé, P., & Wesfreid, J. (1978). Non-Boussinesq convective structures in water near 4°C. *J. Physique* **39**, 1253–1257.

Dubois, M., Normand, C., & Bergé, P. (1978). Wavenumber dependence of velocity field amplitude in convection rolls: Theory and experiments. *Int. J. Heat Mass Transfer* **21**, 999–1002.

Eagles, P.M. (1971). On stability of Taylor vortices by fifth-order amplitude expansions. *J. Fluid Mech.* **49**, 529–550.

Eagles, P.M. (1974). On the torque of wavy vortices. *J. Fluid Mech.* **62**, 1–9.

Eagles, P.M., Stuart, J.T., & DiPrima, R.C. (1978). The effects of eccentricity on torque and load in Taylor vortex flow. *J. Fluid Mech.* **87**, 209–231.

Eckhaus, W. (1965). *Studies in Non-Linear Stability Theory*. Springer, Berlin.

Edwards, B.F. (1988). Crossed rolls at onset of convection in a rigid box. *J. Fluid Mech.* **191**, 583–597.

Edwards, W.S., Beane, S.R., & Varma, S. (1991). Onset of wavy vortices in the finite-length Couette–Taylor problem. *Phys. Fluids A* **3**, 1510–1518.

Edwards, W.S., Tagg, R.P., Dornblaser, B.C., & Swinney, H.L. (1991). Periodic traveling waves with nonperiodic pressure. *Eur. J. Mech. B/Fluids* **10**, 205–210.

Ezersky, A.B., Preobrazhensky, A.D., & Rabinovich, M.I. (1991). Spatial bifurcations of localized structures in Bénard–Marangoni convection. *Eur. J. Mech. B/Fluids* **10**, 211–220.

Farhadieh, R., & Tankin, R.S. (1974). Interferometric study of two-dimensional Bénard convection cells. *J. Fluid Mech.* **66**, 739–752.

Fasel, H., & Booz, O. (1984). Numerical investigation of supercritical Taylor vortex flow for a wide gap. *J. Fluid Mech.* **138**, 21–52.

Fenstermacher, P.R., Swinney, H.L., & Gollub, J.P. (1979). Dynamical instabilities and the transition to chaotic Taylor vortex flow. *J. Fluid Mech.* **94**, 103–128.

Fitzjarrald, D.E. (1976). An experimental study of turbulent convection in air. *J. Fluid Mech.* **73**, 693–719.

Frêne, J., & Godet, M. (1971). Transition from laminar to Taylor vortex flow in journal bearings. *Tribology* **4**, 216–217.

Galdi, G.P., & Straughan, B. (1985). A nonlinear analysis of the stabilizing effect of rotation in the Bénard problem. *Proc. Roy. Soc. London A* **402**, 257–283.

Garazo, A.N., & Velarde, M.G. (1991). Dissipative Korteweg-de Vries description of Marangoni–Bénard oscillatory convection. *Phys. Fluids A* **3**, 2295–2300.

Garcia-Ybarra, P.L., & Velarde, M.G. (1987). Oscillatory Marangoni–Bénard instability and capillary gravity waves in single- and two-component liquid layers with or without Soret thermal diffusion. *Phys. Fluids* **30**, 1649–1655.

Garon, A.M., & Goldstein, R.J. (1973). Velocity and heat transfer measurements in thermal convection. *Phys. Fluids* **16**, 1818–1825.

Georgescu, A. (1985). *Hydrodynamic stability theory.* Nijhoff, Dordrecht.

Gershuni, G.Z., & Zhukhovitskii, E.M. (1976). *Convective Stability of Incompressible Fluids.* Keter, Jerusalem.

Getling, A.V. (1983). Evolution of two-dimensional disturbances in the Rayleigh–Bénard problem and their preferred wavenumbers. *J. Fluid Mech.* **130**, 165–186.

Getling, A.V. (1991). Formation of spatial structures in Rayleigh–Bénard convection. *Usp. Fiz. Nauk* **161**, 1–80; *Sov. Phys. Usp.* **34**, 737–776.

Gilman, P.A. (1986). The solar dynamo: Observations and theories of solar convection, global circulations and magnetic fields. In *Physics of the Sun*, Vol. 1, pp. 95–160. Reidel, Dordrecht.

Ginzburg, V.L., & Landau, L.D. (1950). On the theory of superconductivity. *Zh. Eksp. Teor. Fiz.* **20**, 1064–1082. Translated in *Collected Papers of L. D. Landau*, pp. 546–568. Gordon & Breach, New York, 1965.

Glansdorff, P., & Prigogine, I. (1971). *Thermodynamic Theory of Structure, Stability and Fluctuations.* Wiley-Interscience, New York.

Goldstein, R.J., Chiang, H.D., & See, D.L. (1990). High-Rayleigh-number convection in a horizontal enclosure. *J. Fluid Mech.* **213**, 111–126.

Gollub, J.P., & Benson, S.V. (1978). Chaotic response to periodic perturbation of a convecting fluid. *Phys. Rev. Lett.* **41**, 948–951.

Gollub, J.P., & Benson, S.V. (1980). Many routes to turbulent convection. *J. Fluid Mech.* **100**, 449–470.

Gollub, J.P., & Freilich, M.H. (1976). Optical heterodyne test of perturbation expansions for the Taylor instability. *Phys. Fluids* **19**, 618–626.

Golubitsky, M., & Langford, W.F. (1988). Pattern formation and bistability in flow between counterrotating cylinders. *Physica D* **32**, 362–392.

Golubitsky, M., & Stewart, I. (1986). Symmetry and stability in Taylor–Couette flow. *SIAM J. Math. Anal.* **17**, 249–288.

Gor'kov, L.P. (1957). Steady convection in a plane liquid layer near the critical heat transfer point. *Zh. Eksp. Teor. Fiz.* **33**, 402–407; *Sov. Phys. JETP* **6**, 311–315.

Gorman, M., & Swinney, H.L. (1982). Spatial and temporal characteristics of modulated waves in the circular Couette system. *J. Fluid Mech.* **117**, 123–142.

Gouesbet, G., Maquet, J., Rozé, C., & Darrigo, D. (1990). Surface-tension- and coupled buoyancy-driven instability in a horizontal fluid layer: Overstability and exchange of stability. *Phys. Fluids A* **2**, 903–911.

Gough, D.O. (1977). Stellar convection. In *Problems of Stellar Convection*, Lecture Notes in Physics 71, pp. 348–363, eds. E.A. Spiegel & J.P. Zahn. Springer, Berlin.

Graham, A. (1933). Shear pattern in an unstable layer of air. *Phil. Trans. Roy. Soc. London A* **232**, 285–296.

Graham, R. (1974). Hydrodynamic fluctuations near the convection instability. *Phys. Rev. A* **10**, 1762–1784.

Graham, R., & Domaradski, J.A. (1982). Local amplitude equation of Taylor vortices and its boundary condition. *Phys. Rev. A* **26**, 1572–1579.

Gray, D.D., & Giorgini, A. (1976). The validity of the Boussinesq approximation for liquids and gases. *Int. J. Heat Mass Transfer* **19**, 545–551.

Grodzka, P.G., & Bannister, T. (1972). Heat flow and convection demonstration experiments aboard Apollo 14. *Science* **176**, 506–508.

Grodzka, P.G., & Bannister, T. (1975). Heat flow and convection experiments aboard Apollo 17. *Science* **187**, 165–167.

Haberman, W.L. (1962). Secondary flow about a sphere rotating in a viscous liquid inside a coaxially rotating spherical container. *Phys. Fluids* **5**, 625–626.

Haken, H. (1975). Cooperative phenomena in systems far from thermal equilibrium and in nonphysical systems. *Rev. Mod. Phys.* **47**, 67–121.

Hall, P. (1980a). Centrifugal instabilities of circumferential flows in finite cylinders: Nonlinear theory. *Proc. Roy. Soc. London A* **372**, 317–356.

Hall, P. (1980b). Centrifugal instabilities in finite containers: A periodic model. *J. Fluid Mech.* **99**, 575–596.

Hall, P. (1982). Centrifugal instabilities of circumferential flows in finite cylinders.: The wide gap problem. *Proc. Roy. Soc. London A* **384**, 359–379.

Hall, P., & Walton, I.C. (1977). The smooth transition to a convective regime in a two-dimensional box. *Proc. Roy. Soc. London A* **358**, 199–221.

Harris, D.L., & Reid, W.H. (1964). On the stability of viscous flow between rotating cylinders. Part 2. Numerical analysis. *J. Fluid Mech.* **20**, 95–101.

Hart, J.E., Glatzmaier, G.A., & Toomre, J. (1986). Space laboratory and numerical simulations of thermal convection in a rotating hemi-spherical shell with radial gravity. *J. Fluid Mech.* **173**, 519–544.

Heinrichs, R., Ahlers, G., & Cannell, D.S. (1986). Effects of finite geometry on the wavenumber of Taylor-vortex flow. *Phys. Rev. Lett.* **56**, 1794–1797.

Heutmaker, M.S., & Gollub, J.P. (1987). Wave-vector field of convective flow patterns. *Phys. Rev. A* **35**, 242–260.

Hoard, C.Q., Robertson, C.R., & Acrivos, A. (1970). Experiments on the cellular structure in Bénard convection. *Int. J. Heat Mass Transfer* **13**, 849–856.

Hohenberg, P.C., Kramer, L., & Riecke, H. (1985). Effects of boundaries on one-dimensional reaction-diffusion equations near threshold. *Physica D* **15**, 402–420.

Homsy, G.M., & Hudson, J.L. (1969). Centrifugally driven thermal convection in a rotating cylinder. *J. Fluid Mech.* **35**, 33–52.

Homsy, G.M., & Hudson, J.L. (1971). Centrifugal convection and its effect on the asymptotic stability of a bounded rotating fluid heated from below. *J. Fluid Mech.* **48**, 605–624.

Hopf, E. (1942). Abzweigung einer periodischen Lösung von einer stationären Lösung eines Differentialsystems. *Ber. Verh. Sächs. Akad. Wiss. Leipzig, Math.-Phys. Kl.* **94**, 1–22. Translated in *The Hopf Bifurcation and Its Applications*, pp. 163–193, eds. J.E. Marsden & M. McCracken. Springer, New York, 1976.

Howard, L.N. (1963). Heat transport by turbulent convection. *J. Fluid Mech.* **17**, 405–432.

Inoue, Y., & Ito, R. (1984). Analysis of Bénard convection by the energy-integral method. *Int. Chem. Engr.* **24**, 311–320.

Iooss, G. (1986). Secondary bifurcations of Taylor vortices into wavy inflow or outflow boundaries. *J. Fluid Mech.* **173**, 273–288.

Ivanilov, Iu. P., & Iakovlev, G.N. (1966). The bifurcation of fluid flow between rotating cylinders. *Prikl. Math. Mekh.* **30**, 768–773; *J. Appl. Math. Mech.* **30**, 910–916.

Iudovich, V.I. (1966). Secondary flows and fluid instability between rotating cylinders. *Prikl. Math. Mekh.* **30**, 688–698; *J. Appl. Math. Mech.* **30**, 822–833.

Jeffreys, H. (1926). The stability of a layer of fluid heated from below. *Phil. Mag.* **2**, 833–844.

Jeffreys, H. (1928). Some cases of instability in fluid motion. *Proc. Roy. Soc. London* A **118**, 195–208.

Jeffreys, H. (1951). The surface elevation in cellular convection. *Quart. J. Mech. Appl. Math.* **4**, 283–288.

Jeffreys, H., & Bland, M.E.M. (1951). The instability of a fluid sphere heated within. *Monthly N. Roy. Astron. Soc. London; Geophys. Suppl.* **6**, 148–158.

Jenkins, D.R. (1987). Rolls versus squares in thermal convection of fluids with temperature-dependent viscosity. *J. Fluid Mech.* **178**, 491–506.

Jenkins, D.R. (1988). Interpretation of shadowgraph patterns in Rayleigh–Bénard convection. *J. Fluid Mech.* **190**, 451–469.

Jenssen, O. (1963). Note on the influence of variable viscosity on the critical Rayleigh number. *Acta Polytech. Scand.* **24**, 1–12.

Jhaveri, B., & Homsy, G.M. (1980). Randomly forced Rayleigh–Bénard convection. *J. Fluid Mech.* **98**, 329–348.

Jones, C.A. (1981). Nonlinear Taylor vortices and their stability. *J. Fluid Mech.* **102**, 249–261.

Jones, C.A. (1985). The transition to wavy Taylor vortices. *J. Fluid Mech.* **157**, 135–162.

Jones, C.A., Moore, D.R., & Weiss, N.O. (1976). Axisymmetric convection in a cylinder. *J. Fluid Mech.* **73**, 353–388.

Joseph, D.D. (1966). Nonlinear stability of the Boussinesq equations by the method of energy. *Arch. Rat. Mech. Anal.* **22**, 163–184.

Joseph, D.D. (1976). *Stability of Fluid Motions.* Springer Tracts in Natural Philosophy Vols. 27 & 28. Springer, Berlin.

Kamal, M.M. (1966). Separation in the flow between eccentric rotating cylinders. *J. Basic Engr.* **88**, 717–724.

Kataoka, K. (1986). Taylor vortices and instabilities in circular Couette flow. In *Encyclopedia of Fluid Mechanics*, Vol. 1, pp. 236–274, ed. N.P. Cheremisinoff. Gulf Publ., Houston.

Kelly, R.E., & Pal, D. (1978). Thermal convection with spatially periodic boundary conditions: Resonant wavelength excitation. *J. Fluid Mech.* **86**, 433–456.

Kessler, R. (1987). Nonlinear transition in three-dimensional convection. *J. Fluid Mech.* **174**, 356–379.

Khlebutin, G.N. (1968). Stability of fluid motion between a rotating and a stationary concentric sphere. *Izv. Akad. Nauk SSSR, Mekh. Zh. i Gaza* **3**(6), 53–56; *Fluid Dyn.* **3**(6), 31–32.

Kidachi, H. (1982). Side wall effect on the pattern formation of the Rayleigh–Bénard convection. *Prog. Theor. Phys.* **68**, 49–63.

King, G.P., & Swinney, H.L. (1983). Limits of stability and irregular flow patterns in wavy vortex flow. *Phys. Rev. A* **27**, 1240–1243.

King, G.P., Li, Y., Lee, W., Swinney, H.L., & Marcus, P.S. (1984). Wave speeds in wavy Taylor-vortex flow. *J. Fluid Mech.* **141**, 365–390.

Kirchartz, K.R., Müller, U., Oertel, H., & Zierep, J. (1981). Axisymmetric and nonaxisymmetric convection in a cylindrical container. *Acta Mech.* **40**, 181–194.

Kirchartz, K.R., & Oertel, H. (1988). Three-dimensional thermal cellular convection in rectangular boxes. *J. Fluid Mech.* **192**, 249–286.

Kirchgässner, K. (1961). Die Instabilität der Strömung zwischen zwei rotierenden Zylindern gegenüber Taylor-Wirbeln für beliebige Spaltbreiten. *Z. angew. Math. Phys.* **12**, 14–29.

Kirchgässner, K., & Sorger, P. (1968). Stability analysis of branching solutions of the Navier–Stokes equations. In *Proc. 12th Int Congr. Appl. Mech.*, pp. 257–268, eds. M. Hetényi & W.G. Vincenti. Springer, Berlin.

Kirchgässner, K., & Sorger, P. (1969). Branching analysis for the Taylor problem. *Quart. J. Mech. Appl. Math.* **22**, 183–209.

Kirchner, R.P., & Chen, C.F. (1970). Stability of time-dependent rotational Couette flow. Part 1. Experimental investigation. *J. Fluid Mech.* **40**, 39–47.

Kogelman, S., & DiPrima, R.C. (1970). Stability of spatially periodic supercritical flows in hydrodynamics. *Phys. Fluids* **13**, 1–11.

Kolesov, V.V. (1980). Stability of nonisothermal Couette flow. *Izv. Akad. Nauk SSSR, Mekh. Zh. i Gaza* **15**, 167–170; *Fluid Dyn.* **15**, 137–140.

Kolesov, V.V. (1984). Oscillatory rotationally symmetric loss of stability of nonisothermal Couette flow. *Izv. Akad. Nauk SSSR, Mekh. Zh. i Gaza* **19**, 76–80; *Fluid Dyn.* **19**, 63–67.

Kolodner, P., Walden, R.W., Passner, A., & Surko, C.M. (1986). Rayleigh–Bénard convection in an intermediate-aspect-ratio rectangular container. *J. Fluid Mech.* **163**, 195–226.

Koschmieder, E.L. (1959). Uber Konvektionsströmungen auf einer Kugel. *Beitr. Phys. Atmos.* **32**, 34–42.

Koschmieder, E.L. (1966a). On convection on a uniformly heated plane. *Beitr. Phys. Atmos.* **39**, 1–11.

Koschmieder, E.L. (1966b). On convection on a nonuniformly heated plane. *Beitr. Phys. Atmos.* **39**, 208–216.

Koschmieder, E.L. (1967a). On convection under an air surface. *J. Fluid Mech.* **30**, 9–15.

Koschmieder, E.L. (1967b). On convection on a uniformly heated rotating plane. *Beitr. Phys. Atmos.* **40**, 216–225.

Koschmieder, E.L. (1969). On the wavelength of convective motions. *J. Fluid Mech.* **35**, 527–530.

Koschmieder, E.L. (1974). Bénard convection. *Adv. Chem. Phys.* **26**, 177–212.

Koschmieder, E.L. (1976). Taylor vortices between eccentric cylinders. *Phys. Fluids* **19**, 1–4.

Koschmieder, E.L. (1978). Symmetric circulations of planetary atmospheres. *Adv. Geophys.* **20**, 131–181.

Koschmieder, E.L. (1979). Turbulent Taylor vortex flow. *J. Fluid Mech.* **93**, 515–527, 800.

Koschmieder, E.L. (1991). The wavelength of supercritical surface tension driven Bénard convection. *Eur. J. Mech. B/Fluids* **10**, 233–237.

Koschmieder, E.L., & Biggerstaff, M.I. (1986). Onset of surface-tension-driven Bénard convection. *J. Fluid Mech.* **167**, 49–64.

Koschmieder, E.L., & Pallas, S.G. (1974). Heat transfer through a shallow, horizontal convecting fluid layer. *Int. J. Heat Mass Transfer* **17**, 991–1002.

Koschmieder, E.L., & Prahl, S.A. (1990). Surface tension driven Bénard convection in small containers. *J. Fluid Mech.* **215**, 571–583.

Kramer, L., Ben-Jacob, E., Brand, H., & Cross, M.C. (1982). Wavelength selection in systems far from equilibrium. *Phys. Rev. Lett.* **49**, 1891–1894.

Kramer, L., & Riecke, H. (1985). Wavelength selection in Rayleigh–Bénard convection. *Z. Phys. B* **59**, 245–251.

Kraska, J.R., & Sani, R.L. (1979). Finite amplitude Bénard–Rayleigh convection. *Int. J. Heat Mass Transfer* **22**, 535–546.

Krishnamurti, R. (1968a). Finite amplitude convection with changing mean temperature. Part 1. Theory. *J. Fluid Mech.* **33**, 445–455.

Krishnamurti, R. (1968b). Finite amplitude convection with changing mean temperature. Part 2. An experimental test of the theory. *J. Fluid Mech.* **33**, 457–463.

Krishnamurti, R. (1970a). On the transition to turbulent convection. Part 1. The transition from two- to three-dimensional flow. *J. Fluid Mech.* **42**, 295–307.

Krishnamurti, R. (1970b). On the transition to turbulent convection. Part 2. The transition to time-dependent flow. *J. Fluid Mech.* **42**, 309–320.

Krueger, E.R., Gross, A., & DiPrima, R.C. (1966). On the relative importance of Taylor-vortex and non-axisymmetric modes in flow between rotating cylinders. *J. Fluid Mech.* **24**, 521–538.

Krugljak, Z.B., Kuznetsov, E.A., L'vov, V.S., Nesterikhin, Yu.E., Predtechensky, A.A., Sobolev, V.S., Utkin, E.N., & Zhuravel, F.A. (1980). Laminar-turbulent transition in a circular Couette flow. In *Laminar–Turbulent Transition*, pp. 378–387, eds. R. Eppler & H. Fasel. Springer, Berlin.

Kuo, H.L. (1961). Solution of the non-linear equation of cellular convection and heat transport. *J. Fluid Mech.* **10**, 611–634.

Küppers, G. (1970). The stability of steady finite amplitude convection in a rotating fluid layer. *Phys. Lett. A* **32**, 7–8.

Küppers, G., & Lortz, D. (1969). Transition from laminar convection to thermal turbulence in a rotating fluid layer. *J. Fluid Mech.* **35**, 609–620.

Landau, L.D. (1944). On the problem of turbulence. *C.R. Acad. Sci. USSR* **44**, 311–314. Reprinted in *The Collected Papers of L.D. Landau*, pp. 387–391, Gordon & Breach, New York, 1967.

Landau, L.D., & Lifshitz, E.M. (1957). Hydrodynamic fluctuations. *Zh. Eksp. Teor. Fiz.* **32**, 618–619; *Sov. Phys. JETP*, **5**, 512–513.

Langford, W.F., Tagg, R., Kostelich, E.J., Swinney, H.L., & Golubitsky, M. (1988). Primary instabilities and bicriticality in flow between counter-rotating cylinders. *Phys. Fluids* **31**, 776–785.

Lebon, G., & Cloot, A. (1982). Buoyancy and surface tension driven instabilities in presence of negative Rayleigh and Marangoni numbers. *Acta Mech.* **43**, 141–158.

LeGal, P., Pocheau, A., & Croquette, V. (1985). Square versus roll pattern at convective threshold. *Phys. Rev. Lett.* **54**, 2501–2504.

Legros, J.C., & Platten, J.K. (1983). *Convection in Liquids*. Springer, Berlin.

Leontiev, A.J., & Kirdyashkin, A.G. (1968). Experimental study of flow patterns and temperature fields in horizontal free convection liquid layers. *Int. J. Heat Mass Transfer* **11**, 1461–1466.

Lewandowski, W.M., & Kupski, P. (1983). Methodical investigation of free convection from vertical and horizontal plates. *Wärme Stoffübertr.* **17**, 147–154.

Lewis, J.W. (1928). An experimental study of the motion of a viscous liquid contained between two coaxial cylinders. *Proc. Roy. Soc. London A* **117**, 388–406.

Liang, S.F., Vidal, A., & Acrivos, A. (1969). Buoyancy-driven convection in cylindrical geometries. *J. Fluid Mech.* **36**, 239–256.

Libchaber, A., & Maurer, J. (1978). Local probe in a Rayleigh–Bénard experiment in liquid helium. *J. Physique Lett.* **39**, 369–372.

Lipps, F.B., & Somerville, R.C.J. (1971). Dynamics of variable wavelength in finite-amplitude Bénard convection. *Phys. Fluids* **14**, 759–765.

Long, R.R. (1976). Relation between Nusselt number and Rayleigh number in turbulent thermal convection. *J. Fluid Mech.* **73**, 445–451.

Lorenz, E.N. (1963). Deterministic nonperiodic flow. *J. Atmos. Sci.* **20**, 130–141.

Lorenzen, A., & Mullin, T. (1985). Anomalous modes and finite-length effects in Taylor–Couette flow. *Phys. Rev. A* **31**, 3463–3465.

Lorenzen, A., Pfister, G., & Mullin, T. (1983). End effects on the transition to time-dependent motion in the Taylor experiment. *Phys. Fluids* **26**, 10–13.

Low, A.R. (1929). On the criterion for stability of a layer of viscous fluid heated from below. *Proc. Roy. Soc. London A* **125**, 180–195.

Low, A.R. (1930). Multiple modes of instability of a layer of viscous fluid, heated from below, with an application to Meteorology. In *Proc. 3rd Int. Congr. Appl. Mech.*, ed. C.W. Oseen & W. Weibull, Vol. 1, pp. 109–120. Stockholm.

Lucas, P.G., Pfotenhauer, J.M., & Donnelly, R.J. (1983). Stability and heat transfer of rotating cryogens. Part 1. Influence of rotation on the onset of convection in liquid ^{4}He. *J. Fluid Mech.* **129**, 251–264.

Lücke, M., Mihelcic, M., Wingerath, K., & Pfister, G. (1984). Flow in a small annulus between concentric cylinders. *J. Fluid Mech.* **140**, 343–353.

Luijkx, J.M., & Platten, J.K. (1981). On the onset of free convection in a rectangular container. *J. Nonequilib. Thermodyn.* **6**, 141–158.

L'vov, V.S., Predtechensky, A.A., & Chernykh, A.I. (1981). Bifurcation and chaos in a system of Taylor vortices: A natural and numerical experiment. *Zh. Eksp. Teor. Fiz.* **80**, 1099–1121; *Sov. Phys. JETP* **53**, 562–573.

Machetel, P., & Yuen, D.A. (1988). Infinite Prandtl number convection in spherical shells. In *Mathematical Geophysics*, ed. N.J. Vlaar, G. Nolet, M.J.R. Wortel, & S.A.P.L. Cloetingh, pp. 265–290. Reidel, Dordrecht.

Malkus, W.V.R. (1954a). Discrete transitions in turbulent convection. *Proc. Roy. Soc. London A* **225**, 185–195.

Malkus, W.V.R. (1954b). The heat transport and spectrum of thermal turbulence. *Proc. Roy. Soc. London A* **225**, 196–212.

Malkus, W.V.R., & Veronis, G. (1958). Finite amplitude cellular convection. *J. Fluid Mech.* **4**, 225–260.

Mallock, A. (1896). Experiments on fluid viscosity. *Phil. Trans. Roy. Soc. London A* **187**, 41–56.

Manneville, P., & Piquemal, J.M. (1983). Zigzag instability and axisymmetric rolls in Rayleigh–Bénard convection: The effects of curvature. *Phys. Rev. A* **28**, 1774–1790.

Manneville, P., & Pomeau, Y. (1983). A grain boundary in cellular structures near the onset of convection. *Phil. Mag. A* **48**, 607–621.

Marcus, P.S. (1984a). Simulation of Taylor–Couette flow. Part 1. Numerical methods and comparison with experiment. *J. Fluid Mech.* **146**, 45–64.

Marcus, P.S. (1984b). Simulation of Taylor–Couette flow. Part 2. Numerical results for wavy-vortex flow with one travelling wave. *J. Fluid Mech.* **146**, 65–113.

Marcus, P.S., & Tuckerman, L.S. (1987a). Simulation of flow between concentric rotating spheres. Part 1. Steady states. *J. Fluid Mech.* **185**, 1–30.

Marcus, P.S., & Tuckerman, L.S. (1987b). Simulation of flow between concentric rotating spheres. Part 2. Transitions. *J. Fluid Mech.* **185**, 31–65.

Martinet, B., Haldenwang, P., Labrosse, G., Payan, J.-C., & Payan, R. (1984). Sélection des structures dans l'instabilité de Rayleigh–Bénard. *Compt. Rend. Ser. II* **299**, 755–758.

Marx, K., & Haken, H. (1989). Numerical derivation of the generalized Ginzburg–Landau equations of wavy vortex flow. *Z. Phys. B* **75**, 393–411.

Meksyn, D. (1946a). Stability of viscous flow between rotating cylinders. I. *Proc. Roy. Soc. London A* **187**, 115–128.

Meksyn, D. (1946b). Stability of viscous flow between rotating cylinders. II. Cylinders rotating in opposite directions. *Proc. Roy. Soc. London A* **187**, 480–491.

Metzener, P. (1986). The effect of rigid sidewalls on nonlinear two-dimensional Bénard convection. *Phys. Fluids* **29**, 1373–1377.

Meyer, C.W., Ahlers, G., & Cannell, D.S. (1991). Stochastic influences on pattern formation in Rayleigh–Bénard convection: Ramping experiments. *Phys. Rev. A* **44**, 2514–2537.

Meyer, K.A. (1967). Time-dependent numerical study of Taylor vortex flow. *Phys. Fluids* **10**, 1874–1879.

Meyer-Spasche, R., & Keller, H.B. (1980). Computations of the axisymmetric flow between rotating cylinders. *J. Comput. Phys.* **35**, 100–109.

Mihaljan, J.M. (1962). A rigorous exposition of the Boussinesq approximation applicable to a thin layer of fluid. *Astrophys. J.* **136**, 1126–1133.

Moser, R.D., Moin, P., & Leonard, A. (1983). A spectral numerical method for the Navier–Stokes equations with applications to Taylor–Couette flow. *J. Comput. Phys.* **52**, 524–544.

Mull, W., & Reiher, H. (1930). Der Wärmeschutz von Luftschichten. *Beih. Gesundh.-Ing. Reihe* 1. No. 28.

Müller, U. (1965). Untersuchungen an rotationssymmetrischen Zellularkonvektionsströmungen *I. Beitr. Phys. Atmos.* **38**, 1–8.

Müller, U. (1966). Uber Zellularkonvektionsströmungen in horizontalen Flüssigkeitsschichten mit ungleichmässig erwärmter Bodenfläche. *Beitr. Phys. Atmos.* **39**, 217–234.

Mullin, T. (1982). Mutations of steady cellular flows in the Taylor experiment. *J. Fluid Mech.* **121**, 207–218.

Munson, B.R., & Joseph, D.D. (1971a). Viscous incompressible flow between concentric rotating spheres. Part 1. Basic flow. *J. Fluid Mech.* **49**, 289–303.

Munson, B.R., & Joseph, D.D. (1971b). Viscous incompressible flow between concentric rotating spheres. Part 2. Hydrodynamic stability. *J. Fluid Mech.* **49**, 305–318.

Munson, B.R., & Menguturk, M. (1975). Viscous incompressible flow between concentric rotating spheres. Part 3. Linear stability and experiments. *J. Fluid Mech.* **69**, 705–719.

Nagata, M. (1986). Bifurcations in Couette flow between almost corotating cylinders. *J. Fluid Mech.* **169**, 229–250.

Nagata, W. (1990). Convection in a layer with sidewalls: Bifurcation with reflection symmetries, *J. Appl. Math. Phys.* **41**, 812–828.

Nakabayashi, K. (1983). Transition of Taylor–Görtler vortex flow in spherical Couette flow. *J. Fluid Mech.* **132**, 209–230.

Nakamura, I., Toya, Y., Yamashita, S., & Ueki, Y. (1990). An experiment on a Taylor vortex flow in a gap with a small aspect ratio. *Japan. Soc. Mech. Engr. Int. J.* **33**, 685–691.

Nakaya, C. (1974). Domain of stable periodic vortex flows in a viscous fluid between concentric circular cylinders. *J. Phys. Soc. Japan* **36**, 1164–1173.

Nakaya, C. (1975). The second stability boundary for circular Couette flow. *J. Phys. Soc. Japan* **38**, 576–585.

Neitzel, G.P. (1984). Numerical computation of time-dependent Taylor-vortex flows in finite-length geometries. *J. Fluid Mech.* **141**, 51–66.

Newell, A.C., Lange, C.G., & Aucoin, P.J. (1970). Random convection. *J. Fluid Mech.* **40**, 513–540.

Newell, A.C., Passot, T., & Souli, M. (1990). The phase diffusion and mean drift equations for convection at finite Rayleigh numbers in large containers. *J. Fluid Mech.* **220**, 187–252.

Newell, A.C., & Whitehead, J.A. (1969). Finite bandwidth, finite amplitude convection. *J. Fluid Mech.* **38**, 279–303.

Nield, D.A. (1964). Surface tension and buoyancy effects in cellular convection. *J. Fluid Mech.* **19**, 341–352.

Nield, D.A. (1966). Streamlines in Bénard convection cells induced by surface tension and buoyancy. *Z. angew. Math. Phys.* **17**, 226–232.

Nield, D.A., & Bejan, A. (1991). *Convection in Porous Media.* Springer, Berlin.

Ning, L., Ahlers, G., & Cannell, D.S. (1990). Wave-number selection and traveling vortex waves in spatially ramped Taylor–Couette flow. *Phys. Rev. Lett.* **64**, 1235–1238.

Nissan, A.H., Nardacci, J.L., & Ho, C.Y. (1963). The onset of different modes of instability for flow between rotating cylinders. *A.I.Ch.E.J.* **9**, 620–624.

Normand, C., Pomeau, Y., & Velarde, M.G. (1977). Convective instability: A physicist's approach. *Rev. Mod. Phys.* **49**, 581–624.

Oberbeck, A. (1879). Uber die Wärmeleitung der Flüssigkeiten bei Berücksichtigung der Strömungen infolge von Temperaturdifferenzen. *Ann. Phys. Chem.* **7**, 271–292.

Oliver, D.S., & Booker, J.R. (1983). Planform of convection with strongly temperature-dependent viscosity. *Geophys. Astrophys. Fluid Dyn.* **27**, 73–85.

Ovchinnikova, S.N., & Iudovich, V.I. (1974). Stability and bifurcation of Couette flow in the case of a narrow gap between rotating cylinders. *Prikl. Math. Mekh.* **38**, 1025–1030; *J. Appl. Math. Mech.* **38**, 972–977.

Ovseenko, Iu.G. (1963). On the motion of a viscous fluid between two rotating spheres. *Izv. Vyssh. Ucheb. Zaved, Matematika* (4), 129–139.

Paap, H.-G., & Riecke, H. (1991). Drifting vortices in ramped Taylor vortex flow: Quantitative results from phase equation. *Phys. Fluids A* **3**, 1519–1532.

Pai, S.-I. (1943). Turbulent flow between rotating cylinders. Tech. Note no. 892. National Advisory Commity on Aeronautics, Washington, D.C.

Palm, E. (1960). On the tendency towards hexagonal cells in steady convection. *J. Fluid Mech.* **8**, 183–192.

Palm, E. (1975). Nonlinear thermal convection. *Ann. Rev. Fluid Mech.* **7**, 39–61.

Palm, E., Ellingsen, T., & Gjevik, B. (1967). On the occurrence of cellular motion in Bénard convection. *J. Fluid Mech.* **30**, 651–661.

Palmer, H.J., & Berg, J.C. (1971). Convective instability in liquid pools heated from below. *J. Fluid Mech.* **47**, 779–787.

Park, K., Crawford, G.L., & Donnelly, R.J. (1983). Characteristic lengths in the wavy vortex state of Taylor–Couette flow. *Phys. Rev. Lett.* **51**, 1352–1354.

Park, K., & Donnelly, R.J. (1981). Study of the transition to Taylor vortex flow. *Phys. Rev. A.* **24**, 2277–2279.

Pearson, J.R.A. (1958). On convection cells induced by surface tension. *J. Fluid Mech.* **4**, 489–500.

Pellew, A., & Southwell, R.V. (1940). On maintained convective motions in a fluid heated from below. *Proc. Roy. Soc. London A* **176**, 312–343.

Pérez-Cordón, R., & Velarde, M.G. (1975). On the (non-linear) foundations of Boussinesq approximation applicable to a thin layer of fluid. *J. Physique* **36**, 591–601.

Pérez-Garcia, C., & Carneiro, G. (1991). Linear stability analysis of Bénard–Marangoni convection in fluids with deformable free surface. *Phys. Fluids A* **3**, 292–298.

Pfister, G., & Rehberg, I. (1981). Space-dependent order parameter in circular Couette flow transitions. *Phys. Lett. A* **83**, 19–22.

Pfister, G., Schmidt, H., Cliffe, K.A., & Mullin, T. (1988). Bifurcation phenomena in Taylor–Couette flow in a very short annulus. *J. Fluid Mech.* **191**, 1–18.

Pfotenhauer, J.M., Niemala, J.J., & Donnelly, R.J. (1987). Stability and heat transfer of rotating cryogens. Part 3. Effects of finite cylindrical geometry and rotation on the onset of convection. *J. Fluid Mech.* **175**, 85–96.

Plows, W.H. (1972). UCRL Report 51146, Lawrence Livermore Laboratory, Livermore, Calif.

Pocheau, A., & Croquette, V. (1984). Dislocation motion: A wavenumber selection mechanism in Rayleigh–Bénard convection. *J. Physique* **45**, 35–48.

Pomeau, Y., & Manneville, P. (1979). Stability and fluctuations of a spatially periodic convective flow. *J. Physique Lett.* **40**, 609–612.

Pomeau, Y., & Zaleski, S. (1981). Wavelength selection in one-dimensional cellular structures. *J. Physique* **42**, 515–528.

Proctor, M.R.E. (1981). Planform selection by finite-amplitude thermal convection between poorly conducting slabs. *J. Fluid Mech.* **113**, 469–485.

Rayleigh, Lord (1916a). On convection currents in a horizontal layer of fluid when the higher temperature is on the under side. *Phil. Mag.* **32**, 529–546.

Rayleigh, Lord (1916b). On the dynamics of revolving fluids. *Proc. Roy. Soc. London A* **93**, 148–154.

Reid, W.H., & Harris, D.L. (1958). Some further results on the Bénard problem. *Phys. Fluids* **1**, 102–110.

Reynolds, O. (1886). On the theory of lubrication and its application to Mr Beauchamp Tower's experiments, including an experimental determination of the viscosity of olive oil. *Phil. Trans.* **177**, 157–234.

Riecke, H., & Paap, H.-G. (1986). Stability and wave-vector restriction of axisymmetric Taylor vortex flow. *Phys. Rev. A* **33**, 547–553.

Roberts, P.H. (1965). The solution of the characteristic value problems. *Proc. Roy. Soc. London A* **283**, 550–556.

Roesner, K.G. (1978). Hydrodynamic stability of cylindrical Couette flow. *Arch. Mech.* **30**, 619–627.

Rosenblat, S. (1982). Thermal convection in a vertical circular cylinder. *J. Fluid Mech.* **122**, 395–410.

Rosenblat, S., Davis, S.H., & Homsy, G.M. (1982a). Nonlinear Marangoni convection in bounded layers. Part 1. Circular cylindrical containers. *J. Fluid Mech.* **120**, 91–122.

Rosenblat, S., Homsy, G.M., & Davis, S.H. (1982b). Nonlinear Marangoni convection in bounded layers. Part 2. Rectangular cylindrical containers. *J. Fluid Mech.* **120**, 123–138.

Rossby, H.T. (1969). A study of Bénard convection with and without rotation. *J. Fluid Mech.* **36**, 309–335.

Saltzman, B. (1962a). *Theory of Thermal Convection*. Dover, New York.

Saltzman, B. (1962b). Finite amplitude free convection as an initial value problem. *J. Atmos. Sci.* **19**, 329–341.

San Andres, A., & Szeri, A.Z. (1984). Flow between eccentric rotating cylinders. *J. Appl. Mech.* **51**, 869–878.

Sani, R. (1964). On the non-existence of subcritical instabilities in fluid layers heated from below. *J. Fluid Mech.* **20**, 315–319.

Sawatzki, O., & Zierep, J. (1970). Das Stromfeld im Spalt zwischen zwei konzentrischen Kugelflächen, von denen die innere rotiert. *Acta Mech.* **9**, 13–35.

Scanlon, J.W., & Segel, L.A. (1967). Finite amplitude cellular convection induced by surface tension. *J. Fluid Mech.* **30**, 149–162.

Schaeffer, D.G. (1980). Qualitative analysis of a model for boundary effects in the Taylor problem. *Math. Proc. Camb. Phil. Soc.* **87**, 307–337.

Schlüter, A., Lortz, D., & Busse, F. (1965). On the stability of steady finite amplitude convection. *J. Fluid Mech.* **23**, 129–144.

Schmidt, R.J., & Milverton, S.W. (1935). On the instability of a fluid when heated from below. *Proc. Roy. Soc. London A* **152**, 586–594.

Schrauf, G. (1986). The first instability in spherical Taylor–Couette flow. *J. Fluid Mech.* **166**, 287–303.

Schultz-Grunow, F. (1959). Zur Stabilität der Couette-Strömung. *Z. angew. Math. Mech.* **39**, 101–110.

Schultz-Grunow, F., & Hein, H. (1956). Beitrag zur Couetteströmung. *Z. Flugwiss.* **4**, 28–30.

Schwabe, D., Dupont, O., Queekers, P., & Legros, J.C. (1990). Experiments on Marangoni–Bénard problems under normal and microgravity conditions. *Proc. VIIth Europ. Symp. Materials & Fluid Sci. in Microgravity.* European Space Agency Special Publication 295, pp. 291–298.

Schwarz, K.W., Springett, B.E., & Donnelly, R.J. (1964). Modes of instability in spiral flow between rotating cylinders. *J. Fluid Mech.* **20**, 281–289.

Scriven, L.E., & Sternling, C.V. (1964). On cellular convection driven by surface-tension gradients: Effects of mean surface tension and surface viscosity. *J. Fluid Mech.* **19**, 321–340.

Segel, L.A. (1962). The non-linear interaction of two disturbances in the thermal convection problem. *J. Fluid Mech.* **14**, 97–114.

Segel, L.A. (1965). The nonlinear interaction of a finite number of disturbances to a layer of fluid heated from below. *J. Fluid Mech.* **21**, 359–384.

Segel, L.A. (1966). Non-linear hydrodynamic stability theory and its application to thermal convection and curved flows. In *Non-Equilibrium Thermodynamics, Variational Techniques, and Stability,* pp. 165–197, eds. R.J. Donnelly, R. Herman, & I. Prigogine. University of Chicago Press, Chicago.

Segel, L.A. (1969). Distant side-walls cause slow amplitude modulation of cellular convection. *J. Fluid Mech.* **38**, 203–224.

Segel, L.A., & Stuart, J.T. (1962). On the question of the preferred mode in cellular thermal convection. *J. Fluid Mech.* **13**, 289–306.

Siggia, E.R., & Zippelius, A. (1981). Dynamics of defects in Rayleigh–Bénard convection. *Phys. Rev. A* **24**, 1036–1049.

Silveston, P.L. (1958). Wärmedurchgang in waagerechten Flüssigkeitsschichten. *Forsch. Ing. Wes.* **24**, 29–32, 59–69.

Simpkins, P.G., & Dudderar, T.D. (1978). Laser speckle measurements of transient Bénard convection. *J. Fluid Mech.* **89**, 665–671.

Sirovich, L., & Deane, A.E. (1991). A computational study of Rayleigh–Bénard convection. Part 2. Dimension considerations. *J. Fluid Mech.* **222**, 251–265.

Smith, G.P., & Townsend, A.A. (1982). Turbulent Couette flow between concentric cylinders at large Taylor numbers. *J. Fluid Mech.* **123**, 187–217.

Smith, K.A. (1966). On convective instability induced by surface-tension gradients. *J. Fluid Mech.* **24**, 401–414.

Snyder, H.A. (1965). Experiments on the stability of two types of spiral flow. *Ann. Phys.* **31**, 292–313.

Snyder, H.A. (1969). Wave-number selection at finite amplitude in rotating Couette flow. *J. Fluid Mech.* **35**, 273–298.

Snyder, H.A. (1970). Waveforms in rotating Couette flow. *Int. J. Non-Linear Mech.* **5**, 659–685.

Snyder, H.A., & Lambert, R.B. (1966). Harmonic generation in Taylor vortices between rotating cylinders. *J. Fluid Mech.* **26**, 545–562.

Somerscales, E.F.C., & Dougherty, T.S. (1970). Observed flow patterns at the initiation of convection in a horizontal liquid layer heated from below. *J. Fluid Mech.* **42**, 755–768.

Sommerfeld, A. (1904). Zur hydrodynamischen Theorie der Schmiermittelreibung. *Z. f. Math. Phys.* **50**, 97–155.

Sorokin, V.S. (1953). A variational method in the theory of convection. *Prikl. Math. Mekh.* **17**, 39–48.

Sorokin, V.S. (1954). On steady motions in a fluid heated from below. *Prikl. Math. Mekh.* **18**, 197–204.

Sorour, M.M., & Coney, J.E.R. (1979). The effect of temperature gradient on the stability of flow between vertical, concentric, rotating cylinders. *J. Mech. Eng. Sci.* **21**, 403–409.

Soward, A.M., & Jones, C.A. (1983). The linear stability of the flow in the narrow gap between two concentric rotating spheres. *Q. J. Mech. Appl. Math.* **36**, 19–42.

Sparrow, C. (1982). *The Lorenz Equations: Bifurcations, Chaos, and Strange Attractors.* Springer, New York.

Sparrow, E.M., Munro, W.D., & Jonson, V.K. (1964). Instability of the flow between rotating cylinders: The wide gap problem. *J. Fluid Mech.* **20**, 35–46.

Spiegel, E.A., & Veronis, G. (1960). On the Boussinesq approximation for a compressible fluid. *Astrophys. J.* **131**, 442–447.

Srulijes, J.A. (1979). Zellularkonvektion in Behältern mit horizontalen Temperaturgradienten. Dissertation, Karlsruhe University.

Steinberg, V., Ahlers, G., & Cannell, D.S. (1985). Pattern formation and wavenumber selection by Rayleigh–Bénard convection in a cylindrical container. *Physica Scripta* **32**, 534–547.

Stengel, K.C., Oliver, D.S., & Booker, J.R. (1982). Onset of convection in a variable-viscosity fluid. *J. Fluid Mech.* **120**, 411–431.

Stork, K., & Müller, U. (1972). Convection in boxes: Experiments. *J. Fluid Mech.* **54**, 599–611.

Stork, K., & Müller, U. (1975). Convection in boxes: An experimental investigation in vertical cylinders and annuli. *J. Fluid Mech.* **71**, 231–240.

Stuart, J.T. (1958). On the nonlinear mechanics of hydrodynamic stability. *J. Fluid Mech.* **4**, 1–21.

Stuart, J.T. (1960). On the nonlinear mechanics of wave disturbances in stable and unstable parallel flows. *J. Fluid Mech.* **9**, 353–370.

Stuart, J.T. (1964). On the cellular patterns in thermal convection. *J. Fluid Mech.* **18**, 481–498.

Stuart, J.T. (1971). Nonlinear stability theory. *Ann. Rev. Fluid Mech.* **3**, 347–370.

Stuart, J.T., & DiPrima, R.C. (1978). The Eckhaus and Benjamin–Feir resonance mechanisms. *Proc. Roy. Soc. London A* **362**, 27–41.

Sutton, O.G. (1951). On the stability of a fluid heated from below. *Proc. Roy. Soc. London A* **204**, 297–309.

Sweet, D., Jakeman, E., & Hurle, D.T.J. (1977). Free convection in the presence of both vertical and horizontal temperature gradients. *Phys. Fluids* **20**, 1412–1415.

Swift, J., & Hohenberg, P.C. (1977). Hydrodynamic fluctuations at the convective instability. *Phys. Rev. A* **15**, 319–328.

Swinney, H.L. (1978). Hydrodynamic instabilities and the transition to turbulence. *Prog. Theor. Phys. Suppl.* **64**, 164–175.

Synge, J.L. (1933). The stability of heterogeneous liquids. *Trans. Roy. Soc. Canada* **27**, 1–18.

Tagg, R., Edwards, W.S., Swinney, H.L., & Marcus, P.S. (1989). Nonlinear standing waves in Couette–Taylor flow. *Phys. Rev. A* **39**, 3734–3737.

Takashima, M. (1981). Surface tension driven instability in a horizontal liquid layer with deformable free surface. II. Overstability. *J. Phys. Soc. Japan* **50**, 2751–2756.

Takeda, Y., Kobashi, K., & Fischer, W.E. (1990). Observation of the transient behaviour of Taylor vortex flow between rotating concentric cylinders after sudden starts. *Exp. Fluids* **9**, 317–319.

Tavener, S.J., Mullin, T., & Cliffe, K.A. (1991). Novel bifurcation phenomena in a rotating annulus. *J. Fluid Mech.* **229**, 483–497.

Taylor, G.I. (1921). Experiments with rotating fluids. *Proc. Camb. Phil. Soc.* **20**, 326–329.

Taylor, G.I. (1923). Stability of a viscous liquid contained between two rotating cylinders. *Phil. Trans. Roy. Soc. London A* **223**, 289–343.

Taylor, G.I. (1936a). Fluid friction between rotating cylinders. I. Torque measurements. *Proc. Roy. Soc. London A* **157**, 546–564.

Taylor, G.I. (1936b). Fluid friction between rotating cylinders. II. Distribution of velocity between concentric cylinders when outer one is rotating and inner one is at rest. *Proc. Roy. Soc. London A* **157**, 565–578.

Tesauro, G., & Cross, M.C. (1987). Grain boundaries in models of convective patterns. *Phil. Mag. A* **56**, 703–724.

Thompson, H.A., & Sogin, H.H. (1966). Experiments on the onset of thermal convection in horizontal layers of gases. *J. Fluid Mech.* **24**, 451–479.

Threlfall, D.C. (1975). Free convection in low-temperature gaseous helium. *J. Fluid Mech.* **67**, 17–28.

v. Tippelskirch, H. (1956). Uber Konvektionszellen insbesondere im flüssigen Schwefel. *Beitr. Phys. Atmos.* **29**, 37–54.

v. Tippelskirch, H. (1959). Weitere Konvektionsversuche: Der Nachweis der Ring-zellen und ihrer Verallgemeinerung. *Beitr. Phys. Atmos.* **32**, 23–33.

Torrest, M.A., & Hudson, J.L. (1974). The effect of centrifugal convection on the stability of a rotating fluid heated from below. *Appl. Sci. Res.* **29**, 273–279.

Townsend, A.A. (1984). Axisymmetric Couette flow at large Taylor numbers. *J. Fluid Mech.* **144**, 329–362.

Travis, B., Olson, P., & Schubert, G. (1990). The transition from two-dimensional to three-dimensional planforms in infinite-Prandtl-number thermal convec-tion. *J. Fluid Mech.* **216**, 71–91.

Tuckerman, L.S., & Barkley, D. (1988). A global bifurcation to travelling waves in axisymmetric convection. *Phys. Rev. Lett.* **61**, 408–411.

Turner, J.S. (1973). *Buoyancy Effects in Fluids.* Cambridge University Press.

Ukhovskii, M.R., & Iudovich, V.I. (1963). On the equations of steady-state convec-tion. *Prikl. Math. Mekh.* **27**, 295–300; *J. Appl. Math. Mech.* **27**, 432–440.

Unny, T.E., & Niessen, P. (1969). Thermal instability in fluid layers in the pres-ence of horizontal and vertical temperature gradients. *J. Appl. Mech.* **36**, 121–123.

Van Atta, C. (1966). Exploratory measurements in spiral turbulence. *J. Fluid Mech.* **25**, 495–512.

Velte, W. (1966). Stabilität und Verzweigung stationärer Lösungen der Navier–Stokesschen Gleichungen beim Taylorproblem. *Arch. Rat. Mech. Anal.* **22**, 1–14.

Vernotte, P. (1936). La théorie des tourbillons cellulaires de Bénard. *Compt. Rend.* **202**, 119–121.

Veronis, G. (1959). Cellular convection with finite amplitude in a rotating fluid. *J. Fluid Mech.* **5**, 401–435.

Versteegen, P.L., & Jankowski, D.F. (1969). Experiments on the stability of viscous flow between eccentric rotating cylinders. *Phys. Fluids* **12**, 1138–1143.

Vidal, A., & Acrivos, A. (1966). Nature of the neutral state in surface-tension driven convection. *Phys. Fluids* **9**, 615–616.

Vohr, J.H. (1968). An experimental study of Taylor vortices and turbulence in flow between eccentric rotating cylinders. *J. Lub. Tech.* **90**, 285–296.

Walden, R.W., & Ahlers, G. (1981). Non-Boussinesq and penetrative convection in a cylindrical cell. *J. Fluid Mech.* **109**, 89–114.

Walden, R.W., Kolodner, P., Passner, A., & Surko, C.M. (1987). Heat transport by parallel-roll convection in a rectangular container. *J. Fluid Mech.* **185**, 205–234.

Walgraef, D. (1986). End effects and phase instabilities in a model for Taylor–Couette systems. *Phys. Rev. A* **34**, 3270–3278.

Walgraef, D., Borckmans, P., & Dewel, G. (1982). Fluctuation effects in the transition to Taylor vortex flow in finite geometries. *Phys. Rev. A.* **25**, 2860–2862.

Walgraef, D., Borckmans, P., & Dewel, G. (1984). Onset of wavy vortex flow in fi-nite geometries. *Phys. Rev. A* **29**, 1514–1519.

Walton, I.C. (1978). The linear stability of the flow in a narrow spherical annulus. *J. Fluid Mech.* **86**, 673–693.

Walton, I.C. (1983). The onset of cellular convection in a shallow two-dimensional container of fluid heated non-uniformly from below. *J. Fluid Mech.* **131**, 455–470.

Wannier, G.H. (1950). A contribution to the hydrodynamics of lubrication. *Q. Appl. Math.* **8**, 1–32.

Wasiutynski, J. (1946). Studies in hydrodynamics and structure of stars and planets. *Astrophys. Norvegica* **4**, 1–497.

Watson, J. (1960). On the nonlinear mechanics of wave disturbances in stable and unstable parallel flows. *J. Fluid Mech.* **9**, 371–389.

Weber, J.E. (1973). On thermal convection between non-uniformly heated planes. *Int. J. Heat Mass Transfer* **16**, 961–970.

Weber, J.E. (1978). On the stability of thermally driven shear flow heated from below. *J. Fluid Mech.* **87**, 65–84.

Weinstein, M. (1977a). Wavy vortices in the flow between two long eccentric rotating cylinders. I. Linear theory. *Proc. Roy. Soc. London A* **354**, 441–457.

Weinstein, M. (1977b). Wavy vortices in the flow between two long eccentric rotating cylinders. II. Nonlinear theory. *Proc. Roy. Soc. London A* **354**, 459–489.

Weiss, N.O. (1964). Convection in the presence of constraints. *Phil. Trans. Roy. Soc. London A* **256**, 99–147.

Wendt, F. (1933). Turbulente Strömungen zwischen zwei rotierenden konaxialen Zylindern. *Ing. Arch.* **4**, 577–595.

Wesfreid, J., Pomeau, Y., Dubois, M., Normand, C., & Bergé, P. (1978). Critical effects in Rayleigh–Bénard convection. *J. Phys. Lett.* **39**, 725–731.

White, D.B. (1988). The planform and onset of convection with a temperature-dependent viscosity. *J. Fluid Mech.* **191**, 247–286.

Willis, G.E., & Deardorff, J.W. (1965). Measurements on the development of thermal turbulence in air between horizontal plates. *Phys. Fluids* **8**, 2225–2229.

Willis, G.E., & Deardorff, J.W. (1967a). Development of short-period temperature fluctuations in thermal convection. *Phys. Fluids* **10**, 931–937.

Willis, G.E., & Deardorff, J.W. (1967b). Confirmation and renumbering of the discrete heat flux transitions of Malkus. *Phys. Fluids* **10**, 1861–1866.

Willis, G.E., & Deardorff, J.W. (1970). The oscillatory motions of Rayleigh convection. *J. Fluid Mech.* **44**, 661–672.

Willis, G.E., Deardorff, J.W., & Somerville, R.C.J. (1972). Roll-diameter dependence in Rayleigh convection and its effect upon the heat flux. *J. Fluid Mech.* **54**, 351–367.

Wimmer, M. (1976). Experiments on a viscous fluid flow between concentric rotating spheres. *J. Fluid Mech.* **78**, 317–335.

Wimmer, M. (1981). Experiments on the stability of viscous flow between two concentric rotating spheres. *J. Fluid Mech.* **103**, 117–131.

Wimmer, M. (1983). Die viskose Strömung zwischen rotierenden Kegelflächen. *Z. angew. Math. Mech.* **63**, T 299–301.

Winters, K.H., Plesser, T., & Cliffe, K.A. (1988). The onset of convection in a finite container due to surface tension and buoyancy. *Physica D* **29**, 387–401.

Yahata, H. (1977a). Slowly-varying amplitude of the Taylor vortices near the instability point. *Prog. Theor. Phys.* **57**, 347–360.

Yahata, H. (1977b). Slowly varying amplitude of the Taylor vortices near the instability point. II. Mode-coupling theoretical approach. *Prog. Theor. Phys.* **57**, 1490–1496.

Yahata, H. (1981). Temporal development of the Taylor vortices in a rotating fluid. IV. *Prog. Theor. Phys.* **66**, 879–891.

Yahata, H. (1983). Temporal development of the Taylor vortices in a rotating fluid. V. *Prog. Theor. Phys.* **69,** 396–402.

Yakushin, V.I. (1969). Instability of fluid motion in a thin spherical layer. *Izv. Akad. Nauk. SSSR, Mekh. Zh. i Gaza* **4**(1), 118–119; *Fluid Dyn.* **4**(1), 83–85.

Yakushin, V.I. (1970). Instability of the motion of a liquid between two rotating spherical surfaces. *Izv. Akad. Nauk SSSR, Mekh. Zh. i Gaza* **5**(4), 155–156; *Fluid Dyn.* **5,** 660–661.

Yavorskaya, I.M., Belyaev, Y.N., & Monakhov, A.A. (1975). Experimental study of spherical Couette flow. *Dokl. Akad. Nauk SSSR* **221,** 1059–1062; *Sov. Phys. Dokl.* **20,** 256–258.

Yavorskaya, I.M., Astav'eva, N.M., & Vvedenskaya, N.D. (1978). Stability and non-uniqueness of liquid flow in rotating spherical layers. *Dokl. Akad. Nauk SSSR* **241,** 52–55; *Sov. Phys. Dokl.* **23,** 461–463.

Yavorskaya, I.M., & Belyaev, Yu.N. (1991). Nonuniqueness and multiparametric study of transition to chaos in the spherical Couette system. *Eur. J. Mech. B/Fluids* **10,** 267–274.

Yavorskaya, I.M., Belyaev, Y.N., & Monakhov, A.A. (1977). Stability investigation and secondary flows in rotating spherical layers at arbitrary Rossby numbers. *Dokl. Akad. Nauk SSSR* **237,** 804–807; *Sov. Phys. Dokl.* **22,** 717–719.

Yih, C.-S. (1972). Spectral theory of Taylor vortices. *Arch. Rat. Mech. Anal.* **46,** 218–240.

Yorke, J.A., & Yorke, E.D. (1981). Chaotic behavior and fluid dynamics. In *Hydrodynamic Instabilities and the Transition to Turbulence,* pp. 77–95, eds. H.L. Swinney & J.P. Gollub. Springer, Berlin.

Zaitsev, V.M., & Shliomis, M.I. (1970). Hydrodynamic fluctuations near the convection threshold. *Zh. Eksp. Teor. Fiz.* **59,** 1583–1592; *Soviet Phys. JETP* **32,** 866–870.

Zierep, J. (1958). Eine rotationssymmetrische Zellularkonvektions strömung. *Beitr. Phys. Atmos.* **30,** 215–222.

Zierep, J. (1959). Zur Theorie der Zellularkonvektion III. *Beitr. Phys. Atmos.* **32,** 23–33.

Zierep, J. (1961). Thermokonvektive Zellularströmungen bei inkonstanter Erwärmung der Grundfläche. *Z. angew. Math. Mech.* **41,** 114–125.

Zierep, J. (1963). Zellularkonvektionsströmungen in Gefässen endlicher horizontaler Ausdehnung. *Beitr. Phys. Atmos.* **36,** 70–78.

INDEX